Nutrition, Food Safety and Health

Nutrition, Food Safety and Health

Edited by **Dorothy Green**

Callisto Reference
New York

Published by Callisto Reference,
106 Park Avenue, Suite 200,
New York, NY 10016, USA
www.callistoreference.com

Nutrition, Food Safety and Health
Edited by Dorothy Green

International Standard Book Number: 978-1-63239-707-2 (Hardback)

Printed in China.

Contents

Preface

This book has been a concerted effort by a group of academicians, researchers and scientists, who have contributed their research works for the realization of the book. This book has materialized in the wake of emerging advancements and innovations in this field. Therefore, the need of the hour was to compile all the required researches and disseminate the knowledge to a broad spectrum of people comprising of students, researchers and specialists of the field.

Nutrition deals with studying the interaction of nutrients and their effect on growth and maintenance of the body. Food safety is the branch of science that deals with the effective handling, preparation, and preservation of food and health is the ability of living beings to adapt to the corresponding environment and protect themselves from the infections and diseases. All of these combine to form a synergic effect for the well-being of people. This book attempts to understand and explain various important topics such as human nutrition, food safety and toxicology, public health, nutrition, diet-related non communicable disease, food consumption, food intake, food security, etc. From theories to research to practical applications, case studies related to all contemporary topics of relevance to this field have been included in it. It will help the readers in keeping pace with the rapid changes in this field. This book will prove to be an asset for nutritionists, food scientists, health professionals and students alike.

At the end of the preface, I would like to thank the authors for their brilliant chapters and the publisher for guiding us all-through the making of the book till its final stage. Also, I would like to thank my family for providing the support and encouragement throughout my academic career and research projects.

Editor

Comparative Analysis of the Nutritional Status of Under-five Children and their Mothers in Rural and Urban Areas of Anambra State, Nigeria

Nkiru N. Ezeama[1], Prosper O. U. Adogu[2*], Christian C. Ibeh[2] and Echendu D. Adinma[2]

[1]Department of Community Medicine, Nnamdi Azikiwe University Teaching Hospital, Nnewi, Nigeria.
[2]Department of Community Medicine / Consultant Public Health Physician, Nnamdi Azikiwe University / Nnamdi Azikiwe University Teaching Hospital, Nnewi, Nigeria.

Authors' contributions

This work was carried out in collaboration between all authors. Author NNE headed the project. Author POUA conducted the analysis and interpretation and drafted the manuscript. Author EDA was involved in the early work and assisted with the interpretation of the results. Author CCI was in charge of data management. All authors were involved in the critical revision of the manuscript for important intellectual content and approved the final manuscript.

ABSTRACT

Aim: Malnutrition in the form of under-nutrition is still a major public health problem in developing countries of sub-Saharan Africa including Nigeria. This study compares the nutritional status of under-five children and their mothers in rural and urban areas of Anambra State Nigeria.

Methodology: This was a comparative cross-sectional study carried out in one urban and two rural local government areas of Anambra State, Nigeria namely Awka South, Dunukofia and Anaocha respectively. A total of 657 mother-child pairs were selected from eligible households using the multistage sampling technique. Data on household food security was obtained from the mothers using semi-structured, interviewer-administered questionnaires while anthropometric measurements were carried out on the children and their mothers using weighing scales and height boards.

Results: The overall prevalence of stunting, wasting and underweight among the under-five children in this study were 15.1%, 18.1% and 10.4% respectively, and the proportions were higher in the rural area than in the urban. The prevalence of stunting (height < 152 cm) in the mothers was 7.9% in the rural area compared to 9.1% in the urban area. Majority of the women were overweight (BMI ≥ 25.0), more in the urban (69.3%) than in the rural (59.2%).

**Corresponding author: Email: prosuperhealth50@gmail.com*

Conclusion: This study confirmed that under-five under-nutrition remains a serious public health problem in Anambra State, Nigeria. In order to reduce child morbidity and mortality to which malnutrition contributes significantly, concerted effort must be made by the government to improve child and maternal nutritional status by directing attention to improving household food security through developing and implementing policies that improve the livelihoods of the population.

Keywords: Nutritional status; under-fives; mothers; rural; urban; Nigeria.

1. INTRODUCTION

Malnutrition can be described in several ways; e.g. as a state in which the normal requirements of the body for one or several nutrients are not being met termed under-nutrition or as a state where there is excess intake of energy and /or nutrients termed over-nutrition [1]. Excessive or inadequate intake of macro- and micronutrients could result in either over- or under-nutrition [2]. Under-nutrition is found in every society of the world and children are at the greatest risk. It is the most prevalent form of malnutrition in sub-Saharan Africa (SSA) and is endemic in many developing countries.

Under-nutrition in the form of stunting, wasting and underweight, is endemic in Nigeria, and is clearly seen from results of national surveys such as the NDHS 2008 and MICS 2011 [3,4]. Stunting or low height-for-age (HFA) is indicative of chronic under-nutrition, the result of prolonged episodes of under-nutrition both of the pregnant mother and of the young infant, most importantly during the first 2 years of life [5]. In contrast, wasting or low weight-for-height (WFH) is indicative of acute malnutrition as a result of exposure to a relatively short duration of inadequate food intake often complicated by concurrent infective illness that cause weight loss [3,6]. Underweight is a composite index of HFA and WFH, which takes into account both acute and chronic malnutrition and is measured with the weight-for-age (WFA) index [3].

There is considerable evidence from literature of high levels of child malnutrition found in Nigeria [7-12] and many other developing countries of sub-Saharan Africa [13-15]. Malnutrition increases children's susceptibility to infections; they are more likely to die from common childhood ailments like diarrhoeal disease, malaria and respiratory infection. Those who survive are locked into a vicious cycle of recurring sickness and faltering growth due to frequent illness that cause further decline in their nutritional status [16].

Rural-urban differences have also been observed in the pattern of under-five malnutrition in SSA. Malnutrition is on average, higher in rural areas compared to urban areas; however socioeconomic inequalities are higher in urban areas than in rural areas [17]. The nutritional status of a mother and her child are intricately related. Her nutritional status during pregnancy can affect the body size and composition of her offspring [18]. Maternal nutritional status working alone or synergistically with infections during pregnancy largely determines a child's birth weight in most developing countries [6]. Being born with a low birth weight has serious implications for a child's mental and physical development well into adulthood[6].

Although under-nutrition is the most prevalent form of malnutrition in developing countries especially those in sub-Saharan Africa (SSA), a double burden of persistent under-nutrition with increasing prevalence of over-nutrition in the form of overweight and obesity is being observed in developing countries in transition including Nigeria, while hunger, food insecurity, underweight and other forms of under-nutrition still persist [19,20]. Obesity is a well-recognized risk factor for many non-communicable diseases including cardiovascular disease, type 2 diabetes mellitus, musculoskeletal disorders such as osteoarthritis and respiratory disorders such as sleep apnoea [19,21].

Most studies have associated co-existence of adult obesity and child under-nutrition with urban areas, however in rural Kenya, Mburu and Okello [22] found a prevalence of 22% of households with undernourished children whose mothers were obese [22]. Worldwide, over 10 million children under the age of 5 years die every year from preventable and treatable illnesses despite effective health interventions and malnutrition accounts for at least 50% of these deaths [15]. The situation is no less bleak in Nigeria as results from the 2008 Nigeria Demographic and Health Survey (NDHS) show that 41% of children under five are stunted, 14% are wasted and nearly one in four children is underweight [3].

This study was aimed at determining and comparing the nutritional status of under-five children and their mothers between rural and urban sub-populations in Anambra State, Nigeria.

2. METHODOLOGY

The study areas were Dunukofia, Anaocha and Awka South Local Government Areas, all located in Anambra Central senatorial zone, Anambra State, South Eastern Nigeria. Although Awka South is an urban local government area while Dunukofia and Anaocha are typical rural Local Government Areas (LGAs), they all share some similarities. The predominant ethnic group is Igbo and the language mainly spoken is Igbo. Christianity is the predominant religion while others are traditional religion and Islam. All the communities lie within the tropical rain forest region although most of the original rain forest has been lost due to clearing for commerce, farming and human settlement. Two distinct seasons are observed, a rainy season (annual mean rainfall of 161.9 millimetres) that starts from April to September and a dry season that lasts from October to April. The soil is generally fragile and lacks cohesion such that with poor agricultural practices and bush burning, land in these LGAs and indeed most of Anambra State is especially vulnerable during the rainy season to both sheet and gully erosion and landslides. The topography of these areas is not only undulating but forms a basin surrounded by highlands which drains finally into the Omambala River (from which Anambra State derives its name).

The study design was a community-based comparative cross-sectional survey. The study population comprised mothers aged 15-49 years and their children aged 0-59 months living in households in Ukpo (Dunokofia), Ichida (Anaocha) and Awka (Aka South) for at least one year prior to commencement of the research study.

Inclusion Criteria were households that had non-pregnant women of child-bearing age (15-49 years) with at least one child under the age of five years (0-59 months). Households with pregnant women and other mothers who did not meet the criteria as mentioned above were excluded from the study. Pregnant women were excluded due to physiological conditions of pregnancy that affect weight gain such as water retention which frequently leads to oedema; postural, hormonal and other physiological conditions [23]. This would have affected the comparability of anthropometry between pregnant and non-pregnant women.

Sample Size Determination: The minimum sample size to determine a difference in nutritional status between a rural and an urban community that is significant at the 5% level and with an 80% chance of detecting the difference (power) was calculated using the formula for comparison of two proportions [24]. Making use of figures 30.4% and 20.4% (proportions of food secure rural and urban households, respectively) [25] and anticipating a non-response of 10%, a minimum sample size of 330 per sub-population was calculated. Multiplying by a factor of 2, the total sample size for the study was obtained as 660.

Sampling Technique: A multi-stage sampling technique was used to select 330 households each from the rural and urban sub-populations to obtain a total sample of 660 households. From the 21 LGAs in Anambra state, two LGAs (Dunukofia and Anaocha) were selected from the rural stratum and one (Awka South) from the urban stratum by simple random sampling. Next Ukpo and Ichida communities (both identical in number of wards and population size) were selected from Dunukofia and Anaocha LGAs respectively while Awka community was selected from Awka South all by simple random sampling. Households were then sampled from all the wards in the three study communities as follows: three hundred and thirty (330) households were selected from Awka, 165 from Ukpo and 165 from Ichida to obtain the total sample size of 660 households. The number of households to be sampled in each ward was determined by dividing the total number of households to be sampled in each community by the number of wards in that community. Using the modified cluster sampling technique, neighboring households were successively sampled until the sample size in each ward was achieved. Only one child under five years was recruited per household. If a household had more than one eligible child, one was selected by simple random sampling using the balloting method.

Study instruments: A pre-tested, semi-structured, interviewer-administered questionnaire was used to collect information on socio-demographic characteristics, nutritional intake and anthropometric measurements of each mother and child pair.

Data Collection Methods: Weight and height of the mothers and children were obtained using standardized methods described by Omuemu and Ofili [7]. Height was measured for the mother / child pairs using a ShorrBoard® infant / child / adult portable height-length measuring board [26]. All anthropometric data were used to calculate summary indices for nutritional status – weight-for-age (WFA), weight-for-height / length (WFH), height-for-age (WFA) and body mass index (BMI).

Nutritional status was measured through the use of anthropometric data. WHO child growth standards were used to determine the nutritional status of children and the standard deviation of scores (z-scores) was calculated for the anthropometric measures - weight-for-age (WAZ), weight-for-height / length (WHZ), height-for-age (HAZ) using the following formula [15].

Z-score = (individual value – median value of the reference population)/SD value of the reference population

A cut-off point of –2 standard deviations (SD) below the median of the WHO Multicentre Growth Reference Study population[27] was used for each anthropometric measure of child (0-19years) nutrition status. Based on this, underweight, stunting and wasting were defined as WAZ < -2, HAZ < -2 and WHZ < -2 respectively. Overweight children (<2 years) were defined as having a WHZ >2 SD above the median of the reference population, while overweight children (including mothers<19years) were defined as having BAZ >1. The recommended cut-off criteria for determining overweight and obesity status are different for children birth to five years and five to 19 years of age. In this study, while BAZ was used to assess children 2 to 19 years, WHZ was used to assess children zero to 2 years because there is a lack of convincing evidence that BMI for age is more effective than weight for age or weight for length at assessing adequacy of feeding, and under- and overweight in children younger than two years of age [28]. Maternal (ie >19 years) nutritional status was measured with weight (in kg) and height (in metres) and body mass index (BMI, kg/m^2). A cut-off of 1.52 m was used to indicate short maternal stature [29]. A BMI of < 18.5 kg/m^2 was classified as underweight while ≥25.0 kg/m^2 was classified as overweight/pre-obese and ≥30.0 kg/m^2 / as obese.

2.1 Data Analysis

Out of a total of 660 questionnaires distributed, 657 were returned correctly filled in, giving a response rate of 99.5%. Households were classified under the following socio-economic classes: I, II, III, IV and V and for subsequent analysis these were further collapsed into: upper class (I and II), middlc class (III) and lower class (IV and V) [30]. The social class is derived from occupation and education scores of the father and mother. The average of the 4 scores gives the socio-economic class. Round off to the nearest whole number if a fraction is obtained. An overall score of: 1-2 = high social class (I & II), 3= middle social class (III), while 4-5 = low social class (IV & V). Frequency distributions of the dimensions of child nutritional status namely HFA, WFH and WFA (and BAZ where applicable) were developed. Relevant means and proportions were also calculated while Chi-squared analysis and independent sample t-test were used as tests of statistical significance. Chi-square test was applied in the comparison of discrete nominal variables, while the t-test was used to compare continuous variables and f-test derived as a result of multivariate analyses. A statistically significant finding was set at $p<0.05$.

2.2 Ethical Considerations

Approval for this study was obtained from to the Ethical Committee of the Nnamdi Azikiwe University Teaching Hospital (NAUTHEC). Written informed consent was obtained freely and without coercion from the respondents and respect for the confidentiality of the data obtained from them was ensured.

3. RESULTS

A total of 660 questionnaires were administered, 330 in each of the rural and urban samples, however, 657 questionnaires were analyzed giving a response rate of 99.5%.

Table 1 summarizes the socio-demographic characteristics of the respondents and their children stratified by rural or urban residence. The rural mothers had a mean age of 31.6±6.1 years while the mean age for the urban-dwelling mothers was 30.3±6.7 years. There were 9 (2.7%) teenage mothers (15-19 years) in the urban population but none in the rural sample. A statistically significant difference exists between the mean ages of the rural and urban mothers / respondents (p=0.011) and their age distributions (p=0.015). Majority of the respondents were married at the time of the study, comprising 92.0% and 88.8% of the rural and urban respondents respectively. Statistically significant differences between the rural and urban communities were also observed in the educational status of the respondents (p<0.001) and socioeconomic status (p<0.001). Majority of the households in both sub-populations had more than four members (rural: 254 households, 77.4%; urban: 209 households, 63.5%). This difference in the size of households was also statistically significant (p <0.001). The total study population of under-five children had a mean age of 27.5±17.8 months. The mean age of the children in the rural communities was significantly higher than that of the urban children, 29.1±17.2 vs. 25.9±18.3 months, respectively (p = 0.021). Children in the 12-35 months age group had the highest proportions in the rural sample (42.6%). In contrast, the 36-59 months age group had the highest proportion in urban sample (42.6%). The difference in under-five age distribution between the rural and urban sub-populations was statistically significant (p=0.002). The overall mean birth weight was 3.28±0.52 kg and there was a statistically-significant difference in the mean birth weights between the rural and urban children (p = 0.010).

Table 2 summarizes anthropometric indices for the study sample of under-five children stratified by place of residence. Ninety-nine children (15.1%) in the total sample were stunted (HAZ below -2 SD), 119 (18.1%) were wasted (WHZ below -2 SD), whereas 68 (10.4%) were underweight for age (WAZ below -2 SD). A higher proportion of stunting was found among under-five children in the rural area compared to the urban under-five children. A total of 53 children (16.2%) in the rural area were found to be stunted, of whom 18 (5.5%) were severely stunted (HFA < -3 SD). A higher prevalence of wasting was found among the rural under-five children than those in the urban area. Seventy-three children (22.2%) were wasted, of whom (6.1%) were severely wasted. The urban sample had 46 children (13.9%) wasted out of which 12 (3.6%) were severely wasted. This difference was statistically significant (p < 0.001). More under-five children were found to be underweight in the rural area than in the urban area. Thirty-eight children (11.6%) had low weight-for-age z-scores below -2 SD out, of whom 6 (1.8%) had WFA below -3 SD. Underweight children made up 9.1% of the urban sample, of whom 9 (2.7%) were below -3 SD. These differences were, however, not statistically significant (p = 0.354).

The table also shows the nutritional status of the mothers stratified by place of residence. Using 152 cm as the cutoff point, the difference in the prevalence of stunting between mothers in the rural and urban communities was not statistically significant (7.9% vs. 9.1%; p = 0.584). The mean BMI for the urban women was higher than that of the rural women (urban: 27.6±5.5 kg/m2; rural: 26.3+±4.8 kg/m^2) and this was statistically significant (p = 0.001). The distribution of BMI categories also differed significantly between the mothers in the rural and urban populations (p = 0.024).

Table 3 shows that in the rural sub-sample, stunting was highest among children in the 24-35 months (20.8%) and 48-59 months (20.8%) age groups. Furthermore, the 24-35 months age group was most affected by severe stunting than the rest of the age groups. Among the urban children, the 36-47 months and 48-59 months age groups had higher proportions of stunted children, 23.9% and 23.9%, respectively. Severe stunting was found more among the 36-47 months age group (36.4%) than in the rest of the urban sub-sample. The pattern of the weight-for-height index across the age groups of the rural and urban children under five shows that wasting was highest among the rural children, in the 24-35 months age group (36.9%) and this group also accounted for the highest proportion of severely wasted children in the rural sample (35.0%). In the same vein, the 24-35 months age group also had the highest proportions of wasted children (36.9%) and severely wasted children (50.0%) in the urban sample. The 24-35 months age group in the rural population had the highest number of underweight children, 12 (31.6%) in total. The same age group was also the worst affected in the urban population as 11 (36.7%) children were found to be underweight for their age.

Table 4 shows the nutritional profiles of the sampled households based of the child's weight-for-age and the mother's body mass index stratified by place of residence. Overall, the nutritional profile "normal child-overweight mother" was predominant and this was the same for the rural and urban households (rural: 54.0%; urban: 63.2%) but significantly more prevalent in urban than rural households (p<0.04). The next most common profile was "normal child-normal mother" (rural: 32.0%; urban: 25.2%),

Table 1. Socio demographic characteristics of the mother-child pairs

Socio-demographic characteristic	Rural (n = 328)	Urban (n = 329)	Total (n = 657)	Test statistic	p-value
Age of mother (years)					
15-19	0 (0)	9 (2.7)	9 (1.4)		
20-29	139 (42.4)	155 (47.1)	294 (44.7)	F =15.864	0.015
30-39	146(44.5)	130(39.5)	276 (42)		
40-49	43(13.1)	35(10.7)	78 (11.9)		
Mean±SD	31.6 ±6.1	30.3±6.7	30.9±6.4	t = 2.562	0.011
Marital status					
Single	9 (2.8)	14 (4.3)	23 (3.5)	F = 3.429	0.553
Married	302 (92.0)	292 (88.8)	594 (90.4)		
Others	17(5.2)	23 (7.0)	40(6.1)		
Mothers education					
No formal education	4 (1.2)	2 (0.6)	6 (0.9)	F = 18.19	0.000
Primary	57 (17.4)	30 (9.1)	87 (13.2)		
Secondary	220 (67.1)	216 (65.7)	436 (66.4)		
Tertiary	47 (14.3)	81 (24.6)	128 (19.5)		
Socio-economic class					
Upper class	5 (1.5)	27 (8.2)	32 (4.9)		
Middle class	43 (13.1)	55 (16.7)	98 (14.9)	χ^2 = 18.659	0.000
Lower class	280 (85.4)	247 (75.1)	527 (80.2)		
Size of household					
≤ 4 members	74 (22.6)	120 (36.5)	194 (29.5)	χ^2 = 15.28	0.000
> 4 members	254 (77.4)	209 (63.5)	463 (70.5)		
Mean±SD	6.05±1.87	5.32±1.90	5.68±1.92	t = 5.024	0.000
Sex of child					
Female	163 (49.7)	169 (51.4)	332 (50.5)	χ^2 = 0.184	0.668
Male	165 (50.3)	160 (48.6)	325 (49.5)		
Age of children (in months)					
0– 11	68(20.7)	89(27.1)	157 (23.9)	χ^2 = 19.451	0.002
12 – 35	140(42.6)	100(30.3)	240 (36.5)		
36 – 59	120(36.6)	140(42.6)	260 (23.7)	t = 2.312	
Mean±SD	29.1±17.2	25.9±18.3	27.5 ± 17.8		0.021
Birth weight of child					
<2.5kg	3 (0.9)	9 (2.7)	12 (1.8)	F = 3.177	0.142
≥2.5kg	325 (99.1)	320 (97.3)	645 (98.2)		
Mean±SD	3.33±0.49	3.23±0.53	3.28±0.52	t = 2.596	0.010

Values are number (%) unless otherwise stated. F = Fisher's exact

but markedly higher among rural than urban households (p<0.05). Both the "underweight child-overweight / obese mother" and "underweight child-normal mother" profiles are significantly more prevalent in the rural than the urban households (p<0.03 each).

4. DISCUSSION

There was a significant difference in the mean ages of the respondents in both sub-populations and majority of the respondents were married at the time of the survey. Women in the urban area had higher levels of education and higher socio-economic status compared to the rural women. These findings are consistent with results from the National Demographic and Health Survey (NDHS) of 2008 [3] that showed higher levels of education among urban women than rural women. Larger household sizes were observed more in the rural area than in the urban. The 2008 NDHS also reported larger households (greater than nine members) more among rural dwellers than urban [3]. However, the average household sizes obtained in this study for the rural (6.05 persons) and urban (5.32) samples are higher than the national averages reported for rural and urban areas in the NDHS (4.6 persons and 4.1 persons, respectively) [3]. As seen from the results, majority of the households in both sub-populations had more than four family members (rural: 77.4%; urban: 63.5%). A cross-sectional survey in Lagos, Nigeria also found a higher prevalence of large households of more than four members (52.7%) among its study population [31] although the prevalence is lower than the values obtained in this present study. These results however contrast with the prevalence documented for an urban community in Edo State of Nigeria where majority of the

Table 2. Nutritional status of children aged 0 – 59 months and mothers stratified by place of residence

Nutritional status of children	Rural (n = 328)	Urban (n = 329)	Total (n = 657)	Test Statistic	p-value
Height-for-age Z score					
Below -3 SD	18 (5.5)	11 (3.3)	29 (4.4)	χ^2 = 1.803	0.406
Below -2 SD*	53 (16.2)	46 (14.0)	99 (15.1)		
-2 SD to +2 SD	275 (83.8)	283 (86.0)	558 (84.9)		0.621
Mean Z-score±SD	-0.416±1.3	-0.471±0.43	-0.444±0.9	t = 0.495	
Weight for height z-score					
Below -3 SD	20 (6.1)	12 (3.6)	32 (4.9)	χ^2 = 18.888	0.000
Below -2 SD*	73 (22.2)	46 (13.9)	119 (18.1)		
-2 SD to +2 SD	224 (68.3)	217 (66.0)	441 (67.1)		
Above +2 SD	31 (9.5)	66 (20.1)	97 (14.8)		
Mean Z-score±SD	-0.505±1.96	-1.290±1.38	0.394±1.2	t = -6.571	0.000
Weight-for-age z-score					
Below -3 SD	6 (1.8)	9 (2.7)	15 (2.3)	χ^2 = 3.252	0.354
Below -2 SD*	38 (11.6)	30 (9.1)	68 (10.4)		
-2 SD to +2 SD	284 (86.6)	291 (88.5)	575 (87.5)		
Above +2 SD (BAZ >1	6 (1.8)	8 (2.4)	14 (2.1)		
for children 2-5yrs)	-0.586±1.37	-0.391±1.1	-0.489±1.23	t = -2.025	0.043
Mean Z-score±SD					
Nutritional indices† of mothers	**Rural (n = 328)**	**Urban (n = 329**	**Total (n = 657)**	**Test Statistic**	**P value**
Height of mother					
< 152 cm (HAZ<-2SD for	26 (7.9)	30 (9.1)	56 (8.5)	χ^2 =0.299	0.584
mothers≤19yrs)	302 (92.1)	299 (90.9)	601 (91.5)		
≥ 152 cm					
BMI category (kg/m^2)					
Underweight< 18.5	9 (2.7)	8 (2.4)	17 (2.6)		
Normal 18.5–24.9	125 (38.1)	93 (28.3)	218 (33.2)	χ^2 =7.494	0.024
Overweight/pre-obese					
** 25.0 – 29.9	131 (40.0)	145 (44.1)	276 (42.0)		
Obese ** ≥ 30.0	63 (19.2)	83 (25.2)	146 (22.2)		
Total	328 (100.0)	329 (100.0	657 (100.0)	t = - 3.246	0.001
Mean±SD	26.3±4.8	27.0±5.2	27.6±5.5		

*** Assessed for mothers aged ≤19 years old using BAZ>1 for overweight, pre-obese or obese.*
**Includes children who are below -3 SD from the WHO Child Growth standards population median.*
Values are number (%) unless otherwise stated. † Values are number (%) unless otherwise stated.

households had smaller family sizes of less than four members (74.7%) [7].

This study showed that overall, 15.1% of the under-five children were stunted, 18.1% wasted and 10.4% were underweight for age. The level of low weight-for-age compares well with the 10.0% reported for the South East region of Nigeria in the 2008 NDHS [3] but is much higher than the prevalence of 4.5% documented in the 2011 Nigeria Multiple Indicator Cluster Survey (MICS Nigeria) [4]. A higher level of wasting and lower level of stunting compared to the 2008 NDHS were, however, observed in this study. In comparison, Odunayo and Oyewole[8] obtained much higher levels of stunting and underweight and a lower level of wasting in their study in Ifewara, Nigeria, 26.7%, 23.1% and 9%, respectively[8] , whereas Ajao et al[11] obtained overall prevalence rates of 39.3%, 6.3% and 14.1% for stunting, wasting and underweight, respectively, among rural and urban settlements in Ile-Ife, Nigeria [11]. These findings indicate that there is considerable variability in the nutritional status of children in different parts of the country such that national averages may not clearly depict what really occur at community and household levels. It may also have arisen as a result of differences in methodology employed in both studies. Nevertheless, the lower prevalence of stunting in this study compared to the NDHS findings points to possible improvements in child nutrition that may have occurred since the last national survey was carried out five years ago, and may have reduced the prevalence of chronic malnutrition in Anambra State and possibly Nigeria as a whole. On the one hand, chronic malnutrition in children has long-term consequences; stunted children are likely to accumulate body fat especially central fat

Table 3. WHZ and WAZ of rural and urban children (0-59 months) stratified by age

Age in months	**Weight-for-height z-score (WHZ)**			
Rural (n=328)	**Below -3 SD (n=20)**	**Below -2 SD*(n=73)**	**-2SD to +2SD (n=224)**	**Above +2 SD (n =31)**
<6	2 (10.0)	8 (11.0)	15 (6.7)	6 (19.4)
6-11	2 (10.0)	3 (4.1)	31 (13.9)	5 (16.1)
12-23	2 (10.0)	8 (11.0)	59 (26.3)	3 (9.7)
24-35	7 (35.0)	27 (36.9)	38 (17.0)	5 (16.1)
36-47	4 (20.0)	13 (17.8)	37 (16.5)	3 (9.7)
48-59	3 (15.0)	14 (19.2)	44 (19.6)	9 (29.0)
Urban (n = 329)	**Below-3 SD (n=12)**	**Below -2 SD*(n=46)**	**-2SD to +2SD (n=217)**	**Above +2 SD (n = 66)**
<6	1 (8.3)	7 (15.2)	44 (20.3)	5 (7.6)
6-11	1 (8.3)	4 (8.7)	23 (10.6)	6 (9.1)
12-23	0 (0.0)	2 (4.4)	37 (17.1)	6 (9.1)
24-35	6 (50.0)	17 (36.9)	35 (16.1)	3 (4.5)
36-47	4 (33.4)	13 (28.3)	33 (15.2)	5 (7.6)
48-59	0 (0.0)	3 (6.5)	45 (20.7)	41 (62.1)
	Weight-for-age Z score			
Rural (n = 328)	**Below -3 SD (n = 6)**	**Below -2 SD* (n= 38)**	**-2SD to +2SD (n=284)**	**Above +2 SD (n = 6)#**
<6	1 (16.7)	5 (13.1)	24 (8.5)	0 (0.0)
6-11	1 (16.7)	2 (5.3)	36 (12.7)	1 (16.7)
12-23	0 (0.0)	5 (13.2)	62 (21.8)	3 (50.0)
24-35	0 (0.0)	12 (31.6)	58 (20.4)	0 (0.0)
36-47	2 (33.3)	6 (15.8)	46 (16.2)	1 (16.7)
48-59	2 (33.3)	8 (21.0)	58 (20.4)	1 (16.7)
Urban (n = 329)	**Below -3 SD (n = 9)**	**Below -2 SD* (n= 30)**	**-2SD to +2SD (n=231`)**	**Above +2 SD (n = 8)**
<6	0 (0.0)	5 (16.7)	51 (17.5)	0 (0.0)
6-11	1 (11.1)	1 (3.3)	29 (10.0)	2 (25.0)
12-23	0 (0.0)	0 (0.0)	42 (14.4)	3 (37.5)
24-35	0 (0.0)	11 (36.7)	45 (15.5)	0 (0.0)
36-47	7 (77.8)	9 (30.0)	40 (13.7)	2 (25.0)
48-59	1 (11.1	4 (13.3)	84 (28.9)	1 (12.5)

#Note that BAZ was used to determine cut off point (BAZ>1) for children aged 2-5years
**Includes children who are below -3 SD from the WHO Child Growth standards population median. Values are number (%) unless otherwise stated.*

predisposing them to diseases such as hypertension and diabetes in adulthood[6]. Acute malnutrition, on the other hand, is caused by recent nutritional deficiency working synergistically with other factors such as concurrent infective illness [22] (e.g. malaria) that are still highly endemic in this environment. This may account for the higher level of wasting observed in this study. It remains to be seen what the picture for Nigeria will be in the next round of the NDHS. Nevertheless, the prevalence of under-nutrition still remains high as seen in this study.

Consistent with patterns reported in NDHS 2008 [3], MICS Nigeria [20] and previous studies carried in Lagos and Nepal [6,15] higher levels of stunting, wasting and underweight were observed in the rural areas than the urban area in this study. The rural-urban differences in this study still point to socio-economic and environmental inequalities between rural and urban dwellers. This was confirmed in this study by the higher risk of stunting with decreasing socio-economic class observed more among the rural children than in the urban. However, the prevalence of stunting (14.0%) in the urban area compared to the rural areas (16.2%) suggests that the urban advantage is gradually being eroded due to deterioration in living conditions of urban dwellers as a result of rapid population growth that has surpassed available social amenities [32].

The 24-35 months age group was the worst affected according to all three anthropometric indices. The Ile-Ife study [11] also observed similar findings with the prevalence of wasting and underweight being highest after the age of two years among its study population. Nyahurucha et al. [12] in their study in Tanzania also found this age group to be the most affected

Table 4. Household nutritional profiles based on child's weight-for-age and mother's BMI

Profile	Number of households (%)			Test statistic (z)	p-values	F	p-value
	Rural (n = 328)	Urban (n = 329)	Total (n=657)				
Underweight child-overweight / obese mother	17 (5.2)	20 (6.1)	37 (5.6)	0.22	p>0.05		
Underweight child-normal mother	20 (6.1)	10 (3.1)	30 (4.6)	3.6	P<0.03		
Underweight child-underweight mother	1 (0.3)	0 (0.0)	1 (0.2)	4.0	P<0.03	P=9.577	0.069
Normal child-overweight/obese mother	177 (54.0)	208 (63.2)	385 (58.6)	2.4	P<0.04		
Normal child-normal mother	105 (32.0)	83 (25.2)	188 (28.6)	1.99	P<0.05		
Normal child-underweight mother	8(2.4)	8(2.4)	16(2.4)	0	P=1		
Total	328 (100.0)	329 (100.0)	657 (100.0)				

Statistically significant, F = Fisher's exact test

by under-nutrition [12]. A probable reason for this finding is that most children would have ceased breast feeding by the age of two years, but complementary feeding may be inadequate or inappropriate. In addition, they may be exposed to infectious agents in food, water and the environment resulting in reduced dietary intake and weight loss [4,8,11]. This phenomenon was also observed in another study[33] on worldwide timing of growth faltering involving the three anthropometric indices (WAZ, WHZ and HAZ) which established that WAZ faltered rapidly from 3 to 12 months of age, WHZ declined until 15 months of age while HAZ hitherto close to the reference value at birth, declined sharply until the age of 2 years, after which mean values remained stable between 1.5 and 2 SDs below the reference [33].

Based on the body mass index, the overall prevalence of under-nutrition (BMI < 18.5 kg/m^2) among the mothers in this study was 2.6%. This is much lower than the 6.7% reported in Lagos State, Nigeria [22] and outside the range of 5-20% reported for African women [6]. In addition, this study showed a high prevalence of overweight/obesity (BMI ≥ 25.0 kg/m^2) among the women in the study sample (64.2%) albeit a higher prevalence in the urban community (69.3%) than in the rural (59.2%). The co-existence of child under-nutrition and maternal overweight/obesity is clearly seen from the thirty-seven households overall (5.6%) that had an underweight child and an overweight/obese mother with the prevalence being higher among the urban households than the rural households (6.1% versus 5.2%). Dietary diversity in the household may be affected when there is food and financial shortage and households may resort to consuming limited food choices and food items which are more likely to be high in fat and energy density. The co-existence of child under-nutrition and maternal overweight/obesity should, however, be interpreted with caution as another study has pointed out that stunted child / overweight mother pairs may represent a statistical artifact, not a distinct entity[34]. Nevertheless, it still implies that the households in this study have the twin burden of dealing with the problems of under-nutrition and overweight/obesity in the same household, the so-called Double-Burden phenomenon [22]. This has implications not only for households but for the health system at large. Childhood malnutrition increases the risk and severity of common childhood ailments such as malaria, diarrhoeal disease and pneumonia, whereas adult obesity is linked to hypertension and diabetes [20]. Occurrence of these diseases in the same household could place a heavy financial burden on the family and lead to catastrophic financial losses from health and hospital bills with little available income for adequate nutrition.

5. CONCLUSION

The study showed that child under-nutrition remains a serious public health problem in Nigeria from the results of child malnutrition among households in rural (Ukpo and Ichida) and urban (Awka), Anambra State. Specifically, the "underweight child-overweight / obese mother" and "underweight child-normal mother" profiles are more prevalent among the rural households than the urban ones.

6. RECOMMENDATIONS

In order to reduce child morbidity and mortality to which undernutrition contributes significantly, concerted effort must be made by the government, non-government organizations, the private sector and communities to improve child and maternal nutritional status. A holistic approach targeting all levels of prevention should be taken:

1. Primordial prevention of undernutrition in children should direct attention towards implementing policies that improve the nutrition and livelihoods of the Nigerian population, in both rural and urban areas.
2. Primary prevention aimed at improving the nutritional status of mothers will reduce the risk of a child being born with a low birth weight and being at risk for undernutrition.
3. Secondary preventive measures involving the early diagnosis and treatment of undernourished children need to be emphasized through adequate training of community health workers at the primary health care level.
4. Finally, although stunting cannot be reversed, tertiary measures to prevent children from becoming more malnourished should be explored such as a school breakfast programme in Anambra State [25].

In all these, a lot more attention should be directed at the rural households without necessarily ignoring their urban counterparts.

7. LIMITATIONS OF THE STUDY

Being a household survey, the study effectively excludes residents who do not live within the context of households such as street children and the homeless population. It is also not representative of migrant groups and people who had resided in the study areas for less than a year prior to the start of the study. Furthermore, the result of this study is only generalizable to the state and not the National level because the sample used is only representative of Anambra state. Finally, most of the information was obtained from maternal report with no means of independent verification.

COMPETING INTERESTS

Authors have declared that no competing interests exist.

REFERENCES

1. Manary MJ, Solomons NW. Public health aspects of under-nutrition. In: Gibney MJ Margetts BM, Kearney JM, Arab L, (eds.) Public Health Nutrition. The Nutrition Society. Oxford: Blackwell Publishing. 2004:178-191.
2. Uchendu FN, Atimuo T. The silent and neglected crisis of malnutrition: scientific evidence for taking decisive action. Glob J Health Sci. 2011;3(1):193-202.
3. National Population Commission (NPC) Nigeria, ICF Macro. Nigeria Demographic and Health Survey 2008. Abuja, Nigeria: National Population Commission and ICF Macro. 2009:163-167.
4. National Bureau of Statistics (NBS), United Nations Children's Fund (UNICEF), United Nations Population Fund (UNFPA). Nigeria Multiple Indicator Cluster Survey 2011, Main Report. Abuja, Nigeria: National Bureau of Statistics. April 2013;19-22. Available:http://www.nigerianstat.gov.ng/pages/NBS%20eLibrary (Accessed 30 November 2013)
5. Collins S. Treating severe acute malnutrition seriously. Arch Dis Child. 2007;92:453-461.
6. Baker-Henningham H, Grantham-McGregor S. Nutrition and child development. In: Gibney M.J., Margetts BM, Kearney JM, Arab L, (eds.) Public Health Nutrition. The Nutrition Society. Oxford: Blackwell Publishing. 2004;247-263.
7. Senbanjo IO, Olayiwola IO, Afolabi WA, Senbanjo OC. Maternal and child under-nutrition in rural and urban communities of Lagos state, Nigeria: the relationship and risk factors. BMC Research Notes. 2013; 6:286. Available:http://www.biomedcentral.com/1756-0500/6/286 (Accessed 21 January 2014)
8. Omuemu VO, Ofili AN. Nutritional status of under-five children in an urban community in Benin City. Annals of Biomedical Sciences. 2009;8(1):1–11.
9. Odunayo SI, Oyewole AO. Risk factors for malnutrition among rural Nigerian children. Asia Pac J ClinNutr. 2006;15(14):491-495.

10. Ijarotimi OS, Oyeneyin OO. Effect of economy restructuring on household food security and nutritional status of Nigeria children. J Food Agric Environ. 2005; 3(3&4):27-32.
11. Hamidu, JL, Salami HA, Ekanem AU, Hamman, L. Prevalence of protein-energy malnutrition in Maiduguri, Nigeria. Afr J Biomed Res. 2003;6:123–127.
12. Ajao KO, Ojofeitimi EO, Adebayo AA, Fatusi AO, Afolabi OT. Influence of family size, household food security status, and child care practices on the nutritional status of under-five children in Ile-Ife, Nigeria. Afr J Reprod Health. 2010; 14(4):123
13. Nyaruhucha CNM, Msuya JM, Mamiro PS, Kerengi AJ. Nutritional status and feeding practices of under-five children in Simanjiro district, Tanzania. Tanzan Health Res Bull. 2006;8(3):162-67.
14. Fentaw R, Bogale A, Abebaw D. Prevalence of child malnutrition in agro-pastoral households in Afar Regional State of Ethiopia. Nutr Res Pract. 2013;7(2): 122-131.
15. Medhin G, Hanlon C, Dewey M, Alem A, Tesfaye F, Worku B, et al. Prevalence and predictors of under-nutrition among infants aged six and twelve months in Butajira, Ethiopia: The P-MaMiE Birth Cohort. BMC Public Health. 2010; 10:27. BioMed Central. DOI:10.1186/1471-2458-10-27 (Accessed 9 August 2011).
16. Uthman OA. Using extended concentration and achievement indices to study socioeconomic inequality in chronic childhood malnutrition: The case of Nigeria. Int J Equity Health. 2009;8:22.
17. Fotso J. Child health inequities in developing countries: Differences across urban and rural areas. Int J Equity Health. BioMed Central. 2006;5:9. Available:http://www.equityhealthj.com/content/5/1/9 (Accessed 5 October 2011).
18. Victora CG, Adair L, Fall C, Hallal PC, Martorell R, Richter L, Sachdev HS. Maternal and child under-nutrition: consequences for adult health and human capital. Lancet. 2008;371:340–57. DOI:10.1016/S0140-6736(07)61692-4. (Accessed 5 June 2011).
19. Seidell JC, Visscher TLS Public health aspects of over-nutrition. In: Gibney MJ, Margetts BM, Kearney JM, Arab L (eds.) Public Health Nutrition; the Nutrition Society. Oxford: Blackwell Publishing. 2004;167-177.
20. World Health Organization. Report of the joint WHO/FAO expert consultation on diet, nutrition and the prevention of chronic diseases. Geneva, Switzerland, 2002. WHO Technical Report Series No. 916. Available:http://www.fao.org/docrep/005/ac911e/ac911e01.htm (Accessed 5 June 2012)
21. Food and Agriculture Organisation/World Health Organisation. Rome Declaration on World Food Security and World Food Summit Plan of Action. World Food Summit. November 13-17 1996, Rome, Italy: Food and Agriculture Organisation; 1998.
22. Mburu JG, Okello JJ. The prevalence of under-nourished child obese mother phenomenon in rural areas: evidence from central province of Kenya. African Association of Agricultural Economists (AAAE) Ghana Conference Proceedings. 2007;559-564.
23. Padilha PC, Accioly E, Libera BD, Chagas C, Saunders C. Anthropometric assessment of nutritional status in Brazilian pregnant women. Rev Panam Salud Publica. 2009;25(2):171–8.
24. Kirkwood BR, Sterne JAC. Essential medical statistics 2nd edition. Massachusetts: Blackwell Science. 2003; 420-1.
25. Orewa SI, Iyanbe CO. Household food insecurity in Nigeria: an assessment of the present status of protein – energy malnutrition among rural and low-income urban households. J Appl Sci Res. 2009; 5(10):1615-1621.
26. Weight and Measure LLC. ShorrBoard® infant/child/adult portable height-length measuring board. Weight and Measure LLC; 17802 Shotley Bridge place, Olney, Maryland 20832-1671 USA.
27. World Health Organisation Multicentre Growth Reference Study Group. WHO child growth standards based on length/height, weight and age. Acta Pædiatrica. 2006;Suppl450:76-85.
28. Canadian Paediatric Society. A health professional's guide for using the new WHO growth charts. Paediatr Child Health. 2010;15(2):84–90. PMCID: PMC2865941. Available:www.cps.ca, www.caringforkids.cps.ca

29. World Health Organisation expert committee. Physical status, the use and interpretation of anthropometry. WHO Technical Report Ser. 1995;894:424-438.
30. Ogunlesi TA, Dedeke IOF, Kuponiyi OT. Socio-economic classification of children attending specialist paediatric centres in Ogun state, Nigeria. Niger Med Pract. 2008;54:21-25.
31. Omonona BT, Agoi GA. An analysis of food security situation among Nigerian urban households: Evidence from Lagos State, Nigeria. J Cent EurAgric. 2007;8(3): 397-406.
32. Abuya BA, Clera J, Kimani-Murage E. Effect of mother's education on child's nutritional status in the slums of Nairobi. BMC Paediatrics. 2012;12:80.
33. Victora CG, de Onis M, Hallal PC, Blössner M, Shrimpton R. Worldwide timing of growth faltering: Revisiting implications for interventions. Pediatrics. 2010;125(3):e473-80.
34. Dieffenbach S, Stein AD. Stunted child / overweight mother pairs represent a statistical artifact, not a distinct entity. J Nutr. 2012;142(4):771-3.

2

Review and Perspective on the Composition and Safety of Green Tea Extracts

Jeffrey B. Blumberg[1*], Bradley W. Bolling[2], C. Y. Oliver Chen[1] and Hang Xiao[3]

[1]Friedman School of Nutrition Science and Policy, Tufts University, 711 Washington Street, Boston, MA 02111, USA.
[2]Department of Food Science, University of Wisconsin-Madison, 1605 Linden Drive, Madison, WI 53706, USA.
[3]Department of Food Science, University of Massachusetts Amherst, 338 Chenoweth Laboratory, 100 Holdsworth Way, Amherst, MA 01003, USA.

Authors' contributions

This work was carried out equally in collaboration between all authors. All authors read and approved

ABSTRACT

The growing body of evidence regarding the putative health benefits of green tea (*Camellia sinensis*), including reduced risk of cancer and cardiovascular disease, has led to an increase in the consumption of brewed green tea and the formulation of green tea extracts (GTE) into a variety of food and beverage products and food supplements. The principal bioactive ingredients in green tea beverages and GTE are polyphenols, particularly the flavan-3-ols, which have been shown to act on antioxidant, anti-inflammatory, glucoregulatory, and cell signaling pathways. Some experimental evidence and case reports suggest the use of green tea and GTE is associated with the potential for inducing liver injury. The ability to extrapolate findings from *in vitro* and animal model studies is always limited and the available results on green tea- and GTE-induced liver injury in humans have presented clinical and regulatory challenges due to the difficulty of demonstrating a causal relationship between intake and harm. Attention to the risk for hepatotoxicity has largely been focused on GTE. Existing data are insufficient to identify the causative agent in the preparation or composition of GTE or its dose or duration of use as well as nutrigenetic, medical, and other factors that may contribute to the risk of hepatotoxicity. Responses by different government regulatory agencies regarding the safety of GTE are inconsistent with one another, including the dosage and derivation of its bioactives from aqueous versus hydro-alcoholic extracts. Restrictions on the production of GTE limit the application of innovative extraction

**Corresponding author: Email: jeffrey.blumberg@tufts.edu*

technologies that may improve the efficiency, quality, and cost-effectiveness of production and minimize the presence of endogenous or contaminant heavy metals as well as mycotoxins, pesticides, and other unwanted constituents. The objective of this review is to characterize the phytochemical profile of green tea and GTE, provide an updated description of the methods that could be used to extract green tea constituents, and discuss recent considerations on the potential for harm from consumption of GTE and products containing GTE.

Keywords: Green tea extract; favan-3-ols; hepatotoxicity; cancer; cardiovascular disease; extraction; regulations.

1. INTRODUCTION

Based on archeological evidence, the consumption of tea appears to have first occurred about 5000 years ago during the Paleolithic period, predating the legend of its discovery by the mythical Chinese emperor, Shen Nung. By the fourth century, tea was an important part of Chinese life because of its perceived value as a medicine for the treatment of a variety of ailments [1]. Today, after water, tea is the most widely consumed beverage in the world. Tea is made from leaves of *Camellia sinensis* (L.) O. Kuntzeand is cultivated in more than 30 countries [2,3]. The usual amount of dry green tea leaves used to brew a cup (237 mL) is 1.8-2.5 g. The global average consumption of tea is about 0.5 kg *per capita*, though amounts (in kg *per capita*) are greater where tea drinking is common such as India (0.73), China (0.95), Japan (0.96), Ireland (1.90), United Kingdom (1.97), Turkey (2.04), and Libya (1.90) [4]. Tea is especially rich in flavan-3-ols (also referred to as catechins) but also serves as an important source of flavonols contributing over 90 and 40% of these flavonoids, respectively, to the American diet [5]. Data from the USDA Continuing Survey of Food Intake by Individuals reveals that at the 90th percentile of intake, American men and women consume 3.5 and 3.0 cups (or 830 and 711 mL) daily providing 454 and 393 mg of flavan-3-ols, respectively [6]. These data, together with reports of green tea drinkers in other countries suggest a range of intake of 500-2360 mL daily providing up to 931 mg of flavan-3-ols [7-12].

Numerous research studies suggest that regular consumption of tea is associated with an array of bioactivities, including antibacterial, antioxidant, anti-inflammatory, antiviral, glucoregulatory, and immune-stimulant actions, and putative health benefits, including a reduced risk of chronic conditions like arthritis, some forms of cancer, cardiovascular disease, dental caries, type 2 diabetes mellitus, dyslipidemia, and neurodegenerative diseases [2,13-17]. This diversity of actions appears largely related to the flavonoid constituents of tea via mechanisms partly overlapping with those of other dietary polyphenols, including the modulation of protein kinases, growth factors and transcription factors as well as the modification of intracellular signaling by activation of membrane receptors or molecular targets within cells [15,18,19]. These actions may result directly or indirectly from the antioxidant capacity of green tea *in vivo* to reduce oxidative stress [20-22] and associated DNA damage and lipid peroxidation [23-25].

Given the wide use of tea at 4,624,625 tonnes or 2.3 trillion cups per year, even modest benefits on health could have significant implications for the promotion of public health [4]. An effort to further broaden these benefits beyond drinking tea infusions has led to the extraction of green tea polyphenols for use in functional foods and food supplements. A growing body of research on dried green tea extracts (GTE) suggests similar benefits can be derived from the use of food products containing GTE [26]. Specifications for GTE apply to its constituent polyphenols (especially (-)-epigallocatechin gallate or EGCG) as well as caffeine and L-theanine, other major bioactive components. Green tea consumption has a long, safe history of use [27-30] and green tea preparations such as essential oils, solvent-free oleoresins, and natural extractives (including distillates) are deemed Generally Recognized As Safe (GRAS) [31] for use in foods. GTE are currently formulated into foods, beverages, candy and gums, chocolates, and dietary or food supplements as well as toothpaste. However, concerns have been expressed about the potential for harm of GTE preparations consumed at high doses in supplement form, particularly an idiosyncratic hepatoxicity [32-38]. The objective of this review is to characterize the phytochemical profile of green tea and GTE, provide an updated description of the methods used to extract green tea constituents, and discuss recent considerations on the potential for harm from consumption of GTE and products containing GTE.

2. COMPOSITION OF PHYTOCHEMICALS AND HEAVY METALS IN GREEN TEA LEAVES AND GREEN TEA

On a dry weight basis, the main constituents of tea leaves are polyphenols [39], particularly flavan-3-ols (Table 1). Other constituents of the tea leaf include fiber, carbohydrates (cellulose, pectin, glucose, fructose, sucrose), proteins (enzymes constituting an important fraction), amino acids (L-theanine or 5-N-ethylglutamine, glutamic acid, tryptophan, glycine, serine, aspartic acid, tyrosine, valine, leucine, threonine, arginine, lysine), alkaloids (principally caffeine and smaller amounts of theophylline and theobromine), minerals and trace elements, pigments (chlorophyll and carotenoids), volatile compounds (aldehydes, alcohols, esters, lactones, hydrocarbons), lipids (linoleic and α-linolenic acids), sterols (stigmasterol), and vitamins (vitamins A, C, E, and K as well as several B vitamins)[40-43]. The composition of tea leaves varies widely, being substantially dependent on agricultural, botanical and genetic factors – such as climate, season, growing location, horticultural practice, cultivar, age of the leaves and the plant – and post-harvest factors like the conditions and duration of storage.

The flavan-3-ols in tea leaves, green tea infusions and extracts include EGCG, (-)-epicatechin (EC), (-)-epicatechin gallate (ECG), (-)-epigallocatechin (EGC), (+)-catechin (C), and (+)-gallocatechin (GC) [39] (Table 2). The chemical structures of these compounds are presented in (Fig. 1). Other flavonoids and phenolic acids are present at <1% of dry weight. The major flavonols in tea are glucose or rhamnose conjugates of quercetin and kaempferol [44]. Phenolic acids in tea leaves include quinic acid esters of gallic acid and caffeic acid [44]. The polyphenols are not evenly distributed in the tea plant with the leaf bud and first leaves generally containing the highest concentrations [45]. Flavan-3-ols are concentrated in mesophyll cells of the leaf structure in close proximity to epidermal cells [46].

EGCG is the most abundant flavan-3-ol in tea leaves and green tea infusions, but with its concentration varying about 8-fold between teas and tea preparations [47-54] (Table 2). Though showing less variation in concentration, the next most abundant tea flavan-3-ols are EGC and ECG. In green tea infusions, only EGCG and EC are present consistently. While the content of EGC and ECG in green tea is lower than EGCG, their values in the infusion appear disproportionally larger. The range of caffeine concentrations in green tea infusions is comparable to that of EGCG, suggesting that flavanols and alkaloids leach out the tea matrix in different degrees during steeping or brewing. These variations can be attributed not only to intrinsic factors but also to the methods employed in extraction of flavonoids from tea leaves and green tea for quantification.

Tea trees accumulate significant amounts of trace metals in their leaves [55]. While some of these minerals are essential for human health, some might be associated with toxicity when consumed in large amounts or for long durations if the leaves are consumed directly or dissolved in the infusion. Among the metals and elements listed in the (Table 3) [55], aluminum, arsenic, cadmium, fluoride, and lead can induce acute or chronic toxicity if sufficient exposure is achieved. However, while the amount of these metals in fresh tea leaves and green tea can exceed recommended safe levels for these elements, they are not fully extracted by aqueous or alcoholic solvent, and are present in much lower amounts in infusions [55]. The level of arsenic in green tea infusions is 1000-fold larger than the level generally found in U.S. drinking water (2.0 μg/L, although levels of 0.138 to 1,700 μg/L have been measured in surface water in the U.S.) [56]. As the minimal risk level (MRL) for acute-duration oral exposure (≤14 d) to inorganic arsenic is 5 μg/kg bw/d, daily consumption of 230 mL green tea infusion containing 1.53 μg/L arsenic (the highest identified arsenic content [55]) for 14 d would not attain the MRL for a 70 kg adult. The U.S. Environmental Protection Agency (EPA) requires water suppliers to limit the cadmium concentration in water to <5 μg/L, which is 7- to 47-fold larger than cadmium concentrations reported in green tea infusions. The content of lead in green tea infusions is comparable to levels in surface water and groundwater in the US at 0.005 and 0.03 μg/mL, respectively [57]. Most drinking water supplies in the U.S. contain <5 μg/L of chromium [58]. Green tea infusions may become a significant source of chromium among those who drink a substantial volume daily; the Adequate Intake for chromium established by the Institute of Medicine is 25 and 35 μg/d for women and men, respectively. The majority of chromium in tea is the less-toxic trivalent form, whereas water-soluble hexavalent chromium ranges from 0-10% of total chromium from green tea leaves, depending on its origin; the highest amount of hexavalent chromium found in a cup of green is ~70-fold lower than the maximum acceptable concentration [59]. Importantly, the level of metals and elements in GTE can be readily identified and controlled by parameters in its specification.

Table 1. Approximate composition of tea leaves[1]

Constituents	% dry weight
Polyphenols	30
Fiber	26
Protein	15-20
Carbohydrate	5-7
Minerals	4-9
Alkaloids (caffeine)	2-4
Amino acids	1-4
Pigments	Trace
Lipids and sterols	Trace
Volatile compounds	Trace
Vitamin	Trace

[1]*Chaturvedula and Prakash [39]*

Table 2. Range of flavonoids and alkaloid contents in tea leaves, green tea, and green tea infusion[1,2]

Constituents	Fresh tea leaves	Green tea leaves[3]	Green tea in fusions[4]
	mg/g dry weight		mg/L
Flavonoids			
(-)-EGCG	68 - 134	13 - 113	95 - 712
(-)-EGC	22 - 53	ND - 45	ND - 471
(-)-ECG	5.9 - 18	5.7 - 51	ND - 1508
(-)-EC	1.9 - 11	1.9 - 21.1	25 - 216
(+)-GCG	ND - 2.5	ND - 7.1	ND - 9.3
(+)-CG	ND	ND - 3.7	
(+)-C	ND - 1.2	ND - 1	ND - 688
EGCmG	NR	0.3 - 1.2	NR
Quercetin-rhamnoside-galactoside	trace - 1.0	NR	NR
Quercetin-rhamnoside-glucoside	0.5 - 1.9	NR	NR
Rutin	0.6 - 2.6	NR	NR
Kaempferol	NR	NR	3.6 - 6.4
Myricetin	NR	NR	4.8 - 7.3
Quercetin	NR	NR	13.4 - 20.7
Thearubigins	NR	NR	ND - 221
Alkaloids			
Caffeine	16 - 28	26 - 39	99 - 338
Theobromine	0.2 - 1.6	NR	7.6 - 86

[1]*Abbreviations: C, catechin; CG, catechin gallate; EC, epicatechin; ECG, epicatechin gallate; EGC, epigallocatechin; EGCG, epigallocatechin gallate; EGCmG, epigallocatechin 3-O-methylgallate; GCG, gallocatechin gallate; ND, not detected; NR, not reported*

[2]*Compiled from Bhagwat et al. [47]; Cabrera et al. [48]; Friedman et al. [49]; Komes et al. [50]; Nishitani and Sagesaka [51]; Peterson et al. [52]; Wang et al. [53]; Xu et al. [54]*

[3]*Includes data from loose tea and tea bags*

[4]*Tea infusions made by brewing 2 g tea in 200 mL at 80°C for 3 min*

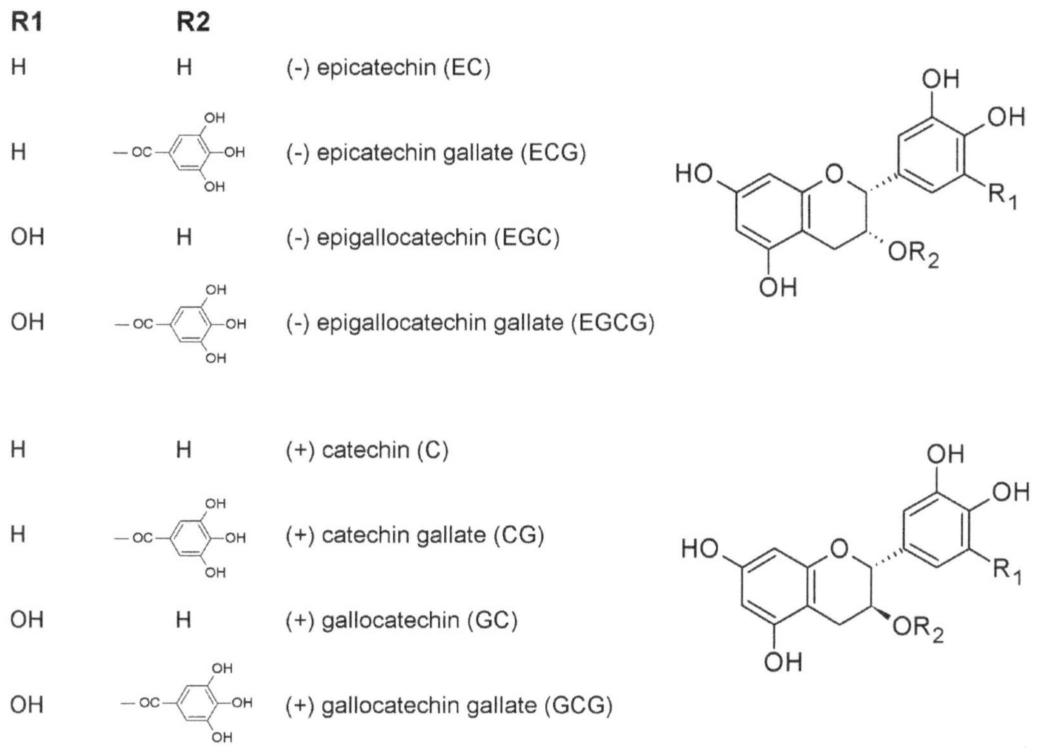

Fig. 1. Chemical structure of main flavan-3-ols in tea leaves

Table 3. Range of trace metal and elemental content in tea leaves, green tea, and green tea infusion[1]

Trace metals	Fresh tea leaves	Green tea	Green tea infusion[2]
	μg/g dry weight		μg/mL
Aluminum	13 - 11,981 (177)[3]	211 - 4074 (41)	0.7 - 6 (24)
Arsenic	0.021 - 0.073 (13)	trace - 1.66 (536)	0.02×10^{-3} - 1.53×10^{-3} (17)
Cadmium	ND[4] - 1.27 (63)	0.013 - 0.114 (141)	0.04×10^{-3} - 0.24×10^{-3} (17)
Chromium	trace - 3.41 (103)	0.45 - 0.99 (2)	ND - 6.91×10^{-3} (17)
Copper	9.4 - 127 (59)	4.7 - 36.5 (169)	0.033 - 0.191 (50)
Fluoride	49 - 808 (33)	49 - 344 (14)	0.59 - 2.52 (20)
Lead	ND - 14.5 (181)	0.11 - 1.93 (141)	0.004 - 0.032 (17)
Manganese	ND - 3154 (56)	211 - 2081 (33)	0.52 - 10.09 (54)
Nickel	6.6 - 14 (43)	ND - 9.35 (28)	0.04 - 0.269 (21)

[1]Summarized from Karak and Bhagat [55]; [2]The range of trace metal and other element contents in green tea infusion is due partly to the different brewing methods, including tea:water ratio, water temperature, and brewing duration; [3]Values in parentheses are the number of samples tested; [4]ND, not detected

3. ADVANCED EXTRACTION TECHNOLOGIES FOR PLANTS/BOTANICALS

To obtain natural products from plant materials, effective extraction techniques are required to remove unwanted constituents and concentrate bioactive compounds. Traditional techniques, such as Soxhlet extraction, have been used for many decades, while novel techniques have been developed more recently to improve extraction efficiency and the cost-effectiveness of the process. It is worthwhile noting that pretreatment of the raw plant material, e.g., breaking, crushing, and milling, can significantly affect the kinetics of solvent extraction [60]. Extraction efficiency is related to the solvent to leaf ratio and depends substantially on polarity of the constituent compounds and the type of solvent. For example, acetone, ethyl acetate, acetonitrile, isopropanol, ethanol, methanol, and water are preferred for extraction of the more amphipathic compounds, such as polyphenols and their glycosides and hexane, petroleum ether, benzene, and chloroform are utilized to maximize extraction of lipophilic constituents, such as fats and essential oils [61]. Extraction solvents can be chosen based on solvent selectivity, capacity, reactivity, stability, solubility, regeneration capability, and toxicity [62]. Several additional novel extraction technologies have emerged that may be useful for the extraction of green tea constituents. For example, ionic liquids have been used for extraction of the stilbene resveratrol from the dried rhizome of *Polygonum cuspidatum* [63]. A surfactant and micelle approach has been used to extract flavonoids from licorice root [64]. Supercritical CO_2 or dimethyl ether with solvent modifiers can be used to decaffeinate tea and extract flavan-3-ols [65]. This section provides a brief overview of novel extraction methods suitable for consideration of plant polyphenols.

3.1 Accelerated Solvent Extraction (ASE)

ASE is also known as pressurized solvent extraction, subcritical solvent extraction or pressurized liquid extraction [66]. ASE is an automated process and widely used in the extraction of functional components from botanicals and plants [67-71]. The high temperature and pressure applied in this protocol not only facilitate extraction efficiency but also reduce solvent use and extraction time by increasing the solubility of target compounds and the rate of solvent diffusion into the plant matrix as well as decreasing solvent viscosity and surface tension [69,72]. Extraction efficiency for thermally labile components can decrease when temperatures are set too high, resulting in a lower quality extract [71]. Since ASE is a relatively fast process, prolonged extraction time does not raise the yield [70-74]. Importantly, ASE often allows for the use of environmentally safe solvents, such as water and aqueous ethanol to achieve more exhaustive extractions of plant materials relative to use of these same solvents under normal extraction conditions. In general, smaller particle sizes of the plant material combined with larger volumes of solvent can increase the amount of functional components dissolved, resulting in greater extraction efficiency [73,75,76].

3.2 Supercritical Fluid Extraction (SFE)

Supercritical fluid is a phase between gas and liquid, and its temperature and pressure are above the critical point. Its density is similar to liquid but its viscosity is similar to gas, resulting in a higher diffusion coefficient than liquid. Due to its high dissolving capacity for many compounds, including phytochemicals, supercritical fluid can be used as an efficient extraction solvent. The principal SFE solvent is CO_2 due to its relatively low critical conditions (30.9ºC and 73.8 bar), profile of safe use, and relatively low cost. Thus, SFE with CO_2 presents a good method for the extraction of thermally labile phytochemicals [77,78]. Supercritical CO_2 has a very high diffusion coefficient, leading to a much faster extraction rate than conventional solvent extraction [79]. Due to its non-polarity, CO_2 may not be an efficient solvent for extraction of polar bioactives such as phenolic compounds. However, the efficiency of SFE for polar compounds is significantly increased when combined with organic solvent modifiers like ethanol [80,81]. While extracted phytochemicals can be easily recovered by decreasing the pressure of the supercritical CO_2 [78], the high cost and sophisticated operation conditions limit the broad utilization of SFE. However, it is noteworthy that SFE is the method of choice for the commercial production of quality decaffeinated tea.

3.3 Ultrasound-Assisted Extraction (UAE)

Ultrasound waves can change the physical and chemical properties of plant material and induce cavitations causing an increase in pressure and temperature close to the plant surface, disrupting cell walls and enhancing the release of intracellular compounds into the extraction solvent [82,83]. UAE can also cause swelling of the plant material that can facilitate the extraction of phytochemicals. UAE is a simple and cost-effective method which has been shown to improve the extraction of polyphenols, though extraction efficiency is dependent on the nature of the plant matrix [66,72]. UAE is also suitable for thermally labile compounds because satisfactory extraction can be achieved at relatively low temperature [72].

3.4 Enzyme-Assisted Extraction (EAE)

The principle of EAE is use of enzymes, such as cellulases, pectinases and hemicellulase, to break the plant cell wall to facilitate the extraction of targeted constituents from plantmatrices [84]. EAE has been used for extraction of proteins, phenolics, oils, and carotenoids [85-89]. Although EAE may offer better extraction efficiency in comparison with conventional extraction methods, this method extends considerably the time for completion of conventional extraction protocols and may not be better than other novel techniques using pre-extraction treatment, e.g., UAE [89,90]. During EAE, specific enzymes are needed for different raw materials, which require the information on the exact composition of the cell walls. In one study, UAE pretreatment led to a better and faster extraction of phenolic compounds from acerola than an EAE-added protocol [90]. Additional factors that limit the application of EAE to green tea include the relatively high cost of enzymes and environmental sensitivity regarding the use of these enzymes.

3.5 Pulsed Electric Field (PEF) Assisted Extraction

PEF assisted extraction uses an electrical field to induce electroporation of cell membranes which results in an increased permeability of extraction solvent into the plant matrix and extraction efficiency [66,91,92]. PEF treatment improves the extraction of phytochemicals, including polyphenols like anthocyanins and tannins [91]. Compared to other pre-treatments such as UAE, PEF assisted extraction requires less time and lower energy costs[93]. PEF has been used as a pre-treatment before pressing to enhance the yield and quality (turbidity and odor intensity) of juices from different fruits and vegetables [94,95]. PEF may yield comparable or better extraction efficiency than UAE and requires less time (seconds vs. minutes) and energy than UAE [91,96].

3.6 Microwave Assisted Extraction (MAE)

As electromagnetic radiation, microwaves can penetrate plant materials and interact with polar molecules such as water to generate heat [72]. The heat generated internally in the plant materials disrupts cell structure and facilitates the dissolution of phytochemicals from the plant matrices [95]. Moreover, homogenously increased temperature of extraction solvent and plant materials by MAE generally leads to enhanced extraction efficacy. MAE also reduces extraction time in comparison to conventional extraction methods [97-99]. However, in certain cases, unwanted products can be generated during MAE, especially when the temperature is too high [100]. The major advantage of MAE is that it offers significant reductions in extraction time and solvent usage with a similar or even better extraction efficacy in comparison to the conventional extraction methods [66].

4. TECHNOLOGIES FOR GREEN TEA EXTRACTION

On a commercial scale, extractions of green tea for beverages or food supplements are typically conducted using aqueous, ethanolic or supercritical fluid solvents with either batch or continuous processes [101]. A variety of laboratory-scale methods for GTE preparation have been reported and demonstrate that various solvents, temperatures, and other process variables contribute to a wide range of constituents in the final product. Market studies of commercial GTE have reported a marked variation in the content of EGCG (11.9-fold), polyphenols (2.6-fold), and caffeine (6.5-fold) [102-104] (Table 4). It is noteworthy that the variation in the phytochemical content of GTE in these studies is equivalent or less than that reported in surveys of commercial green teas used for infusions.

4.1 Continuous or Batch Aqueous Extraction

A single aqueous batch extraction for GTE most closely mimics the typical home preparation of green tea. Phytochemical yields have been reported for both batch and continuous extraction processes, including reverse flow continuous extraction. For batch processes, time, temperature, and tea:water ratios have been manipulated in laboratory studies. Total flavan-3-ol content plateaus from 30 to 120 min during aqueous batch extractions of green tea [50,105]. Vuong et al. [105] reported that EC content plateaued at 30 min, but catechin content continued to increase to 120 min in a batch aqueous process. Similarly, L-theanine aqueous batch extraction was maximized at 80°C for 30 min [106]. Aqueous shaking extraction of flavan-3-ols from dry green tea leaves at 25°C peaked at 2 h, but had only minor differences from 0.5 to 24 h with a maximum 8% difference between 2 and 6 h [107]. Reflux extraction of green tea leaves beyond 30 min up to 3 h at 80°C did not yield significantly more flavan-3-ols or caffeine [107]. Longer aqueous extraction resulted in lower caffeine:flavan-3-ol ratios [108].

Table 4. Fold-variation in phytochemical content of GTE and commercial green tea samples on a weight/weight basis

GTE constituent	GTE dietary supplements, fold variation (n=10)[1]	Commercial green teas, fold variation (n=23)[2]
Caffeine	6.5	79.7
EGCG	11.9	24.1
EGC	18	142
ECG	16.9	22.7
EC	7	26
Theophylline	-[3]	47.5
Total Polyphenols	2.6	-

[1]*Seeram et al. [103]*; [2]*Friedman et al. [104]*; [3]*not analyzed*

The water:tea ratio also impacts the yield of flavan-3-ols and L-theanine. L-theanine yield is maximized at water:tea ratios >20:1 and may introduce differences of≥25% [106]. In contrast, water:tea ratios of ≥50:1 yield a maximum content of flavan-3-ols, 80% greater than 10:1 ratios [105].

A continuous or semi-continuous aqueous extraction of green tea leaves increases polyphenol yields beyond that obtained with batch extraction. Following 3 min steeping of green tea leaves, re-extraction slightly increased the proportion of flavan-3-ols, as a third 3 min extraction had 33, 22, and 15% of flavan-3-ols, phenolic acids, and methylxanthines, respectively, of the first extract [50]. Similarly, successive aqueous extraction yielded 11% more caffeine and theobromine at the third extract of loose green tea leaves relative to the first [109]. Flavan-3-ols had a cumulative 91% extraction efficiency compared to an 80% efficiency for caffeine at 95°C following four sequential 10 min extractions at 95°C [110]. This result is significantly greater than the 65% and 59% yield of flavan-3-ols and caffeine, respectively, after the first 10 min of steeping [110]. While fluorosis from tea consumption is extremely rare, a semi-continuous extraction process can be utilized to reduce fluorine content of GTE. Discarding the initial 20 s infusion of green tea at 50°C and then extracting in boiling water for 5 min reduced the fluorine content by 26% [111]. Successive 20 min boiling aqueous extractions of green tea yielded ~30% less aluminum after each additional extract [112].

Since the flavan-3-ol content peaks at 0.5 to 2 h in batch extractions, increasing yields of semi-continuous extraction are expected. Although not compared in the reports discussed above, re-extraction likely provides an advantage of increasing rates of diffusion of solutes from leaves to water. Ultimately, time and temperature of aqueous extractions can be manipulated in either continuous or batch processes to change the yield of the constituent phytochemicals and minerals.

4.2 Extraction Temperature

Extraction temperature plays an important role in the yields offlavan-3-olsfrom green tea with efficiency varying dependent on the individual compound [113]. Perva-Uzunalić [110] obtained maximum aqueous extraction efficiency of flavan-3-ols after 20 min at 80°C and 10 min at 95°C, suggesting an interactive effect of temperature and time on extraction efficiency. Response surface methodology, which is generally employed to generate an optimal response by conducting a sequence of designated experiments, has been used characterize the yield of bioactive components of GTE. Zhang et al. [114] reported that steeping 1 g tea powder in 16 mL water for 40 min at 96°C extracted the most EGCG, theanine, and total phenols. A potential degradation of flavan-3-ols may occur when a protocol includes both high temperature and extended extraction. For example, aqueous extraction of green tea for 7 h at 98°C reduced the flavan-3-ol content of GTE by 20% [115]. Increasing temperature from 80 to 95°C augmented epimerization of EGCG to (-) gallocatechin gallate (GCG) by 52% from 0.25 g/100 g green tea leaves [115]. The caffeine content of aqueous extracts reaches a plateau at 80°C from 20-40 min extraction, and then decreases slightly at 90°C [113]. Thus, temperature employed during extraction not only affects extraction efficiency of flavan-3-ols but can also contribute to an altered flavonoid profile through epimerization.

4.3 Extraction Solvent

Laboratory-scale preparations of GTE have utilized a variety of solvents with various effects on extract yields [107,110,116] (Table 5). Friedman et al. [104] compared an 80% ethanol extraction to an aqueous extraction of 24 green teas and found that aqueous alcohol extraction yielded 36% more flavan-3-ols, 12% more caffeine, ~50% more theobromine, and 30-200% more theophylline than water alone. As shown in (Table 5), the percentage of ethanol in the extraction solvent is crucial to yield of flavan-3-ols, but not necessarily caffeine, with 40% ethanol in water being the best as compared to all other aqueous ethanol solvents and methanol, acetonitrile, and acetone [107,110,116]. Similarly, Rusak et al. [117] noted the yield offlavan-3-ols from loose green tea extracted for 30 min was highest with 40% ethanol compared to 10 and 70% ethanol, and water. The 40% ethanol extract recovered 18% EGC, 74% EGCG, 40% GCG, and 97% ECG more than water extraction. Thus, aqueous ethanol provides a larger ratio of flavan-3-ols to caffeine than other extraction solvents.

In contrast, the flavonol and proanthocyanin yield from green tea leaves appear only marginally affected by ethanol or aqueous ethanol extraction [110]. However, acetonitrile and acetone extraction provides 33 to 83% flavonols and proanthocyanidins compared to water extraction. Recovery of caffeine and theobromine was greatest with 70 and 15% methanol in water, respectively, following three 10-min sequential extractions of green tea leaves [118]. Theophylline was only recovered in the 70% methanol extract, but not in methanol alone, water, acetonitrile or acetone [118].

Less information is available about how solvents affect other green tea constituents. Mossion et al. [119] reported that mineral content of water decreased extraction of aluminum, total organic carbon, and total polyphenols. Kanrar et al. [120] demonstrated an ethyl acetate-cyclohexane mixture (9:1, v/v) was effective at extracting pesticide residues from green tea. Pollution associated with industrialization near tea cultivation areas can increase the presence of heavy metals, particularly lead, in raw tea leaves. Different solvent extractions vary in their capacity for extracting lead from tea plants [121,122]. Sullivan et al. [123] have noted that while hydroethanolic solvents are not completely efficient in the extraction of heavy metal accumulations, they can be an effective decontamination step in herbal product processing.

4.4 pH

The pH of aqueous extraction solvents varies from ~5.5 to 7.2 based on temperature and the quantity of green tea leaves utilized for extraction [124]. Yoshida et al. [124] reported that the aqueous pH was nearly constant at ~6.5 when the ratio of green tea leaves to water was between 0.5 and 3%. Stability of tea flavan-3-olsdecreases with increasing pH above pH 3, at which they are quite stable [107]. Thus, the pH of extraction solvent strongly determines extraction efficiency of these flavonoids from green tea leaves. The effect of pH varies during extraction, with a pH of 4.5 providing a maximum recovery of gallic acid, EGC, and EGCG, whereas ECG and EC are stable up to pH 6.5 [102]. Extraction at pH 7 results in greater proportions of EC and ECG to other flavan-3-ols [102]. GTE extraction held at 80°C for 20 min at pH 8, led to twice the epimerization of EGC to GC compared to pH 6 [124]. The mineral content of GTE is also affected by the acidity of extraction solvents. Water extraction with 0.12 µmol/L HCl provides a yield with 44% aluminum, 19% calcium, and 44% magnesium of the original amount in the green tea leaves [112].

Table 5. Effect of extraction solvent on green tea extract phytochemical yield, as fold-difference of water extract[1]

Phytochemical	Aqueous ethanol				Methanol	Acetonitrile	Acetone
	20%	40%	50%	80%			
EGCG	2.76	3.32	3.26	1.92	1.3-3.2	-	-
Catechins	1.46	1.61	1.46	0.79 - 1.15	1.17	1.24	0.63
Flavonols	-[2]	-	0.91	1.02	0.97	0.39	0.33
Proanthocyanidins	-	-	0.97	0.96	0.91	0.83	0.83
Caffeine	1.03	1.03	0.97	-	-	-	-

[1]*Compiled results from Perva-Uzunalić et al. [110]; Meterc et al. [116]; Choung and Lee [107].*
[2]*not determined*

4.5 Other Extraction Technologies

Other extraction technologies for preparing GTE may modulate the yield offlavan-3-ols. In a semi-continuous SFE using CO_2, Sosa et al. [125] tested whether encapsulation with a poly-ε-caprolactone, a biodegradable particle, applied during the solvent removal step would affect the constituent profile of the final GTE. They found that total phenols varied 60-100% of the maximum theoretical composition, but caffeine content was reduced to 13%. Ultrasonic extraction of green tea leaves for 1 h had 5% less flavan-3-ols, mainly as EGC and ECG, and the same caffeine content relative to a 2 h aqueous extraction [107]. Conventional hot-water steeping of green tea yielded 55% magnesium, 260% calcium, and a similar percentage of aluminum compared to focused microwave oven heating [112]. Ionic liquids have also been used to extract caffeine and theophylline from commercial tea drinks [126].

4.6 Procedures for Solvent Removal

GTE is dried prior to distribution. Spray drying, drum drying, or lyophilization can be used to process GTE. Excessive heat during a high pressure spray drying process of 145°C pre-expansion temperature and 79°C resulted in 88% less polyphenols relative to optimized conditions at ≤130°C [116]. Temperatures during spray drying between 33 and 65°C did not affect the amount of total polyphenols but the residual water content increased with lower temperatures, a condition which is unacceptable for commercial applications.

5. POSTPRODUCTION HANDLING OF GREEN TEA

5.1 Grinding

The size of the dry tea leaf particles also affects the yield of extracted compounds as smaller particle sizes increase surface area available for extraction. Aqueous extraction of green tea leaves with a particle size ≤1 mm have ~10% greater yields of flavan-3-ols [105]. Gains in fluoride yields with smaller particle sizes relative to flavonoids maybe exploited to limit fluorine in GTE [111]. Caffeine and total flavonoids, but not EGCG, are dependent on particle size for supercritical CO_2 decaffeination [65].

5.2 Storage

The duration that green tea leaves are stored before use may affect GTE composition. Friedman et al. [104] found that after 6 mo storage at 20°C, EGCG concentration in green tea leaves decreased by 28% and total flavan-3-ols decreased by 32%. Storing tea leaves under vacuum or nitrogen and/or low humidity and temperature can minimize flavonoid oxidation.

5.3 Decaffeination

Green tea may be decaffeinated prior to extraction. At optimal conditions for supercritical CO_2 decaffeination of ground green tea leaves, 99.6% caffeine, 40.5% flavan-3-ols, and 43.1% chlorophyll are extracted using ethanol as a co-solvent [127]. Ethanol is superior to water in this regard, and an 50% aqueous ethanol modifier for supercritical CO_2 decaffeination resulted in 70% caffeine and 6% flavan-3-ols being removed from green tea powder [65]. Perva-Uzunalic et al. [110] noted that supercritical CO_2 reducesflavan-3-ols by 17% and flavonols by 10% from the starting material. Up to 43% of chlorophylls are removed after 120 min of supercritical CO_2 decaffeination with an ethanol co-solvent [128]. While the loss of polyphenols can be minimized during the decaffeination process, commercial decaffeinated green teas generally have a lower content of flavan-3-ols than non-decaffeinated green tea [102].

6. POTENTIAL HEALTH BENEFITS OF GREEN TEA INFUSIONS AND EXTRACTS

Before discussing the reported potential for adverse effects of GTE, it is worth briefly noting the potential health benefits associated with green tea and GTE. A wealth of basic research as well as evidence from observational studies and randomized clinical trials indicate the consumption of green tea infusions or GTE to be associated with an improvement in intermediary biomarkers of chronic disease and/or a reduced risk of some forms of cancer, cardiovascular disease, type 2 diabetes mellitus, neurodegenerative conditions like Alzheimer's disease and Parkinson's disease, and other age-associated diseases. These putative benefits appear due principally to the array of bioactivity of the flavonoid constituents of green tea, particularly the flavan-3-ols, including antioxidation, anti-inflammation, glucoregulation, anti-proliferation, energy regulation, and modulation of signal transduction pathways [16,129-135]. A brief summary of selected systematic reviews and meta-analyses of the association between green tea and health outcomes are presented in (Tables 6 and 7). In addition, green tea infusions and GTE have a modest effect on weight loss and/or weight management, though this benefit appears due principally to an interaction between the constituent flavonoids and caffeine. Randomized clinical trials indicate green tea flavonoids have a positive impact on blood glucose, suggesting a potential benefit to people with impaired glucoregulation and an increased risk of type 2 diabetes. Evidence from clinical studies also indicates a modest effect of green tea in decreasing serum total cholesterol and LDL-cholesterol [155]. Data from observational studies showing green tea is associated with a reduced risk of stroke is consistent with results from clinical trials. While rodent model studies reveal a marked chemopreventive effect of green tea on several forms of cancer, results from observational studies and from clinical trials examining intermediary biomarkers are mixed and dependent on cancer type, sex, and study design.

7. SAFETY PROFILE OF GREEN TEA EXTRACTS

Green tea consumption has a long, safe history of use as a beverage and green tea preparations are deemed GRAS for use in foods; however, concerns about the potential for hepatic toxicity of GTE prepared by aqueous solvent extractions, particularly hydro-alcoholic extracts, have been raised. As discussed above, the polyphenol profile of alcohol extractions of green tea leaves appears to affect overall GTE composition, and the extraction solvent causes a wider effect on more than processing steps or raw materials. Based on laboratory analysis of 17 to 54 samples of quality grade commercial green tea infusions (Table 3), aqueous and hydro-alcoholic extracts of green tea are expected meet guidelines for the suggested safe levels of metals in herbals, food ingredients, and dietary supplements (Table 8). For example, the aluminum content reported in green tea infusions (0.7 - 6 mg/mL) is ~33-fold less than CODEX standards. Depending on the location that a tea is grown, the content of arsenic, cadmium, copper, and lead in dry tea may exceed recommendations for exposure to these elements, but are not fully extracted by aqueous or alcoholic extraction, and are present in much lower amounts in infusions [55]. Reports on the heavy metal content of commercial or laboratory-prepared GTE are lacking in the literature, however, commercial products are commonly controlled for contaminants inclusive of heavy meals through established quality specifications in accordance with applicable current Good Manufacturing Practices (cGMP). Upon drying, residual solvents are not expected to be at concentrations above toxic threshold due to the drying process (Table 9). While tea leaves and infusions are unlikely to have residue solvents as their processing typically does not include water or organic solvents, data from these sources and GTE are not readily available from literature. While decaffeinating tea with methylene chloride may leave residual traces, it is excluded from import into the United States and some other countries. In contrast, ethanol is regarded as a solvent with low toxic potential. There is no evidence that hydro-alcoholic processes extract significantly different proportions of bioactive or toxic compounds from green tea that would lead to toxicity. However, components of GTE other than polyphenols and caffeine are rarely reported in animal experiments or case studies of toxicity, so data on these constituents of GTE are quite limited.

Green tea flavonoids consumed in a phospholipid complex are better absorbed than in a free form but do not appear to be associated with toxicity [160]. Nanoparticles and other technologies for enhancing the bioavailability and/or distribution of tea flavonoids have been tested in animal models without evidence of harm but remain untested in humans. In humans, green tea flavan-3-ols appear similarly bioavailable from infusions and GTE. Importantly, foods co-consumed with tea may impair maximum plasma flavonoid status and the time required to achieve maximal concentrations [161]; such actions would likely reduce any potential toxicity associated with tea polyphenols [162]. Indeed, consumption of GTE in a fed state resulted in lower and less variable toxicity than found under fasted conditions in dogs fed GTE at 50 to 1000 mg/kg body weight (bw)/d [163,164]. Further, clinical toxicokinetic studies by Ullmann et al. [165,166] confirmed that GTE under fasting conditions or following a single bolus led to more marked increases in bioavailability and plasma concentrations of EGCG than administration with split doses or with food. Rarely, moderate consumption of aqueous green tea infusions, aqueous GTE, and hydro-ethanolic GTE have been associated with liver injury in humans but confounding factors and effect modifiers make it impossible to assign a direct causal relationship to individual or combined tea flavonoids or related polyphenols [167]. Evidence regarding toxicity from tea and tea polyphenols is described below.

Table 6. Meta-analyses and systematic reviews of randomized clinical trials on the health benefits of green tea or green tea catechins in anthropometry, glucoregulation, lipid profile, blood pressure, and coronary heart disease risk[1]

Authors	Number of study (n)	Subject (n)	Duration	Dosage in RCT	Objectives	Findings
Anthropometry						
Jurgens et al. [132]	RCT (14)	1562	12-13 wk	140.85-1206.9 mg GTC/d	BW in overweight or obese adults	BW: ↔ outside Japan; ↓ in Japan. WC: ↔
Hursel et al. [136]	RCT (11)	1188	12-13 wk	270-1207 mg GTC/d	BW	BW: ↓ Effect modification by caffeine and ethnicity
Phung et al. [137]	RCT (15)	1243	8-24 wk	141-1207 mg GTC/d	BMI, BW, WC, WHR	GTC with caffeine BMI: ↓ BW: ↓ WC: ↓ WHR: ↔ GTC alone: ↔
Glucoregulation						
Zheng et al. [138]	RCT (22)	1548	3-24 wk	236-1207 mg GTC/d	Glucoregulation	FG: ↓ FI: ↔ HbA1C: ↔ HOMA: ↔
Liu et al. [139]	RCT (17)	1133	4-14 wk	208-1207 mg GTC/d	Glucoregulation	FG: ↓ FI: ↓ HbA1C: ↓
Lipid profile						
Kim et al. [133]	RCT (20)	1415	3-24 wk	145-3000 mg GTC/d	Lipid profile	TC: ↓ LDL-C: ↓ HDL-C: ↔ TG: ↔
Zheng et al. [140]	RCT (14)	1136	3-24 wk	150-2500 GTC mg/d	Lipid profile	TC: ↓ LDL-C: ↓ HDL-C: ↔ TG: ↔
Coronary heart disease						
Hooper et al. [141]	RCT (170) [12 green tea]	6557 (515 green tea)	0-52 wk (0-12 wk for green tea)	Not available	CHD risk	LDL-C: ↓ BP: ↔
Taubert et al. [142]	RCT (10) [1 green tea]	516 (343 green tea)	2 mo	1 tea bag/d, 544 mg phenols	BP	BP: ↔

[1]Abbreviations: BMI, body mass index; BP, blood pressure; BW, body weight; CHD, coronary heart disease; FG, fasting blood glucose; FI fasting insulin; GTC, green tea catechins; HbA1C, glycated hemoglobin; HDL-C, high density lipoprotein cholesterol; HOMA, homeostatic model assessment; LDL-C, low density lipoprotein cholesterol; RCT, randomized clinical trial; TC, total cholesterol; TG, triglyceride; WC, waist circumference; WHR, waist and hip ratio; ↓, decrease; ↔, no change

7.1 Evidence from Cell Studies

Primary hepatocytes from adult male Wistar rats were used to investigate the potential toxicity of hydro-ethanolic GTE constituents [168]. Hepatocellular necrosis was observed at doses of 1 to 3 mg GTE/mL medium. However, individual flavan-3-ols, caffeine, and theanine were not cytotoxic up to 3 mg/mL medium except for the observation that EGCG reduced cell viability at the 1 mg/mL dose only. The lipid-soluble fraction did not appear to account for increased cytotoxicity. With an LD_{50} of 200 μM, Galati et al. [169] reported that EGCG was the most potent of green tea phenolics in inducing mitochondrial membrane potential collapse in isolated rat hepatocytes. In contrast, Nishikawa et al. [170] found a benefit of EGCG at 50-100 μg/mL in stimulating apoptosis of human hepatocellular carcinoma cell lines. However, the direct relevance and ability to extrapolate these findings is limited as only sub-micromolar concentrations are observed *in vivo* following purified flavan-3-ol and flavonol administration to rodents and humans [171].

Table 7. Meta-analyses and systematic reviews of observational studies on the health benefits of green tea or green tea catechins in coronary heart disease, stroke, and cancer risk[1]

Authors	Number of study (n)	Subject (n)	Duration	Objectives	Findings[2]
Coronary heart disease					
Wang et al. [143]	Cohort (12), Case-control (6) [5 green tea]	416,676 (13,978 cases)	5-20 y follow up	CAD (CHD and MI)	RR=0.72 (0.58-0.89) 1 cup/d: ↓ 10% CAD risk
Stroke					
Arab et al. [144]	Cohort (8), Case-control (2), Cross-sectional (1)	213,897 (5,537 cases)	4-15 y	Total stroke	RR=0.79 (0.73-0.85, ≥3 vs.<1 cup/d)
Cancer					
Zheng et al. [145]	Case-control (20), Cohort (4)	7,376 cases, 487,894 control in case-control; 8,874,734 person-y in cohort	NA[3]	Esophageal cancer	Cohort: RR=0.77 (0.57-1.04, highest vs. non/lowest) Case-control: RR=0.70 (0.51-0.96)
Zheng et al. [146]	Case-control (8), Cohort (2)	3,731 (3,557 cases)	0-9 y	Esophageal cancer	Overall: ↔ Case control: RR=0.32 (0.10-0.54, high), 0.43 (0.21-0.66, medium), 0.45 (0.10-0.79, low) vs. non-drinker in female only
Kang et al. [147]	Cohort (7), Case-control (11)	223,044 (6814 cases)	NA	Stomach cancer	Overall: RR=0.86 (0.74-1.00) subgroup analysis (6 studies): RR=0.68 (0.53-0.87, ≥5 cup/d vs. lowest)
Myung et al. [148]	Case-control (8), Cohort (5)	254,869 (6,636 cases)	NA	Stomach cancer	RR=1.10 (0.92-1.32, highest vs. lowest consumption)
Wang et al. [149]	Cohort (6)	352,275 (1,675 cases)	NA	Colorectal cancer	Overall RR=0.90 (0.72-1.08) Shanghai: RR=0.70 (0.55-0.85, highest vs. lowest) Singapore: RR=1.36 (1.06-1.74)
Tang et al. [150]	Cohort (5), Case-control (7)	106,069 (5,495 cases)	NA	Lung cancer	RR=0.78 (0.61-1.00) in highest intake

Table 7 Continued............

Ogunleye et al. [151]	Case-control (9)	164,943 (5,617 cases)	NA	Breast cancer	RR=0.81, (0.75-0.88, >4 vs. ≤1 cup/d)
Tang et al. [152]	Cohort (2), Case-control (5)	107,764 (3,487 cases)	NA	Endometrial cancer	Overall: RR=0.85 (0.77-0.94) Cohort: RR=1.05 (0.85-1.28) Case control: RR=0.81 (0.71-0.93)
Zheng et al. [153]	Cohort (6), Case-control (7)	111,499 (3,608 cases)	NA	Prostate cancer	Overall: RR=0.72 (0.45-1.15) Case control: RR=0.43 (0.25-0.73) Cohort: RR=1.00 (0.66-1.53)
Nagle et al. [154]	Case-control (3)	6,092 (2,567 cases)	NA	Ovarian cancer	RR=0.58 (0.33-1.01), ≥1 cup/d vs. never/seldom

[1]Abbreviations: CAD; coronary artery disease; CHD, coronary heart disease; MI, myocardial infarction; RR, relative risk; [2]Values in parenthesis is the 95% confidence interval; [3] NA: not available

Table 8. Established safe levels of metals in herbals, foods, and food ingredients, and as provided by dietary supplements

Metal	Herbal products (w/w)	Foods and ingredients (w/w)	Dietary supplements (dose)
Aluminum	[1]	0.2 mg/mL[3]	-
Arsenic	-	0.1-0.5 mg/kg[4]	15 µg/d(inorganic)[5]
Cadmium	0.3 mg/kg[2]	0.3-400 µg/kg[4]	5 µg/d[5]
Copper	-	0.1-0.4 mg/kg[4]	-
Lead	10 mg/kg[2]	0.1-2 mg/kg[4]	10 µg/d[5,]
Mercury	-	1-100 µg/kg[4]	15 µg/d[5,6]

[1]Abbreviations: -, no established guidelines; [2]World Health Organization (WHO) guidelines for assessing quality of herbal medicines with reference to contaminants and residues, 2007 [156]; [3] EPA 40 CFR part 143 [157]; [4] Joint FAO/WHO Food Standards Programme, CODEX Committee on Contaminants in Foods, Sixth Session, Maastricht, The Netherlands. Working document for information and use in discussions related to contaminants and toxins in the GSCFTT (General Standard for Contaminants and Toxins in Foods and Feeds), March 26-30, 2012 [158]; [5]United States Pharmacopeial Convention Chapter 2232 Elemental Contaminants in Dietary Supplements, USP 36-NF31 S1 [159]; [6] Hg in total as methylmercury, 2 µg/d

Table 9. Recommended limits for residual solvents in U.S. dietary supplements with relevance to GTE[1]

Solvent category	Residual solvent	Recommended limits† (mg/d)
Class 2 (solvents to be limited)	acetonitrile	4.1
	chloroform	0.6
	dichloromethane	6
	hexane	2.9
	methanol	30.
Class 3 (solvents with low toxic potential)	acetic acid‡	50
	acetone	50
	ethanol‡	50
	ethyl acetate	50

[1]American Herbal Product Association Guidance Policies [159]. †Based on a 10 g dose; ‡ Higher levels permitted when intended as part of the product formulation

7.2 Evidence from Animal Models

In rodents, the No Observed Adverse Effect Level (NOAEL) with oral ingestion ranges from 65.5 to 2500 mg GTE/kg body weight (bw), providing intakes of flavan-3-ols of 120 to 3542 mg/kg bw/d and EGCG of 50 to 1500 mg/kg bw/d over periods ranging from 4 to 26 weeks

[163,172-178]. Applying an uncertainty factor of 100, Isbrucher et al. [175] suggested an acceptable daily intake (ADI) for EGCG, when adjusted to 100% purity, of 5.3 mg/kg bw, which would be approximately equivalent to 322 mg/d for the average individual. The lowest level of GTE (72% flavan-3-ols) causing liver toxicity in female F344/NTac rats was 1000 mg/kg bw/d following14 wk of consumption *via* gavage [172]. However, at 65.5 mg/kg bw/d, nasal toxicity was observed in B6C3F1mice, manifested bychronic inflammation of the olfactory epithelium, nerve atrophy, and metaplasia [172], though the relevance to humans of these signs is not clear. In contrast, consumption of 2500 mg/kgbw/d of an aqueous (but not fully characterized) GTE *via*gavage for 28 d did not cause any adverse effects on body or organ weights, serum enzymes, or histopathology in male or female ICR mice [173]. Chengelis et al. [179] also reported that male and female Sprague Dawley rats consuming up to 2 g/kg bw/d GTE *via* gavage for 28 d did not exhibit any adverse liver effects. They also reported no adverse effects of heat- and non-heat sterilized green tea flavan-3-ols, but noted that decaffeinated purified green tea flavan-3-ols reduced body weight. This animal model contrasts with human studies indicating that caffeine is an important contributor to weight loss associated with green tea and GTE [180]. Toxicological studies using rodents have not investigated the difference between hydrophilic and lipophilic components of GTE.

Highly purified flavan-3-olshave been observed to induce hepatotoxicity in rodents at lower doses than GTE. For example, in an acute study using male CF-1 mice, Lambert et al. [181] administered a single oral dose of EGCG at 1500 mg/kg bwand observed liver injury and death. The observed hepatocellular necrosis and apoptosis was associated with increased biomarkers of oxidative stress in plasma and liver. Plasma alanine transaminase was found to be increased 24 and 48 h after EGCG administration. Based on allometric scaling, Lambert et al. [181] calculated the minimum equivalency to this dose to be the EGCG amount from 10.5 cups of green tea, and noted this amount in humans would be typically consumed over the course of a day without any apparent adverse effect. Consistent with an apparent greater sensitivity of mice to flavan-3-ol toxicity, Takami et al. [178] fed male and female F344 rats a semi-purified GTE (81% polyphenols) at 5% (w/w) of the diet for 90 d and observed only an increase in plasma alkaline phosphatase and aspartate transaminase but no histopathological changes in the liver or other tissues.

Bun et al. [182] directly tested the effect of GTE derived using different extraction solvents on liver function in rats in two experiments. A preliminary 6-wk trial compared administration by gavage of 2500 mg/kg bw/d of GTE prepared with methylene chloride (from which potential non-polar hepatotoxins might be concentrated) and its distilled water vehicle (with 3% Tween 80) (control). A second 12-wk experiment compared the effects following gavage administration of an aqueous green tea extract (1400 mg/kg bw), an 80% hydro-ethanolic green tea extract (2000 mg/kg bw), or the vehicle (control).Signs characteristic of hepatotoxicity, including serum levels of lactate dehydrogenase, aspartate aminotransferase, alanine aminotransferase, alkaline phosphatase, and γ-glutamyl transferase, were not significantly affected by either aqueous or hydro-ethanolic extracts.

Long-term exposure to a concentrated GTE material (*c.a.* 76% catechins) has also been examined in rodents in a 2-y carcinogenesis study conducted by National Toxicology Program [183]. Only at the highest dose tested for each species (300 and 1000 mg/kg bw/d dose for B6C3F1 mice and Wistar rats, respectively), non-neoplastic lesions in the liver were observed, including hepatocyte necrosis, chronic inflammation, oval cell hyperplasia and hematopoietic cell proliferation. However, it is noteworthy that no increase in liver tumors was observed after chronic exposure to GTE in either species and that the manifestation of liver damage did not appear to progress to a more severe level when compared to those observed in the 14-wk study by Chan et al. [172].

Two studies of GTE toxicity studies have been conducted in dogs. Isbrucker et al. [174] fed Beagles doses of 50, 150 or 500 mg/kg bw/d of GTE (80% EGCG) and observed no adverse effects when administered over 13 wk in divided doses during a pre-fed state. Mortality resulted when the two higher doses were administered in a single bolus dose to fasted dogs; however, this model was considered an unrealistic comparison to the human condition. Kapetanovic et al. [163] also tested oral doses of GTE (56-72% EGCG) in Beagles at 200, 500, and 1000 mg/kg bw/d for 6.5 mo. Similar to the findings of Isbrucker et al. [174], they found that dosing in a fed state resulted in lower and less variable exposure as well as less frequent and lesser severity than found under fasting conditions. Importantly, Chow et al. [162] reported the same relationship in humans wherein volunteers consuming an acute dose of 400 to 12000 mg of GTE showed a >3.5-fold increase in the average maximum plasma concentration of EGCG when taken in a fasting condition than when taken with food.

In contrast to evidence from animal models of hepatotoxicity, several reports have shown a hepatoprotective action of green tea preparations. For example, Chen et al. [184] found a protective effect of intraperitoneal EGCG for 3 d at 50 and 75 mg/kg bw against acute carbon tetrachloride hepatoxicity in mice and Bruno et al. [185] found GTE reduced serum alanine transaminase and aspartate transaminase activity as well as hepatic steatosis in obese mice. Fiorini et al. [186] found EGCG at 85 mg/kg bw administered via intraperitoneal injection or drinking water protected against hepatic ischemia/reperfusion injury in steatotic mice. Dobrzynska et al. [187] and Zhang et al. [188] both reported that feeding green tea polyphenols to rats reduced the adverse effects of ethanol toxicity in the liver.

7.3 Evidence from Human Studies

As it is unethical to attempt to induce harm in human studies, there is no direct evidence of toxicity from green tea or GTE. Clinical toxicity from GTE or purified tea polyphenols can only be deduced from animal model data and/or observational evidence. Similar to rodents, humans extensively metabolize flavonoids to glucuronidated, methylated, and sulfated forms for relatively rapid excretion through bile, urine or feces and do not appear to accumulate these compounds in plasma or tissues for long periods [189]. In general, EGCG, EGC, and EC reach their maximum plasma concentrations within 1.5-2.5 h after consumption and return to the baseline values within 24 h [190]. Flavan-3-ol bioavailability is also dose-dependent in acute oral administrations, e.g., at 50, 100, 200, 400, 800 or 1600 mg EGCG [165]. Plasma and urine concentrations of flavan-3-ols are absent 24 h after an acute dose and their accumulation is essentially undetectable even after repeated doses of 200, 400, or 800 mg EGCG daily for ≥10 d [166,191]. However, these data do not imply an absence of bioactivity, as repeated provision of 800 mg/d green tea polyphenols resulted in an inhibition in the capacity of phase II enzymes in plasma and a consequent increase in the maximum of free EGCG concentration in plasma, but the kinetics of circulating EGCG were not altered.

Only a few, minor adverse side effects have been reported from randomized clinical trials of GTE, though most of these studies have had durations of less than 6 mo. Reports of minor to moderate side effects of GTE consumption include gastric irritation, diarrhea, nausea, and vomiting [192]. These untoward reactions appear unrelated to the chemical composition of GTE [28,130,132].

In 2008, up to 34 cases of liver damage associated with GTE had been reported world-wide and were summarized by Sarma et al. [28]. For most case reports, it is usually unclear whether the toxicity was due to GTE *per se* or related to other chemicals introduced during the extraction process, to concomitant medications, to other intentional or adulterant ingredients in the product or other factors. Though of variable quality, 27 of these reports were categorized according to the Naranjo causality algorithm scale as possible causality and 7 as probable causality. These cases were associated with the use of 6 different products varying markedly in product composition, solvent extraction, dose, and duration of use. The reported hepatotoxicity was coincident with increases in the activity of serum alanine transaminase and/or alkaline phosphatase or acute hepatitis-like syndrome [193]. However, the Naranjo scale has been found to lack validity and reproducibility when evaluating hepatotoxicity and is no longer recommended for such assessments [194]. The NIH Liver Tox database includes case reports of GTE extract and indicates no dose-response relationship between GTE intake and hepatotoxicity [193]. Many of the cases of liver toxicity associated with GTE were identified in Europe and linked to the use of alcohol extracts; however, hepatotoxicity has been reported for both aqueous and alcoholic GTE [28]. Liver injury has also been reported in at least 3 individuals drinking green tea infusions: an adult consuming 4-6 cups/d for 6 mo; an individual drinking an unspecified amount of green tea daily; and an individual consuming 6 cups/d of a commercial green tea infusion for 4 mo [33,194-196]. More recently, Teschke et al. [197] analyzed 185 case reports, spontaneous reports, review articles, and comments regarding liver injury associated with the use of 60 herbs, herbal drugs, and herbal dietary supplements, including GTE, and concluded that convincing causality assessment was rarely provided. The most robust current approach for structured, quantitative, and validated hepatotoxicity cases is the Council for International Organizations of Medical Sciences (CIOMS) scale for initial causality assessment [198]. While not without limitations, the CIOMS scale compiles liver specific criteria for challenge, de-challenge, risk factors, exclusion of unrelated diseases, and co-medication [199,200].

In contrast to case reports, randomized clinical trials of GTE do not provide evidence of hepatotoxicity. For example, in a 3-wk randomized clinical trial testing a GTE daily containing ~670 mg flavan-3-ols in healthy men, Frank et al. [201] found no untoward effects on γ-glutamyl-transpeptidase, alanine transaminase or aspartate transaminase. In a 12-wk phase II randomized clinical trial of patients with oral premalignant lesions, Tsao et al. [202] found no adverse effects, including on hematological or blood chemistry outcomes, from GTE supplementation at 500 to 1000 mg/m^2 (~14.7 to 29.4 mg/kg bw or ~1034 to 2068 mg/d). Similarly, Bettuzzi et al. [203] reported no adverse effects following the administration of green tea flavanols at 600 mg/d for 1 y in a safety assessment study among men with high-grade prostate intraepithelial neoplasia. In a randomized clinical trial of 40 overweight and obese Japanese children, 6 to 16 y, Matsuyama et al. [204] tested the effect of 575.9 vs 75.4 mg daily of green tea flavanols and found no significant effects on γ-glutamyl-transpeptidase, alanine transaminase, aspartate transaminase or albumin. In contrast, supplementation with a GTE at 1300 mg/d for a mean of 34.5 d in men with prostate cancer and scheduled for a radical prostatectomy was associated with statistically significant decreases in aspartate transaminase and alkaline phosphatase; non-significant decreases in alanine transaminase, γ-glutamyl-transpeptidase, and lipase were also reported [205]. While these results indicate no adverse effect on liver function and suggest a favorable trend, it is important to note that all these values were within normal limits throughout the study.

Hepatoxicity from herbal products and pharmaceuticals is not uncommon and has often raised concerns about product quality [206-208]. Navarro et al. [38] assayed 97 herbal dietary supplements and found that 29 of 73 products that did not identify GTE or any of its constituents did contain flavan-3-ols, suggesting that packaging and labels of such products seem to be unreliable as regards GTE content. In their study matching clinical histories with the analysis of tea flavonoids, among patients with confirmed hepatotoxicity, no statistically significant association was observed between the presence of flavan-3-ols or the dose consumed and liver injury causality score, severity, or pattern of liver injury [38]. Liver injury associated with green tea infusions and GTE is consistent with the definition of idiosyncratic drug-induced hepatotoxicity based on its rare occurrence and lack of dose-dependency [38,209]. While green tea infusions and GTE appear well tolerated in most all individuals, they have been associated with liver injury in rare circumstances, possibly as a result of specific environmental, genetic, and/or medical conditions.

8. AUTHORITATIVE REVIEWS OF GREEN TEA EXTRACTS

In 2003, the French Health Product Safety Agency (AFSSAPS) and the Spanish Food Safety and Nutrition Agency suspended the marketing authorization of an 80% ethanolic GTE (Exolise®) due to reports of hepatic disorders [210]. This product was indicated at 375 mg/d of GTE, standardized to 25% EGCG, and obtained using 80% ethanol as the solvent. More recent cases were reported to AFSSAPS in 2006 with aqueous GTE and AFSSA, the French Food Safety Agency, in 2009 [32,211,212]. In 2007, the U.S. Pharmacopeia (USP) determined GTE was a Class 2 substance and issued a press release suggesting specific cautionary wording be added to any dietary supplements containing GTE as an ingredient: "Take with food. Discontinue use and consult a healthcare practitioner if you have a liver disorder or develop symptoms of liver trouble such as abdominal pain, dark urine, or jaundice" [213]. However, two years later, the USP undertook a follow-up comprehensive review and, based on updated information, reclassified GTE as an article falling within Class A (Admitted into the Compendia – does not require a caution/warning statement in labeling) and canceled the published proposal for a cautionary labeling statement [214]. Additional expert reviews based on case reports, observational studies, and clinical trials confirm the general safety of green tea and GTE consumption [28,30,130,132,215-217] (Table 10).

In 2009, the European Food Safety Authority Scientific (EFSA) Cooperation Group (ESCO) published guidance for the safety assessment of botanicals in which green tea was considered based upon knowledge of the risk of hepatotoxicity and the worldwide, long-term consumption of traditional tea infusions; intakes of brewed green tea from the mean to 95th percentile of intake at 362 to 1097 g/d, respectively, were noted in China and Japan and were associated with mean and 95th percentile intakes of EGCG at 95 to 289 and 148 to 448 mg/d and of total flavan-3-ols at 186 to 565 and 307 to 931 mg/d, respectively [216]. The ESCO concluded that traditional green tea infusion is assumed to be safe on level A (safety presumed based upon available knowledge derived from long-term use). Dried aqueous GTE which have a similar composition and do not exceed the concentration of polyphenols compared to a traditional green tea infusion were also classified to be on level A when consistent to EFSA guidance on the safety assessment of botanicals intended for use in the preparation of beverages [216,218]. In marked contrast, this EFSA report [216] questions the safety of use of GTE for use in food supplements and recommends application of a margin of safety (MOS) – the ratio between the NOAEL (determined from animal studies) and daily intake for EGCG – of 100. Though absent specific information, the Norwegian Food Safety Authority recently issued a warning about using highly concentrated GTE supplements based upon some cases of hepatotoxicity reported by the Norwegian Medicines Agency [219].

Similar to EFSA guidelines, France has published restrictions on the use of preparations of green tea in food supplements limiting them to traditional extracts (i.e., aqueous extracts at a maximal amount equivalent to 30 mg/d of EGCG

Table 10. Summary of systematic reviews that include safety assessment of green tea or GTE

Review	Date	Conclusions
US Pharmacopeia, Safety of Green Tea Extract[1]	2008	Assigned Class 2 safety rating, suggested labeling: "Take with food. Discontinue use and consult a healthcare practitioner if you have a liver disorder or develop symptoms of liver trouble such as abdominal pain, dark urine, or jaundice"
Green tea consumption and liver disease; interventional and observational studies with liver cancer, cirrhosis, and fatty liver disease as outcomes[2]	2008	Increased green tea consumption may reduce the risk of chronic liver diseases.
US Pharmacopeia, Green Tea Extract Monograph[3]	2009	Class A safety rating, no cautionary label statement
Green tea for the prevention of cancer[4]	2009	Green tea consumption appears safe at moderate, regular, and habitual use (3-5 cups/d, providing ~250 mg catechins/d)
EFSA ESCO advise on the EFSA guidance for the safety assessment of botanicals[5]	2009	Aqueous GTE for food supplement use given A except for weight reduction purposes which received grade B, other GTE (e.g. hydro-alcoholic preparations) given grade B
Green tea for weight loss and weight maintenance in overweight or obese adults[6]	2012	Assessed randomized control trials of at least 12 wk duration of green tea preparations. Most adverse effects reported appeared unrelated to green tea or control intervention.
French Agency for Food, Environmental and Occupational Health Safety (ANSES)[7]	2012	Based on case reports, hepatitis severity following GTE consumption could not be conclusively associated with GTE consumption, dose response could not be achieved.

[1]*Sarma et al. [28];* [2]*Jin et al. [215];* [3]*Sarma. [30];* [4]*Boehm et al. [130];* [5]*EFSA J. [215];* [6]*Jurgens et al. [132];* [7]*ANSES, 21011-SA-0108 [217]*

for a person weighing 60 kg) and excludes hydro-alcoholic extracts and powders [220]. In some contrast, the French Agency for Food, Environmental and Occupational Health Safety recently evaluated 17 case reports of hepatitis associated with GTE consumption and found no conclusive evidence for a causal relationship because the data on quantity of green tea and GTE consumed were lacking so that no dose-response relationship could be established between intake and severity of hepatitis [217]. Belgium allows a maximum dose of EGCG at 1600 mg/d and a maximum amount of 25% ethanol permitted in the extraction process [221]. Currently, the Plant LIBRA project, co-financed in the context of the 7th European Union Framework Program, is evaluating the safety of GTE using several assessment approaches, including the Observed Safe Level [222], and Minimum Anticipated Biological Effect Level [223]and Estimation of the Optimal Range [224,225].

More recently, Cochrane database reviews have confirmed the general safety of consuming green tea infusions at 3-5 cups/d [130] and the safety of infusions and GTE associated with weight loss studies [132]. The safety of GTE was reviewed by the US Pharmacopeia and assigned as a Class A substance that does not require a caution or warning statement [28,30]. GTE prepared as essential oils, solvent-free oleoresins, and other natural extractives are currently classified as GRAS for food and ingredient use by the U.S. Food and Drug Administration [226]. Further, purified tea flavan-3-ols, an aqueous GTE, and L-theanine from green tea have self-determined GRAS status (Table 11).

Table 11. Industry-reported GRAS status of GTE and green-tea derived ingredients

Product	Manufacturer	Date
Teavigo, green tea extract standardized to 94% EGCG[1]	DSM	2005
Purified green tea catechins[2]	Kao, Inc.	2007
L-theanine extracted from tea leaves[3]	Blue California	2010
AssuriTEA aqueous green tea extract[1]	Kemin	2012

[1]AIBMR Life Sciences, GRAS Self-determination Inventory Database [227]; [2]Assuming maximum use of 540 mg catechins/350 mL, with up to 1 to 1.5 g green tea catechins presumed safe; FDA GRN No. 225 & 259 [228]; [3]FDA GRN No. 338 [229]

9. SUMMARY AND PERSPECTIVE

The interest in the putative health benefits of green tea has led to an increase in the consumption of brewed green tea and the application of GTE to a variety of food and beverage products and food supplements. With some experimental evidence of toxicity from green tea and GTE and a number of case reports of hepatotoxicity, the concept of a safety threshold for green tea and GTE has been raised. While there are several approaches to determining safety thresholds [222,224,225, 230,231,232], some suggest a conservative threshold for nutrients and food supplements should be equivalent to at least 10-fold less than the lowest dose associated with toxicity observed in humans. In the case reports of GTE-associated hepatotoxicity summarized by Bonkovsky [34], the reported range of intake was 10-29 mg GTE/kg bw/d. using the lower 10 mg/kg bw dose, an estimated acceptable intake would be 1.0 mg/kg bw/d or 70 mg EGCG daily for a reference 70 kg person. However, such a calculation is in marked contrast to usual intakes and the GRAS status of green tea and GTE and the expert and systematic reviews on the safety of consuming moderate levels of GTE and green tea infusions [28,30,130,132,215]. For example, the 90th percentile of EGCG intake associated with green tea consumption in the U.S. is 189 and 218 mg/d for women and men, respectively; the 95th percentile of EGCG intake associated with green tea consumption in China and Japan is 288 and 447 mg/d [4,8]. These values are about 3- to 6-times higher than a 10-fold safety threshold of 1.0 mg/kg bw/d and suggest an inappropriate application of the precautionary principle, particularly in the absence of established causality. Although most individual studies have been relatively short-term, recent Cochrane database reviews confirm the general safety of green tea and GTE [130,132]. Nonetheless, it is noteworthy that Bettuzzi et al. [203] found no untoward responses in a year-long randomized, placebo-controlled trial of a GTE providing 311 mg/d EGCG and 142 mg/d of other flavan-3-ols. The noted gap could be attributed to the factors such as dosing frequencies and conditions, as well as the medical conditions, concomitant use of other products and sensitivities of individuals.

Responses by government regulatory agencies to the experimental data and case reports of hepatotoxicity associated with use of GTE are contradictory, e.g., with EFSA stating that GTE for food supplement use cannot be considered safe and suggesting a MOS (calculated with data from animals) of at least 100 [216] and the USP admitting GTE into its Compendia without a requirement for any caution/warning label(though USP's Dietary Supplement Information Expert Committee continues to monitor clinical case reports) [213]. After an accountability assessment of 17 cases of hepatitis linked to green tea and GTE reported between 2009 and 2011, the French Agency for Food, Environmental and Occupational Health Safety concluded that these cases could not be clearly associated with green tea due to confounding by other likely causes and an inability to obtain adequate information on intake [217]. This contrast between approaches and recommendations arises in part from the rare and idiosyncratic nature of the injury and the apparent absence of a dose-response relationship between GTE and hepatotoxicity [36,37,193]. The sensitivity and extremely low occurrence of hepatotoxicity with GTE may have a nutrigenetic basis, e.g., polymorphism(s) in pathways for flavonoid metabolism [35,233].This relationship is highlighted by the recent report of Navarro et al. [38] which indicated no statistically significant association between GTE dose or content of flavan-3-ols and liver injury causality score, severity, or pattern of liver injury. Despite concern raised in some reports about a great risk for harm induced by GTE derived from hydro-alcoholic versus aqueous extractions, specifically hepatotoxicity, there is no evidence that the former provides significantly different proportions of bioactive or toxic compounds than the latter. This work is essentially confirmed by the observation that hepatotoxicity associated with GTE have been reported from cases following use of variety of products varying markedly in composition, solvent extraction, dose, and duration of use. Although the pathogenesis of hepatotoxicity associated with GTE is unknown, it is important to note that EGCG, the component of GTE most often identified as the cause of hepatotoxicity, is readily soluble in both water and alcohol-water mixtures. Regulations ranking the potential hepatotoxicity of GTE via its extraction with water versus hydro-alcoholic solvents are not based on the available evidence. Rather than an extraction process, it appears that the greater number of cases linked to a limited number of GTE products may be largely due to their greater sales and indication for chronic use, such as for weight loss.

The evidence derived from observational studies and clinical trials over the last three decades suggest that flavan-3-ols from green tea may contribute importantly to reducing the risk of prevalent age-related chronic diseases. The biological plausibility underlying this association is derived from a large body of experimental studies revealing several relevant molecular mechanisms of action of these flavonoids. However, the application of this information, e.g., through the development of innovative dietary products containing GTE, requires careful consideration of the safety profile of its constituent phytochemicals. Clinical case reports of hepatotoxicity associated with use of GTE and products containing GTE have raised concern and generated responses from regulatory agencies. In addition to the usual limitations associated with the extrapolation of evidence from *in vitro* and animal model experiments, data on liver injury induced by GTE and its specific flavan-3-ols constituents revealed by these approaches are mixed in that some demonstrate hepatotoxicity while others show hepatoprotective actions. Case reports are important but also limited, especially with rare and idiosyncratic reactions, in their ability to reveal causality due to confounding medical and lifestyle factors of the patients.

10. CONCLUSION

The scientific evidence regarding the potential benefits and risks of green tea polyphenols and GTE continues to emerge. Existing data are insufficient to identify in the preparation or composition of GTE the causative agent, its dose or duration of use as well as nutrigenetic, medical, and other factors contributing to the risk of hepatotoxicity. While reference to quantitative equivalents associated with traditional dietary intakes of green tea, its constituent flavonoids, and related dietary polyphenols is important in evaluating GTE products, opportunities should be allowed for innovation to improve the efficiency, quality, and cost-effectiveness of their production through extraction technologies. This approach can also prove useful in minimizing the presence of endogenous or contaminant heavy metals as well as mycotoxins, pesticides, and other unwanted constituents.

ACKNOWLEDGEMENTS

This work was supported by an educational grant from the International Alliance of Dietary/Food Supplement Associations (IADSA) to the

Friedman School of Nutrition Science and Policy at Tufts University. IADSA had no role in the preparation or review of the manuscript. We appreciate the assistance of Jo Obelsky and Jennifer O'Leary-Chen in formatting the manuscript.

COMPETING INTERESTS

Authors have declared that no competing interests exist.

REFERENCES

1. MacFarlane A, MacFarlane I, editors. The empire of tea: The remarkable history of the plant that took over the world. Woodstock, NY: The Overlook Press; 2003.
2. Hodgson JM, Croft KD. Tea flavonoids and cardiovascular health. Mol Aspects Med. 2010;31:495-502.
3. Mukhtar H, Ahmad N. Tea polyphenols: Prevention of cancer and optimizing health. Am J Clin Nutr. 2000;71:1698-702S.
4. International Tea Committee. Annual Bulletin of Statistics. 2013. Accessed 18 October 2013. Available: http://www.inttea.com/ publications.asp.
5. Chun OK, Chung SJ, Song WO. Estimated dietary flavonoid intake and major food sources in U.S. adults. J Nutr. 2007;137:1244-52.
6. CFSII. Continuing food survey of intake by individuals. 2003. Accessed 8 January 2014. Available: http://www.ars.usda.gov/Services/ docs.htm?docid=14392.
7. Tomata Y, Kakizaki M, Nakaya N, Tsuboya T, Sone T, Kuriyama S, et al. Green tea consumption and the risk of incident functional disability in elderly Japanese: The Ohsaki Cohort 2006 Study. Am J Clin Nutr. 2012;95:732-9.
8. Robert Koch Institute. Verzehr von grünem tee. Bundes-Gesundheitssurvey. 1998. Accessed 30 October 2013. Available: http://www. rki.de/EN/Content/ Health_ Monitoring/Health_Reporting/GBEDownloadsB/was_essen_wir_heute.pdf?_ blob=publicationFile.
9. Khokhar S, Magnusdottir SG. Total phenol, catechin, and caffeine contents of teas commonly consumed in the United Kingdom. J Agric Food Chem. 2002;50:565-70.
10. Kaegi E. Unconventional therapies for cancer: Green tea. Canadian Med Assoc J. 1998;158:1033-5.
11. Henderson L, Gregory J, Swan G. National Dietary and Nutrition Survey: Adults aged 19 to 64 years. Food Standards Agency. 2002. Accessed 8 October 2013. Available:http://food.gov.uk/ multimedia/pdfs/ ndnsprintedreport.pdf.
12. Smith A. Effects of caffeine on human behavior. Food Chem Toxicol. 2002;40:1243-55.
13. Blumberg JB. Introduction to the proceedings of the fifth international scientific symposium on tea and human health. Am J Clin Nutr. 2013;98:1607-10S.
14. Bogdanski P, Suliburska J, Szulinska M, Stepien M, Pepek-Musialik D, Jablecka A. Green tea extract reduces blood pressure, inflammatory biomarkers, and oxidative stress and improves parameters associated with insulin resistance in obese, hypertensive patients. Nutr Res. 2012;32:421-7.
15. Del Rio D, Rodriguez-Mateos A, Spencer JP, Tognolini M, Borges G, Crozier A. Dietary (poly) phenolics in human health: Structures, bioavailability, and evidence of protective effects against chronic diseases. Antiox Redox Signal. 2013;18:1818-92.
16. Liu K, Zhou R, Wang B, Chen K, Shi LY, Zhu JD, et al. Effect of green tea on glucose control and insulin sensitivity: A meta-analysis of 17 randomized controlled trials. Am J Clin Nutr. 2013;98:340-8.
17. Mak JC. Potential role of green tea catechins in various disease therapies: progress and promise. Clin Exp Pharmacol Physiol. 2012;39:265-73.
18. Kim H-S, Quon MJ, Kim J-A. New insights into the mechanisms of polyphenols beyond antioxidant properties: Lessons from the green tea polyphenol, epigallocatechin-3-gallate. Redox Biol. 2014;2:187-95.
19. Lorenz M. Cellular targets for the beneficial actions of tea polyphenols. Am J Clin Nutr. 2013;90:1642-50S.
20. Katiyar SK, Afaq F, Perez A, Muhkhtar H. Green ptea polyphenol (-)-epigallocatechin-3-gallate treatment of human skin inhibits ultraviolet radiation-induced oxidative stress. Carcinogenesis. 2001;22:287-94.
21. Meng Q, Velalar CN, Ruan R. Regulating the age-related oxidative damage, mitochondrial integrity, and antioxidative

enzyme activity in Fischer 344 rats by supplementation of the antioxidant epigallocatechin-3-gallate. Rejuvenation Res. 2008;11:649-60.
22. Yan J, Zhao Y, Suo S, Liu Y, Zhao B. Green tea catechins ameliorate adipose insulin resistance by improving oxidative stress. Free Radic Biol Med. 2012;52:1648-57.
23. Basu A, Sanchez K, Leyva MJ, Wu M, Betts NM, Aston CE, Lyons TJ. Green tea supplementation affects body weight, lipids, and lipid peroxidation in obese subjects with metabolic syndrome. J Am Coll Nutr. 2010;29:31-40.
24. Nakagawa K, Ninomiya M, Okubo T, Aoi N, Juneja LR, Kim M, et al. Tea catechin supplementation increase antioxidant capacity and prevents phospholipid hydroperoxidation in plasma of humans. J Agric Food Chem. 1999;47:3967-73.
25. Orsolic N, Sirovina D, Gajski G, Garaj-Vrhovac V, Jembrek MJ, Kosalec I. Assessment of DNA damage and lipid peroxidation in diabetic mice: effects of propolis and epigallocatechin gallate (EGCG). Mutat Res. 2013;757:36-44.
26. Namal Senanayake SPJ. Green tea extract: Chemistry, antioxidant properties and food applications – a review. J Funct Food. 2013;5:1529-41.
27. Blumenthal M, editor. The ABC clinical guide to herbs. (1st ed.) American Botanical Council. Austin, TX. Thieme New York. New York. 2003;335-50.
28. Sarma DN, Barrett ML, Chavez ML, Gardiner P, Ko R, Mahady GB, et al. Safety of green tea extracts: A systematic review by the US Pharmacopeia. Drug Saf. 2008;31:469-84.
29. Shivanand P. Tea: effects on cardiovascular system, blood pressure, EEG and many other system of human being. Int J Pharm Sci Rev Res. 2010;2:31-4
30. Sarma DN. Update on the USP green tea extract monograph. U.S. Pharmacopeia. 2009. Accessed 26 October 2013. Available:http://www.usp.org/usp-nf/notices/retired-compendial-notices/update-usp-green-tea-extract-monograph.
31. GRAS. 21 CFR Ch. I (4-1-12 Edition) § 182.20. 2010. Substances generally recognized as safe, Subpart A – General Provisions Sec. Essential oils, oleoresins (solvent-free), and natural extractives (including distillates). Food and Drug Administration, Dept. Health and Human Services. Washington, DC. Accessed 8 October 2013. Available: http://www.gpo.gov/fdsys/pkg/ CFR-2012-title21-vol3/ pdf/CFR-2012 -title21-vol3-sec182-20.pdf.
32. Gloro R, Hourmand-Ollivier I, Mosquet B, Mosquet L. Rousselot P, Salamé E, et al. Fulminant hepatitis during self-medication with hydroalcoholic extract of green tea. Eur J Gastroenterol Hepatol. 2005;17: 1135-7.
33. Jimenez-Saenz M, Martinez-Sanchez MdC. Acute hepatitis associated with the use of green tea infusions. J Hepatol. 2006;44:616-7.
34. Bonkovsky HL. Hepatotoxicity associated with supplements containing Chinese green tea (*Camellia sinensis*). Ann Intern Med. 2006;144:68-71.
35. Javaid A, Bonkovsky HL. Hepatotoxicity due to extracts of Chinese green tea (*Camellia sinensis*): A growing concern. J Hepatol. 2006;45:334-5.
36. Mazzanti G, Menniti-Ippolito FM, Moro PA, Cassetti F, Raschetti R, Santuccio C, et al. Hepatoxicity form green tea: A review of the literature and two unpublished cases. Eur J Clin Pharmacol .2009;65:331-4.
37. Fong TL, Klontz KC, Canas-Coto A, Casper SJ, Durazo FA, Davern TJ, et al. Hepatotoxicity due to Hydroxycut: A case series. Am J Gastroenterol. 2010;105: 1561-6.
38. Navarro VJ, Bonkovsky HL, Hwang S-I, Vega M, Barnhart H, Serrano J. Catechins in dietary supplements and hepatotoxicity. Dig Dis Sci. 2013;58:2682-90.
39. Chaturvedula VSP, Prakash I. The aroma, taste, color and bioactive constituents of tea. J Med Plants Res. 2011;5:2110-24.
40. Chacko SM, Thambi PT, Kuttan R, Nishigaki I. Beneficial effects of green tea: a literature review. Chin Med. 2010;5:13. DOI: 10.1186/1749-8546-5-13.
41. Cabrera C, Artacho R, Giménez R. Beneficial effects of green tea – a review. J Am Coll Nutr. 2006;25:79-99.
42. Wang Y, Ho CT. Polyphenolic chemistry of tea and coffee: A century of progress. J Agric Food Chem. 2009;57:8109-14.
43. Butt MS, Sultan MT. Green tea: Nature's defense against malignancies. Crit Rev Food Sci Nutr. 2009;49:463-73.
44. Zhao Y, Chen P, Lin L, Harnly JM, Yu L, Li Z. Tentative identification, quantitation, and

principal component analysis of green pu-erh, green, and white teas using UPLC/DAD/MS. Food Chem. 2011;126: 126-77.
45. Saijo R, Kato M, Takeda Y. Changes in the contents of catechins and caffeine during leaf development. Proceedings International Conference on Ocha (Tea) Culture and Science, Shizuoka. Japan. 2004;251-4.
46. Suzuki T, Yamazaki N, Sada Y, Oguni I, Moriyasu Y. Tissue distribution and intracellular localization of catechins in tea leaves. Biosci Biotechnol Biochem. 2003;67:2683-6.
47. Bhagwat S, Haytowitz DB, Holden JM. (Ret.). USDA database for the flavonoid content of selected foods, Release 3.1. U.S. Department of Agriculture, Agricultural Research Service. Accessed 14 March 2014. Available: http://www.ars.usda.gov/ nutrientdata/flavonoid.
48. Cabrera C, Giménez R, López MC. Determination of tea components with antioxidant activity. J Agric Food Chem. 2003;51:4427-35.
49. Friedman M, Levin CE, Lee S-U, Kozukue N. Stability of green tea catechins in commercial tea leaves during storage for 6 months. J Food Sci. 2009;74:H47-51
50. Komes D, Horžić D, Belščak A, Ganić KK, Vulić I. Green tea preparation and its influence on the content of bioactive compounds. Food Res Int. 2010;43:167-76.
51. Nishitani E, Sagesaka YM. Simultaneous determination of catechins, caffeine and other phenolic compounds in tea using new HPLC method. J Food Compos Anal. 2004;17:675-85.
52. Peterson J, Dwyera J, Bhagwat S, Haytowitz D, Holden J, Eldridge AL, Beecherd G, et al. Major flavonoids in dry tea. J Food Compos Anal. 2005;18:487-501.
53. Wang D, Lu J, Miao A, Xie Z, Yang D. HPLC-DAD-ESI-MS/MS analysis of polyphenols and purine alkaloids in leaves of 22 tea cultivars in China. J Food Compos Anal. 2008;21:361-9.
54. Xu W, Song Q, Li D, Wan X. Discrimination of the production season of Chinese green tea by chemical analysis in combination with supervised pattern recognition. J Agric Food Chem. 2012;60:7064-70.
55. Karak T, Bhagat RM. Trace elements in tea leaves, made tea and tea infusion: A review. Food Res Int. 2010;43:2234-52.
56. ATSDR (Agency for Toxic Substances and Disease Registry). Toxicological profile for arsenic. Atlanta, GA: U.S. Department of Health and Human Services, Public Health Services. 2007a. Accessed 14 July 2014. Available: http://www.atsdr.cdc.gov/tox guides/toxguide-2.pdf.
57. ATSDR (Agency for Toxic Substances and Disease). Toxicological profile for lead. Atlanta, GA: U.S. Department of Health and Human Services, Public Health Services. 2007b.Accessed 1 June 2013. Available:http://www.atsdr.cdc.gov/toxguid es/ toxguide-13.pdf.
58. ATSDR (Agency for Toxic Substances and Disease). Toxicological profile for chromium. Atlanta, GA: U.S. Department of Health and Human Services, Public Health Services. 2012. Accessed 4 July 2013. Available: http://www.atsdr.cdc.gov/tox guides/toxguide-7.pdf.
59. Mandiwana KL, Panichev N, Panicheva S. Determination of chromium (VI) in black, green and herbal teas. Food Chem. 2011;129:1839-43.
60. Eggers R, Pilz S. High pressure processing. In: Bart H-J, Pilz S, editors. Industrial Scale Natural Products Extraction. Weinheim, Germany: Wiley-VCH. 2011;87-122.
61. Kassing M, Jenelten U, Schenk J, Strube J. A new approach for process development of plant-based extraction processes. Chem Eng Tech. 2010;33:377-87.
62. Strube J, Backer W, Schulte M. Process engineering and mini-plant technology. In: Bart HJ, Pilz S, editors. Industrial scale natural products extraction. Weinheim, Germany: Wiley-VCH. 2011;123-80.
63. Du FY, Xiao XH, Li GK. Application of ionic liquids in the microwave-assisted extraction of trans-resveratrol from Rhizma Polygoni Cuspidati. J Chrom A. 2007;1140:56-62.
64. Jordan V, Muller U. Alternative solvents in plant extraction. In: Bart H-J, Pilz S, editors. Industrial Scale Natural Products Extraction. Weinheim, Germany: Wiley-VCH. 2011;55-86.
65. Sun QL, Hua S, Ye JH, Lu JL, Zhen XQ, Liang YR. Decaffeination of green tea by supercritical carbon dioxide. J Med Plant Res. 2010;4:1161-8.

66. Wijngaard H, Hossain MB, Rai DK, Brunton N. Techniques to extract bioactive compounds from food by-products of plant origin. Food Res Int. 2012;46:505-13.
67. Klein AP, Beach ES, Emerson JW, Zimmerman JB. Accelerated solvent extraction of lignin from *Aleurites moluccana* (Candlenut) nutshells. J Agr Food Chem. 2010;58:10045-8.
68. Ibanez E, Herrero M, Martin-Alvarez PJ, Senorans FJ, Reglero G, Cifuentes A. Accelerated solvent extraction: A new procedure to obtain functional ingredients from natural sources. ACS Sym Ser. 2006;926:65-78.
69. Hossain MB, Barry-Ryan C, Martin-Diana AB, Brunton NP. Optimisation of accelerated solvent extraction of antioxidant compounds from rosemary (*Rosmarinus officinalis* L.), marjoram (*Origanum majorana* L.) and oregano (*Origanum vulgare* L.) using response surface methodology. Food Chem. 2011;126:339-46.
70. Herrero M, Martin-Alvarez PJ, Senorans FJ, Cifuentes A, Ibanez E. Optimization of accelerated solvent extraction of antioxidants from *Spirulina platensis* microalga. Food Chem. 2005;93:417-23.
71. Wijngaard HH, Ballay M, Brunton N. The optimisation of extraction of antioxidants from potato peel by pressurised liquids. Food Chem. 2012;133:1123-30.
72. Wang LJ, Weller CL. Recent advances in extraction of nutraceuticals from plants. Trends in Food Sci Tech. 2006;17:300-12.
73. Luthria DL. Influence of experimental conditions on the extraction of phenolic compounds from parsley (*Petroselinum crispum*) flakes using a pressurized liquid extractor. Food Chem. 2008;107:745-52.
74. Rostagno MA, Palma M, Barroso CG. Pressurized liquid extraction of isoflavones from soybeans. Anal Chim Acta. 2004;522:169-77.
75. Denery JR, Dragull K, Tang CS, Li QX. Pressurized fluid extraction of carotenoids from *Haematococcus pluvialis* and *Dunaliella salina* and kavalactones from Piper methysticum. Anal Chim Acta. 2004;501:175-81.
76. Pompeu DR, Silva EM, Rogez H. Optimisation of the solvent extraction of phenolic antioxidants from fruits of *Euterpe oleracea* using response surface methodology. Bioresource Technol. 2009;100:6076-82.
77. Herrero M, Mendiola JA, Cifuentes A, Ibanez E. Supercritical fluid extraction: Recent advances and applications. J Chromatogr A. 2010;1217:2495-511.
78. Poiana M, Sicari V, Mincione B. Supercritical carbon dioxide ($SC\text{-}CO_2$) extraction of grape fruit flavedo. Flavour Frag J. 1998;13:125-30.
79. Roy BC, Goto M, Kodama A, Hirose T. Supercritical CO_2 extraction of essential oils and cuticular waxes from peppermint leaves. J Chem Tech Biotech. 1996;67:21-6.
80. Santos SAO, Villaverde JJ, Silva CM, Neto CP, Silvestre AJD. Supercritical fluid extraction of phenolic compounds from *Eucalyptus globulus* Labill bark. J Supercrit Fluid. 2012;71:71-9.
81. Casas L, Mantell C, Rodríguez M, Ossa EJMDL, Roldán A, Ory ID, et al. Extraction of resveratrol from the pomace of *Palomino fino* grapes by supercritical carbon dioxide. J Food Eng. 2010;96:304-8.
82. Chen BY, Kuo CH, Liu YC, Ye LY, Chen JH, Shieh CJ. Ultrasonic-assisted extraction of the botanical dietary supplement resveratrol and other constituents of *Polygonum cuspidatum*. J Nat Prod. 2012;75:1810-3.
83. Chemat F, Zill-e-Huma, Khan MK. Applications of ultrasound in food technology: Processing, preservation and extraction. Ultrason Sonochem. 2011;18: 813-35.
84. Puri M, Sharma D, Barrow CJ. Enzyme-assisted extraction of bioactives from plants. Trends Biotechnol. 2012;30:37-44.
85. Chandran J, Amma KPP, Menon N, Purushothaman J, Nisha P. Effect of enzyme assisted extraction on quality and yield of volatile oil from black pepper and cardamom. Food Sci Biotech. 2012;21:1611-7.
86. Choudhari SM, Ananthanarayan L. Enzyme aided extraction of lycopene from tomato tissues. Food Chem. 2007;102:77-81.
87. Gibbins RD, Aksoy HA, Ustun G. Enzyme-assisted aqueous extraction of safflower oil: Optimisation by response surface methodology. Int J Food Sci Tech. 2012;47:1055-62.
88. Li BB, Smith AB, Hossain MM. Extraction of phenolics from citrus peels II. Enzyme-assisted extraction method. Sep Purif Technol. 2006;48:189-96.

89. Sari YW, Bruins ME, Sanders JPM. Enzyme assisted protein extraction from rapeseed, soybean, and microalgae meals. Industr Crops Products. 2013;43:78-83.
90. Le HV, Van VML. Comparison of enzyme-assisted and ultrasound-assisted extraction of vitamin C and phenolic compounds from acerola (*Malpighia emarginata* DC.) fruit. Int J Food Sci Tech. 2012;47:1206-14.
91. El Darra N, Grimi N, Maroun RG, Louka N, Vorobiev E. Pulsed electric field, ultrasound, and thermal pretreatments for better phenolic extraction during red fermentation. Eur Food Res Technol. 2013;236:47-56.
92. Puertolas E, Cregenzan O, Luengo E, Alvarez I, Raso J. Pulsed-electric-field-assisted extraction of anthocyanins from purple-fleshed potato. Food Chem. 2013;136:1330-6.
93. Huang W, Wue A, Niu H, Jia Z, Wang J. Optimised ultrasonic-assisted extraction of flavonoids from Folium eucommiae and evaluation of antioxidant activity in multi-test systems in vitro. Food Chem. 2009;114:1147-54.
94. Chittapalo T, Noomhorm A. Ultrasonic assisted alkali extraction of protein from defatted rice bran and properties of the protein concentrates. Int J Food Sci Tech. 2009;44:1843-9.
95. Turk MF, Vorobiev E, Baron A. Improving apple juice expression and quality by pulsed electric field on an industrial scale. Lwt-Food Sci Technol. 2012;49:245-50.
96. Corrales M, Toepfl S, Butz P, Knorr D, Tauscher B. Extraction of anthocyanins from grape by-products assisted by ultrasonics, high hydrostatic pressure or pulsed electric fields: a comparison. Innovat Food Sci Emerg Tech. 2008;9:85-91.
97. Kaufmann B, Christen P, Veuthey JL. Parameters affecting microwave-assisted extraction of with anolides. Phytochem Anal. 2001;12:327-31.
98. Liazid A, Guerrero RF, Cantos E, Palma M, Barroso CG. Microwave assisted extraction of anthocyanins from grape skins. Food Chem. 2011;124:1238-43.
99. Perez-Serradilla JA, de Castro MDL. Microwave-assisted extraction of phenolic compounds from wine lees and spray-drying of the extract. Food Chem. 2011;124:1652-9.
100. Tsubaki S, Sakamoto M, Azuma J. Microwave-assisted extraction of phenolic compounds from tea residues under autohydrolytic conditions. Food Chem. 2010;123:1255-8.
101. Pfennig A, Deliniski D, Johannisbauer W, Josten H. Extraction technology. In: Bart H-J, Pilz S, editors. Industrial Scale Natural Products Extraction. Weinheim: Wiley-VCH. 2011;181-218.
102. Henning SM, Fajardo-Lira C, Lee HW, Youssefian AA, Go VL, Heber D. Catechin content of 18 teas and a green tea extract supplement correlates with the antioxidant capacity. Nutr Cancer. 2003;45:226-35.
103. Seeram NP, Henning SM, Niu Y, Lee R, Scheuller HS, Heber D. Catechin and caffeine content of green tea dietary supplements and correlation with antioxidant capacity. J Agric Food Chem. 2006;54:1599-603.
104. Friedman M, Levin CE, Choi S-H, Kozukue E, Kozukue N. HPLC Analysis of catechins, theaflavins, and alkaloids in commercial teas and green tea dietary supplements: Comparison of water and 80% ethanol/water extracts. J Food Sci. 2006;71:C328-37.
105. Vuong QV, Golding JB, Stathopoulos CE, Nguyen MH, Roach PD. Optimizing conditions for the extraction of catechins from green tea using hot water. J Sep Sci. 2011;34:3099-106.
106. Vuong QV, Stathopoulos CE, Golding JB, Nguyen MH, Roach PD. Optimum conditions for the water extraction of L-theanine from green tea. J Sep Sci. 2011;34:2468-74.
107. Choung M-G, Lee M-S. Optimal extraction conditions for simultaneous determination of catechins and caffeine in green tea leaves. Food Sci Biotech. 2011;20:327-33.
108. Astill C, Birch MR, Dacombe C, Humphrey PG, Martin PT. Factors affecting the caffeine and polyphenol contents of black and green tea infusions. J Agric Food Chem. 2001;49:5340-7.
109. Hicks MB, Hsieh YHP, Bell LN. Tea preparation and its influence on methylxanthine concentration. Food Res Int. 1996;29:325-30.
110. Perva-Uzunalić A, Škerget M, Knez Ž, Weinreich B, Otto F, Grüner S. Extraction of active ingredients from green tea (*Camellia sinensis*): Extraction efficiency of major catechins and caffeine. Food Chem. 2006;96:597-605.
111. Ma SC, Wang XL, Liang YR. Kinetic study

on leaching of fluorine from green tea. Afr J Pharm Pharmaco. 2012;6:1410-6.
112. Costa LM, Gouveia ST, Nobrega JA. Comparison of heating extraction procedures for Al, Ca, Mg, and Mn in tea samples. Anal Sci. 2002;18:313-8.
113. Ziaedini A, Jafari A, Zakeri A. Extraction of antioxidants and caffeine from green tea (*Camelia sinensis*) leaves: kinetics and modeling. Food Sci Technol Int. 2010;16:505-10.
114. Zhang X, Xu F, Gao Y, Wu J, Sun Y, Zeng X. Optimising the extraction of tea polyphenols, (-)-epigallocatechin gallate and theanine from summer green tea by using response surface methodology. Int J Food Sci Tech. 2012;47:2151-7.
115. Chen Z-Y, Zhu QY, Tsang D, Huang Y. Degradation of green tea catechins in tea drinks. J Agric Food Chem. 2000;49:477-82.
116. Meterc D, Petermann M, Weidner E. Extraction of green tea and drying with a high pressure spray process. Hem Ind. 2007;61:222-8.
117. Rusak G, Komes D, Likić S, Horžić D, Kovač M. Phenolic content and antioxidative capacity of green and white tea extracts depending on extraction conditions and the solvent used. Food Chem. 2008;110:852-8.
118. Sharma V, Gulati A, Ravindranath SD, Kumar V. A simple and convenient method for analysis of tea biochemicals by reverse phase HPLC. J Food Comp Anal. 2005;18:583-94.
119. Mossion A, Potin-Gautier M, Delerue S, Le Hécho I, Behra P. Effect of water composition on aluminum, calcium and organic carbon extraction in tea infusions. Food Chem. 2008;106:1467-75.
120. Kanrar B, Mandal S, Bhattacharyya A. Validation and uncertainty analysis of a multiresidue method for 67 pesticides in made tea, tea infusion, and spent leaves using ethyl acetate extraction and gas chromatography/mass spectrometry. J AOAC Int. 2010;93:411-24.
121. Wan CX. Primary speciation analysis of lead in tea. J Jiamusi Univ. 2012;30:798-800.
122. Xu J, Yu MG, Chen YX, Fu XP, Guan DC. Characteristic of distribution and chemical forms of Pb in tea plant varieties. Chinese J Appl Ecol. 2011;22:891-6.
123. Sullivan J, Greenfield J, Cumberford G, Grant J, Stewart J. Extraction efficiencies of heavy metals in hydroethanolic solvent from herbs of commerce. J AOAC Intl. 2010;2:496-8.
124. Yoshida Y, Kiso M, Goto T. Efficiency of the extraction of catechins from green tea. Food Chem. 1999;67:429-33.
125. Sosa MV, Rodríguez-Rojo S, Mattea F, Cismondi M, Cocero MJ. Green tea encapsulation by means of high pressure antisolvent coprecipitation. J Supercrit Fluid. 2011;56:304-11.
126. Tian M, Yan H, Row KH. Solid-phase extraction of caffeine and theophylline from green tea by a new ionic liquid-modified functional polymer sorbent. Anal Lett. 2009;43:110-8.
127. Park HS, Im NG, Kim KH. Response surface analysis of caffeine and chlorophyll extraction in supercritical decaffeination of green tea. J Biotechnol. 2008;136:477-8S.
128. Park HS, Im NG, Kim KH. Extraction behaviors of caffeine and chlorophylls in supercritical decaffeination of green tea leaves. LWT - Food Sci Technol. 2012;45:73-8.
129. Arab L, Khan F, Lam H. Tea consumption and cardiovascular disease risk. Am J Clin Nutr. 2013; 98:1651-59S.
130. Boehm K, Borrelli F, Ernst E, Habacher G, Hung SK, Milazzo S, et al. Green tea (*Camellia sinensis*) for the prevention of cancer. Cochrane Database Syst Rev. 2009;3: CD005004.
131. Grassi D, Desideri G, Di Giosia P, De Feo M, Fellini E, Cheli P, Ferri L, Ferri Cet al. Tea, flavonoids, and cardiovascular health; Endothelial protection. Am J Clin Nutr. 2013;98:1660-66S.
132. Jurgens TM, Whelan AM, Killian L, Doucette S, Kirk S, Foy E. Green tea for weight loss and weight maintenance in overweight or obese adults. Cochrane Database Syst Rev. 2012;12:CD008650.
133. Kim A, Chiu A, Barone MK, Avino D, Wang F, Coleman CI, et al. Green tea catechins decrease total and low-density lipoprotein cholesterol: A systematic review and meta-analysis. J Am Diet Assoc. 2011;111:1720-9.
134. Shen C-L, Chyu M-C, Wang J-S. Tea and bone health: Steps forward in translational nutrition. Am J Clin Nutr. 2013;98:1694-9S.
135. Yuan J-M. Cancer prevention by green tea: Evidence from epidemiologic studies. Am J Clin Nutr. 2013;98:1676-81S.
136. Hursel R, Viechtbauer W, Westerterp-Plantenga MS. The effects of green tea on

weight loss and weight maintenance: a meta-analysis. Int J Obes (Lond). 2009;33:956-61.

137. Phung OJ, Baker WL, Matthews LJ, Lanosa M, Thorne A, Coleman CI. Effect of green tea catechins with or without caffeine on anthropometric measures: A systematic review and meta-analysis. Am J Clin Nutr. 2010;91:73-81.
138. Zheng XX, Xu YL, Li SH, Hui R, Wu YJ, Huang XH. Effects of green tea catechins with or without caffeine on glycemic control in adults: A meta-analysis of randomized controlled trials. Am J Clin Nutr. 2013;97:750-62.
139. Liu K, Zhou R, Wang B, Chen K, Shi LY, Zhu JD, Mi MT. Effect of green tea on glucose control and insulin sensitivity: A meta-analysis of 17 randomized controlled trials. Am J Clin Nutr. 2013:98:340-8.
140. Zheng XX, Xu YL, Li SH, Liu XX, Hui R, Huang XH. Green tea intake lowers fasting serum total and LDL cholesterol in adults: A meta-analysis of 14 randomized controlled trials. Am J Clin Nutr. 2011;94:601-10.
141. Hooper L, Kroon PA, Rimm EB, Cohn JS, Harvey I, Le Cornu KA, et al. Flavonoids, flavonoid-rich foods, and cardiovascular risk: A meta-analysis of randomized controlled trials. Am J Clin Nutr. 2008;88:38-50.
142. Taubert D, Roesen R, Schömig E. Effect of cocoa and tea intake on blood pressure: A meta-analysis. Arch Intern Med. 2007;167:626-34.
143. Wang ZM, Zhou B, Wang YS, Gong QY, Wang QM, Yan JJ, et al. Black and green tea consumption and the risk of coronary artery disease: A meta-analysis. Am J Clin Nutr. 2011;93:506-15.
144. Arab L, Liu W, Elashoff D. Green and black tea consumption and risk of stroke: A meta-analysis. Stroke. 2009;40:1786-92.
145. Zheng JS, Yang J, Fu YQ, Huang T, Huang YJ, Li D. Effects of green tea, black tea, and coffee consumption on the risk of esophageal cancer: A systematic review and meta-analysis of observational studies. Nutr Cancer. 2013;65:1-16.
146. Zheng P, Zheng HM, Deng XM, Zhang YD. Green tea consumption and risk of esophageal cancer: a meta-analysis of epidemiologic studies. BMC Gastroenterol. 2012;12:165.
147. Kang H, Rha SY, Oh KW, Nam CM. Green tea consumption and stomach cancer risk: A meta-analysis. Epidemiol Health. 2010;32:e2010001.
148. Myung SK, Bae WK, Oh SM, Kim Y, Ju W, Sung J, et al. Green tea consumption and risk of stomach cancer: A meta-analysis of epidemiologic studies. Int J Cancer. 2009;124:670-7.
149. Wang ZH, Gao QY, Fang JY. Green tea and incidence of colorectal cancer: Evidence from prospective cohort studies. Nutr Cancer. 2012;64:1143-52.
150. Tang N, Wu Y, Zhou B, Wang B, Yu R. Green tea, black tea consumption and risk of lung cancer: A meta-analysis. Lung Cancer. 2009;65:274-83.
151. Ogunleye AA, Xue F, Michels KB. Green tea consumption and breast cancer risk or recurrence: A meta-analysis. Breast Cancer Res Treat. 2010;119:477-84.
152. Tang NP, Li H, Qiu YL, Zhou GM, Ma J. Tea consumption and risk of endometrial cancer: a meta-analysis. Am J Obstet Gynecol. 2009;201:605.e1-8.
153. Zheng J, Yang B, Huang T, Yu Y, Yang J, Li D. Green tea and black tea consumption and prostate cancer risk: An exploratory meta-analysis of observational studies. Nutr Cancer. 2011;63:663-72.
154. Nagle CM, Olsen CM, Bain CJ, Whiteman DC, Green AC, Webb PM. Tea consumption and risk of ovarian cancer. Cancer Causes Control. 2010;21:1485-91.
155. Onakpoya I, Spencer E, Heneghan C, Thompson M. The effect of green tea on blood pressure and lipid profile: A systematic review and meta-analysis of randomized clinical trials. Nutr Metabol Cardiovasc Dis; 2014. DOI:10.1016/j.numecd.2014.01.016.
156. World Health Organization (WHO) guidelines for assessing quality of herbal medicines with reference to contaminants and residues, 2007. Accessed 1 March 2014. Available: http://www.who.int/medicinedocs/index/assoc/s14878e/s14878e.pdf.
157. EPA 40 CFR part 143. Accessed 1 March 2014. Available: http://www.ecfr.gov/cgi-bin/text-idx?tpl=/ecfrbrowse/Title40/40cfr143_main_02.tpl.
158. Joint FAO/WHO Food Standards Programme, CODEX Committee on Contaminants in Foods, Sixth Session, Maastricht, The Netherlands. Working document for information and use in discussions related to contaminants and

toxins in the GSCFTT (General Standard for Contaminants and Toxins in Foods and Feeds), March 26-30, 2012. Accessed 1 March 2014. Available:ftp://ftp.fao.org/codex/meetings/cccf/cccf6/cf06_INFe.pdf.

159. United States Pharmacopeial Convention Chapter 2232 Elemental Contaminants in Dietary Supplements, USP 36-NF31 S1.

160. American Herbal Product Association Guidance Policies. Accessed 2 December 2013. Available:http://www.ahpa.org/default.aspx?tabid=223.

161. Pietta P, Simonetti P, Gardana C, Brusamolino A, Morazzoni P, Bombardelli E. Relationship between rate and extent of catechin absorption and plasma antioxidant status. Biochem Mol Biol Int. 1998;46:895-903.

162. Chow HH, Hakim IA, Vining DR, Crowell JA, Ranger-Moore J, Chew WM, et al. Effects of dosing condition on the oral bioavailability of green tea catechins after single-dose administration of Polyphenon E in healthy individuals. Clin Cancer Res. 2005;11:4627-33.

163. Kapetanovic IM, Crowell JA, Krishnaraj R, Lindeblad M, Lyubimov A. Exposure and toxicity of green tea polyphenols in fasted and non-fasted dogs. Toxicol. 2009;260:28-36.

164. Wu KM, Yao J, Boring D. Green tea extract-idncued lethal toxicity in fasted but not in nonfasted dogs. Intl J Toxicol. 2011;30:19-20.

165. Ullman U, Haller J, Decourt JP, Girault N, Girault J, Richard-Caudron AS, et al. A single ascending dose study of epigallocatechin gallate in healthy volunteers. J Int Med Res. 2003;31:88-101.

166. Ullmann U, Haller J, Decourt JD, Girault J, Spitzer V, Weber P. Plasma-kinetic characteristics of purified and isolated green tea catechin epigallocatechin gallate (EGCG) after 10 days repeated dosing in healthy volunteers. Int J Vitam Nutr Res. 2004;74:269-78.

167. Yarnell E, Abascal K. Hepatotoxicity of botanicals. Alt Comp Ther. 2014;20:136-44.

168. Schmidt M, Schmitz HJ, Baumgart A, Guedon D, Netsch MI, Kreuter MH, et al. Toxicity of green tea extracts and their constituents in rat hepatocytes in primary culture. Food Chem Toxicol. 2005;43:307-14.

169. Galati G, Lin A, Sultan AM, O'Brien PJ. Cellular and *In vivo* hepatotoxicity caused by green tea phenolic acids and catechins. Free Radic Biol Med. 2006;40:570-80.

170. Nishikawa T, Nakajima T, Moriguchi M, Jo M, Sekoguchi S, Ishii M, et al. A green tea polyphenol, epigalocatchin-3-gallate, induces apoptosis of human hepatocellular carcinoma, possibly through inhibition of Bcl-2 family proteins. J Hepatol. 2006;44:1074-82.

171. Lee MJ, Maliakal P, Chen L, Meng X, Bondoc FY, Prabhu S, et al. Pharmacokinetics of tea catechins after ingestion of green tea and (-)-epigallocatechin-3-gallate by humans: formation of different metabolites and individual variability. Cancer Epidemiol Biomarkers Prev. 2002;11:1025-32.

172. Chan PC, Ramot Y, Malarkey DE, Blackshear P, Kissling GE, Travlos G, et al. Fourteen-week toxicity study of green tea extract in rats and mice. Toxicol Pathol. 2010;38:1070-84.

173. Hsu YW, Tsai CF, Chen WK, Huang CF, Yen CC. A subacute toxicity evaluation of green tea (*Camellia sinensis*) extract in mice. Food Chem Toxicol. 2011;49:2624-30.

174. Isbrucker RA, Edwards JA, Wolz E, Davidovich A ,Bausch J. Safety studies on epigallocatechin gallate (EGCG) preparations. Part 2: Dermal, acute, and short-term toxicity studies. Food Chem Toxicol. 2006;44:636-50.

175. Isbrucker RA, Edwards JA, Wolz E, Davidovich A, Bausch J. Safety studies on epigallocatechin gallate (EGCG) preparations. Part 3: Teratogenicity and reproductive toxicity studies in rats. Food Chem Toxicol. 2006;44:651-61.

176. Morita O, Knapp JF, Tamaki Y, Stump DG, Moore JS, Memec MD. Effects of green tea catechin on embryo-fetal development in rats. Food Chem Toxicol. 2009;47:1296-1303.

177. Morita O, Kirkpatrick JB, Tamaki Y, Chengelis CP, Beck MJ, Bruner RH. Safety assessment of heatp-sterilized green tea catechin preparation: A 6-month repeat-dose study in rats. Food Chem Toxicol. 2009;1760-70.

178. Takami S, Imai T, Hasumura M, Cho YM, Onose J, Hirose M. Evaluation of toxicity of green tea catechins with 90-day dietary administration to F344 rats. Food Chem

Toxicol. 2008;46:2224-9.
179. Chengelis CP, Kirkpatrick JB, Regan KS, Radovsky AE, Beck MJ, Morita O, et al. 28-Day oral (gavage) toxicity studies of green tea catechins prepared for beverages in rats. Food Chem Toxicol. 2008;46:978-89.
180. Hursel R, Westerterp-Plantenga. Catechin- and caffeine-rich teas for control of body weight in humans. Am J Clin Nutr. 2013;98:1682-93S.
181. Lambert JD, Kennett MJ, Sang S, Reuhl KR, Ju J, Yang CS. Hepatotoxicity of high oral dose (-)-epigallocatechin-3-gallate in mice. Food Chem Toxicol. 2010;48:409-16.
182. Bun SS, Bun H, Guédon D, Rosier C, Ollivier E. Effect of green tea extracts on liver functions in Wistar rats. Food Chem Toxicol. 2006;44:1108-13.
183. NTP Technical Report on the Toxicology of Green Tea Extract in F344/NTac Rats and B6C3F1/N Mice and Toxicology and Carcinogenesis Studies of Green Tea Extract in Wistar HAN[Crl:WI(Han)] Rats and B6C3Fa/N Mice (Gavage Studies). NTP TR 585. NIH Publication No. 14-5927; 2014.
184. Chen JH, Tipoe GL, Liong EC, So HS, Leung KM, Tom WM, et al. Green tea polyphenols prevent toxin-induced hepatotoxicity in mice by down-regulating inducible nitric oxide-derived prooxidants. Am J Clin Nutr. 2004;80:742-51.
185. Bruno RS, Dugan CE, Smyth JA, DiNatale DA, Koo SI. Green tea extract protects leptin-deficient, spontaneously obese mice from hepatic steatosis and injury. J Nutr. 2008;138:323-31.
186. Fiorini RN, Donovan JL, Rodwell D, Evans Z, Cheng G, May HD, Milliken et al. Short-term administration of (-)-epigallocatechin gallate reduces hepatic steatosis and protects against warm hepatic ischemia/reperfusion injury in steatotic mice. Liver Transpl. 2005;11:298-308.
187. Dobrzynska I, Sniecinska A, Skrzydlewska E, Figaszewski Z. Green tea modulation of the biochemical and electric properties of rat liver cells that were affected by ethanol and aging. Cell Mol Biol Lett. 2004;9:709-21.
188. Zhang XG, Xu P, Liu Q, Yu CH, Zhang Y, Chen SH, et al. Effect of tea polyphenols on cytokine gene expression in rats with alcoholic liver disease. Hepatobiliary Pancreat Dis Int. 2006;5:268-72.
189. Lee MJ, Wang ZY, Li H, Chen L, Sun Y, Gobbo S, et al. Analysis of plasma and urinary tea polyphenols in human subjects. Cancer Epidemiol Biomarkers Prev. 1995;4:393-9.
190. Yang CS, Chen L, Lee MJ, Balentine D, Kuo MC, Schantz SP. Blood and urine levels of tea catechins after ingestion of different amounts of green tea by human volunteers. Cancer Epidemiol Biomarkers Prev. 1998;7:351-4.
191. Chow HH, Cai Y, Hakim IA, Crowell JA, Shahi F, Brooks CA, et al. Pharmacokinetics and safety of green tea polyphenols after multiple-dose administration of epigallocatechin gallate and polyphenon E in healthy individuals. Clin Cancer Res. 2003;9:3312-9.
192. Gruenwald J, Brendler T, Jaenike C, editors. PDR for herbal medicines. 1st ed. Medical Economics Co. Inc. Montvale, NJ. 1998.
193. NIH. Green Tea (*Camellia sinesis*). LiverTox. 2013a. Accessed 26 October 2013. Available: http://livertox.nlm.nih.gov/GreenTea.htm.
194. García-Cortés M, Stephens C, Lucena MI, Fernández-Castañer A, Andrade RJ. Causality assessment methods in drug induced liver injury: Strengths and weaknesses. J Hepatol. 2011;55:583-91.
195. Federico A, Tiso A, Loguercio C. A case of hepatotoxicity caused by green tea. Free Radic Biol Med. 2007;43:474.
196. Rohde J, Jacobsen C, Kromann-Andersen H. Toxic hepatitis triggered by green tea. Ugeskr Laeger. 2011;173:205-6.
197. Teschke R, Wolff A, Frenzel C, Schulze J, Eickhoff A. Herbal hepatotoxicity: A tablular compilation of reported cases. Liver Int. 2012;32:1543-56.
198. Teschke R, Schwarzenboeck A, Hennermann KH. Causality assessment in hepatotoxicity by drugs and dietary supplements. Br J Clin Pharmacol. 2008;66:758-66.
199. NIH. Drug record. Herbals and dietary supplements. 2013b. Accessed 26 October 2013. Available: http://livertox.nih.gov/Herbals and Dietary Supplments.htm.
200. Teschke R, Frenzel C, Schulze J, Eickhoff A. Herbal hepatotoxicity: challenges and pitfalls of causality assessment methods. World J Gastroenterol. 2013;19:2864-82.
201. Frank J, George TW, Lodge JK, Rodriguez-Mateos AM, Spencer JPE,

Minihane AM, et al. Daily consumption of an aqueous greet tea extract supplement does not impair liver function or alter cardiovascular disease risk biomarkers in healthy men. J Nutr. 2009;139:58-62.
202. Tsao AS, Martin J, TangX-M, Lee JJ, El-Naggar AK, Wistuba I, Culotta et al. Phase II randomized, placebo-controlled trial of green tea extract in patients with high-risk oral premalignant lesions. Cancer Prev Res. 2009;931-41.
203. Bettuzzi S, Brausi M, Rizzi F, Castagnetti G, Peracchia G, Corti A. Chemoprevention of human prostate cancer by oral administration of green tea catechins in volunteers with high-grade prostate intraepithelial neoplasia: a preliminary report from a one-year proof-of-principle study. Cancer Res. 2006;66:1234-40.
204. Matsuyama T, Tanaka Y, Kaminmaki I, Nagao T, Tokimitsu I. Catechin safely improved higher levels of fatness, blood pressure, and cholesterol in children. Obesity. 2008;16:1338-48.
205. McLarty J, Bigelow RLH, Smith M, Elmajian D, Ankem M, Cardilli JA. Tea polyphenols decrease serum levels of prostate-specific antigen, hepatocyte growth factor, and vascular endothelial growth factor in prostate cancer patients and inhibit production of hepatocyte growth factor and vascular endothelial growth factor *In vitro*. Cancer Prev Res. 2009;2:673-82.
206. Fontana RJ, Seeff LB, Andrade RJ, Björnsson E, Day CP, Serrano J, et al. Standardization of nomenclature and causality assessment in drug-induced liver injury: summary of a clinical research workshop. Hepatol. 2010;52:730-42.
207. Larrey D, Faure S. Herbal medicine hepatotoxicity: A new step with development of specific biomarkers. J Hepatol. 2011;54:599-601.
208. Appelhans K, Frankos V. Herbal medicine hepatotoxicity revisited. J Hepatol. 2012;56:504-5.
209. Corsini A, Ganey P, Ju C, Kaplowitz N, Pessayre D, Roth R, Watkins PB, et al. Current challenges and controversies in drug-induced liver injury. Drug Saf. 2012;35:1099-117.
210. AFSSAPS. French Agency for Sanitary Safety of Health Products. Press release. Suspension of the authorization for placing on the market of the medicinal product EXOLISE® (epigallocatechol gallate); 2003. Accessed 14 July 2014. Available: http://www.ansm.sante.fr/S-informer/Presse-Communiques-Points-presse/Suspension-de-l-autorisation-de-mise-sur-le-marche-de-la-specialite-pharmaceutique-EXOLISE-R-gallate-d-epigallocatechol
211. García-Cortés M, Borraz Y, Lucena MI, Peláez G, Salmerón J, Diago M, et al. Liver injury induced by "natural remedies": An analysis of cases submitted to the Spanish Liver Toxicity Registry. Rev Esp Enferm Dig. 2008;100:688-95.
212. AFSSA. French Agency for Food Safety. Communication to the Working Group; 2009.
213. USP (U.S. Pharmacopeia). Powdered decaffeinated green tea extract: In-process revision. Pharmacopeial Forum. 2007;33:1220-3.
214. USP (U.S. Pharmacopeia). Update on the USP green tea extract monograph. 10 April 2009. Accessed 26 December 2013. Available: http://www.usp.org/usp-nf/notices/ retired-compendial-notices/update-usp-green-tea-extract-monograph.
215. Jin X, Zheng RH, Li YM. Green tea consumption and liver disease: A systematic review. Liver Int. 2008;28:990-6.
216. EFSA. EFSA Scientific Cooperation (ESCO) Report. Advice on the EFSA guidance document for the safety assessment of botanicals and botanical preparations intended for use as food supplements, based on real case studies EFSA J. 2009;7:280.
217. ANSES. French Agency for Food, Environmental and Occupational Health Safety. [The evaluation of risk of hepatotoxicity related to consumption of foods containing green tea.] 21011-SA-0108. 2012. Accessed 14 July 2014. Available:http://www.anses.fr/Documents/ NUT2011sa 0108.pdf.
218. EFSA Guidance. Guidance on the safety assessment of botanicals and botanical preparations intended for us as ingredients in food supplements. EFSA J. 2009;7:1249.
219. NFSA. 2014. Accessed 26 October 2013. Available:http://www.matportalen.no/ verktoy/ advarsler/gronn te ekstrakt kan gi lever bivirkninger.
220. DGCCRF. Directorate General for

Competition and Consumer del Fraud. Re: Restriction on the use of green tea preparations in food supplements. Note of information no. 2011-2060; 2011.
221. Belgisch Staatsblad. [Royal Order for botanicals.] 2012. Accessed 14 July 2014. Available: 15 August 2013. Available: http://www.ejustice.just.fgov.be/mopdf/2012/ 04/04_1.pdf.
222. Hathcock JN, Shao A. Expanded approach to Tolerable Upper Intake guidelines for nutrients and bioactive substances. J Nutr. 2008;138:1992-5S.
223. EMA. European Medicines Agency, Committee for Medicinal Products for Human Use. 22 March 2007.Accessed 18 January 2014. Available: http://www.ema.europa.eu/docs/en_GB/document library/Scientific_guideline/2009/09/WC500002989.pdf.
224. Renwick AG. Risk characterization of chemicals in food. Toxicol Lett. 2004;149:163-76.
225. PlantLIBRA. EC Project No. 245199. 2013. Accessed 26 October 2013. Available: http://www.plantlibra.eu/web/node/44.
226. FDA. 21 CFR 182.20 Part 182 -- Substances generally recognized as safe, Subpart A--General Provisions. Sec. 182.20 Essential oils, oleoresins (solvent-free), and natural extractives (including distillates). In: Services DoHaH, editor. Washington, DC: United States Food and Drug Administration; 2012. Accessed 8 November 2013. Available: http://www.accessdata.fda.gov/scripts/cdrh/cfdocs/cfcfr/CFRSearch.cfm?fr=182.20.
227. AIBMR Life Sciences, GRAS Self-determination Inventory Database. Accessed 1 March 2014. Available: http://www.aibmr.com/resources/GRAS-database.
228. Assuming maximum use of 540 mg catechins/350 mL, with up to 1 to 1.5 g green tea catechins presumed safe; FDA GRN No. 225 & 259. Accessed 1 March 2014. Available: http://www.accessdata.fda.gov/scripts/fcn/gras_notices/grn000225.pdf.
229. FDA GRN No. 338. Accessed 1 March 2014. Available: http://www.accessdata.fda.gov/scripts/fcn/gras_notices/GRN000338.pdf.
230. FAO/WHO. A model for establishing upper levels of intake for nutrients and related substances. FAO/WHO Technical Workshop on Nutrient Assessment. Geneva. 2005. Accessed 22 April 2014. Available:http://www.who.int/ipcs/highlights/full_report.pdf.
231. Food and Nutrition Board, Institute of Medicine. Dietary reference intakes: A risk assessment model for establishing upper intake levels for nutrients. National Academy Press. Washington, DC; 1998.
232. Taylor CL, Tetley EA. Nutrient risk assessment as a tool for providing scientific assessments to regulators. J Nutr. 2008;138:1996-2002.
233. Lambert JD, Sang S, Yang CS. Possible controversy over dietary polyphenols: Benefits vs. risks. Chem Res Toxicol. 2007;20:583-5.

3

Public Health Risk-benefit Assessment Associated with Food Consumption–A Review

Géraldine Boué[1,2], Sandrine Guillou[2,1], Jean-Philippe Antignac[3,4], Bruno Le Bizec[3] and Jeanne-Marie Membré[1,2*]

[1]INRA, Secalim UMR 1014, F-44026 Nantes, France.
[2]LUNAM University, Oniris, Nantes, France.
[3]LUNAM University, Oniris, Laboratory Study of Residues and Contaminants in Food (LABERCA), F-44307 Nantes, France.
[4]INRA, F-44307 Nantes, France.

Author's contribution

This work was carried out in collaboration between all authors. Authors GB, SG and JMM designed the structure of review. Author GB managed the literature search, synthesized the information and wrote the first draft of the manuscript. Authors SG, JPA, BLB and JMM revised the manuscript and contributed ideas throughout the review. All authors read and approved the final manuscript.

ABSTRACT

Background: In the food safety field, risk assessment, including microbial and chemical components, has been applied for many years. However, a whole and integrated public health assessment also depends on the nutritional composition of food. While the fact that foods and diets can be a source of both risks and benefits now appears undisputed, carrying out a risk-benefit assessment (RBA) is still an emerging and challenging scientific subject.
Aims: The purpose of the present review was to synthesize RBA studies associated with food consumption and to summarize the current methodological options and/or tendencies carried out in this field.
Methods: The different data sources explored included around 20 accessible databases using the main terms "risk", "benefit" and "food" as keyword enquiries in article title and full-text. The initial research process led to 3293 screened papers, 160 of which were examined in detail.
Results: There were 126 articles dealing with RBA studies and 34 with the RBA methodological framework. Most of the available papers dealt with the comparison of nutritional beneficial effects and chemical adverse effects related to fish consumption. The majority of studies undertook a comparison of consumer exposure to risks and benefits with regard to reference safety values. However, more varied studies have emerged during the last 15 years, contributing to the

**Corresponding author: Email: Jeanne-marie.membre@oniris-nantes.fr*

diversification and the development of this issue.
Conclusion: RBA appears to be a promising scientific discipline and should be the next step in assessing the overall impact of food on health.

Keywords: Risk-benefit assessment; food; chemistry; microbiology; nutrition.

1. INTRODUCTION

Food safety management has adopted a risk-based approach in both the microbiological and chemical fields. In this context, the impact of more and more hazards associated with food consumption is evaluated by a risk assessment framework. In the nutritional field, food is recognized as having beneficial effects on health but also adverse effects. As a result, the concept of an integrated risk-benefit assessment has emerged in the last decade.

The risk can be defined as the probability that an adverse health effect affecting an organism, a system, or a sub-population will occur, as a consequence of an exposure to a hazard in food [1]. In contrast, the benefit is defined as the probability that a positive health effect will occur. Risk and benefit can be simultaneously related to the consumption of most foods that are commonly associated with various types of microbial (e.g. pathogens), chemical (e.g. acute toxic or endocrine-disrupting substances), and/or nutritional (e.g. saturated fatty acids) hazards, together with beneficial nutritional components (e.g. unsaturated fatty acids).

Risk-benefit assessment (RBA) falls with in the concept of risk-benefit analysis, which is an integrative approach associating three interconnected and complementary parts: risk-benefit management, risk-benefit assessment, and risk-benefit communication. The EFSA agency [2] advises mirroring the traditional risk analysis process to undertake a risk-benefit analysis, while considering some differences like the addition of a benefit assessment and a risk-benefit comparison as illustrated in Fig. 1.

The objective of RBA is to assess risk and benefit scientifically and objectively in the same integrative methodology. Then, risk-benefit management sets up two kinds of public health action: modification of food standards, reconsidering legislation to improve the quality of food available, and establishing recommendations for consumers to change their food habits into a healthier diet and lifestyle (food choice, consumption habits and cooking practices).

Several studies of RBA have already been undertaken and methodological developments in this field were first carried out by the European Food Safety Authority (EFSA) [2,4] and the Netherlands National Institute for Public Health and the Environment (RIVM) [5]. Then, some European collaborative research projects have worked on the RBA framework through these programs: "Benefit Risk Assessment For Food" (BRAFO)[6], "Best PRActices of Risk-BEnefit Analysis" (BEPRARIBEAN)[7], "Quality of Life – Integrated Benefit and Risk Analysis" (QALIBRA) [8] and "Benefit-Risk Assessment for Food: an Iterative Value-of-Information Approach"(BENERIS) [9].

In this context, RBA is becoming an established discipline. The aim of our work was to synthesize RBA studies associated with food consumption and to summarize the methodologies in a common framework.

2. METHODS

The research of articles aimed to collect RBA studies associated with food consumption and information on the RBA methodology. We followed the PRISMA data search process advised by Moher et al. [10] to organize the research of articles.

Databases explored included Web of Science, PubMed, MEDLINE, CABI, FSTA, ScieLO, Science Direct, EBSCO HOST, ACS Publications, Annual Reviews, edp Sciences, Endocrine Society, Cambridge Journals, NRC, High wire Press, World Cat, Science.gov and Google Scholar. Other sources were explored like Google, citation tracking, key journal search etc.

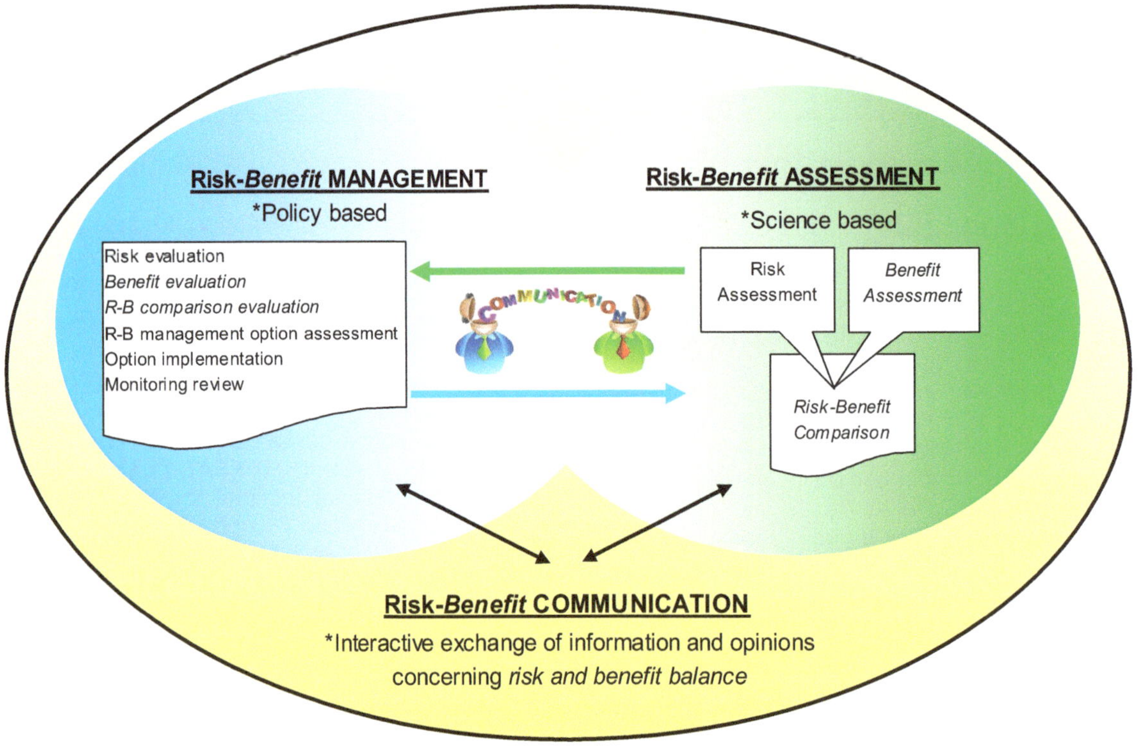

Fig. 1. Evolution of the conventional risk assessment conceptual framework toward an integrated risk-benefit paradigm, adapted from WHO [3]. The new elements are indicated in italics

These data searches were restricted to articles introducing RBA in terms of public health associated with food consumption in the fields of nutrition, chemistry, and microbiology. Only studies written in English or French without a publication date restriction were considered. The latest research was undertaken on 20th May 2014.

The same research was done on all the databases mentioned above. First, the search was based on the keywords "Food", "Risk* AND (Benefit* OR Beneficial*)" in the title but this did not provide all the relevant articles. Therefore, the search criteria were extended to the topic. Unfortunately, some databases did not have the option to search by topic. In this case, the nearest available option was used or, if there was none, we looked for the word "food" in the whole article. Below are the keywords used when the topic option was available, and when it was not.

When the topic option was available (e.g. for Web of Science):

- TITLE: (risk* AND (benefit* OR beneficial*)) AND TOPIC: (food)
- TITLE: ((chemi* OR toxicolo* OR microbi* OR nutrition) AND (risk* AND (benefit* OR beneficial*))) and TOPIC: (food)
- TITLE: ((risk* AND (benefit* OR beneficial*)) AND (health)) and TOPIC: (food)
- TITLE: ((risk* AND (benefit* OR beneficial*)) AND (public health)) and TOPIC: (food)
- TITLE: ((risk* AND (Benefit* OR Beneficial*) AND (review)) and TOPIC: (food)
- TITLE: ((risk* AND (benefit* OR beneficial*)) AND (state of the art)) and TOPIC: (food)

When the topic option was not available (e.g. for Science Direct):

- TITLE(risk* AND (benefit* OR beneficial*)) and FULL-TEXT(food)
- TITLE((chemi* OR toxicolo* OR microbi* OR nutrition) AND (risk* AND (benefit* OR beneficial*))) and FULL-TEXT(food)
- TITLE((risk* AND (benefit* OR beneficial*)) AND (health)) and FULL-TEXT(food)

- TITLE((risk* AND (benefit* OR beneficial*)) AND (public health)) and FULL-TEXT(food)
- TITLE((risk* AND (Benefit* OR Beneficial*) AND (review)) and FULL-TEXT(food)
- TITLE((risk* AND (benefit* OR beneficial*)) AND (state of the art)) and FULL-TEXT(food)

The article screening was carried out in three consecutive steps. The first selection of articles was based on the title accordance with the terms searched, then the abstract was explored, and finally the full article was screened. Articles that met the following criteria were selected for inclusion:

- The full article was written in English or French.
- The article was specific to the food sector.
- The main subject was a study of RBA introducing a comparison of risk and benefit or was about the methodology of RBA.
- The RBA assessment was specific to the field(s) of nutrition and/or microbiology and/or chemistry. Other subjects, like economy and sociology, were excluded.
- Reviews dealing with risks and benefits of food, like a review of the positive and adverse health effects due to the consumption of a specific food, were also selected to identify potential RBA studies.

Regarding articles dealing with the RBA methodology, the different steps recommended to undertake an RBA and the terminologies used were identified in order to summarize a common framework, which is presented in the Results section. The RBA studies identified were classified into two groups: performed and potential studies. For each study undertaken, the topic, the scientific field (microbiology, nutrition and chemistry), the type of comparison and the main results are presented in Table S1 (Appendix). Potential studies were investigated to compile a non-exhaustive list of future research needed in RBA.

3. RESULTS

Based on the research process, 3293 papers were identified comprising 2896 peer-reviewed articles found through databases and 397 from other sources corresponding to the grey literature (mainly scientific reports and theses) or from on-line documents (website pages, electronic articles, web-seminars). The results and process are summarized in Fig. 2. The screening step excluded 1819 papers by title checking and 182 by abstract reading. The screening was extensive because RBA is also an important topic in medicine, with the aim of balancing the beneficial effects of drugs against their potential adverse effects.

At the end of the query process, 160 articles were included in the review, 126 dealing with RBA studies (70 applications with recommendations and 56 studies on positive and negative health effects), and 34 with the RBA methodological framework.

3.1 Studies of Risk-benefit Assessment

There were 70 articles reporting RBA applied to food. In this section, these are presented chronologically, by scientific discipline (microbiology, chemistry, and/or nutrition), by comparison criteria and by category of applications. Beside RBA studies in the strictest sense, there were also 56 studies on positive and negative health effects, which could potentially be used in RBA.

3.1.1 History of RBA studies

The first RBA study appeared in 1999. Since then, the number has increased gradually Fig. 3. The first case study undertaken concerned the assessment of fish consumption, which is still by far the most studied topic (70% of RBAs). Fish consumption is a well-known source of both health benefits provided by omega-3 and risks due to environmental pollutants (dioxins, PCBs and methyl mercury). These RBAs have often been conducted at the level of a specific country by food safety agencies or various scientific groups.

Beside RBA on fish, many other case studies have emerged: supplementation or fortification of foods, assessment of nitrates and nitrites in fruits and vegetables, food-specific molecules such as acrylamide created during the manufacturing process, water and milk treatment, replacement of sugar by intense sweeteners, consumption of trans-unsaturated fatty acids, fish cooking practices, etc. Fig. 3.

3.1.2 Scientific fields of RBA studies

All these studies fall within the fields of nutrition and/or microbiology and/or chemistry. However, only a few studies have performed an integrated approach including these three disciplines Fig. 4.

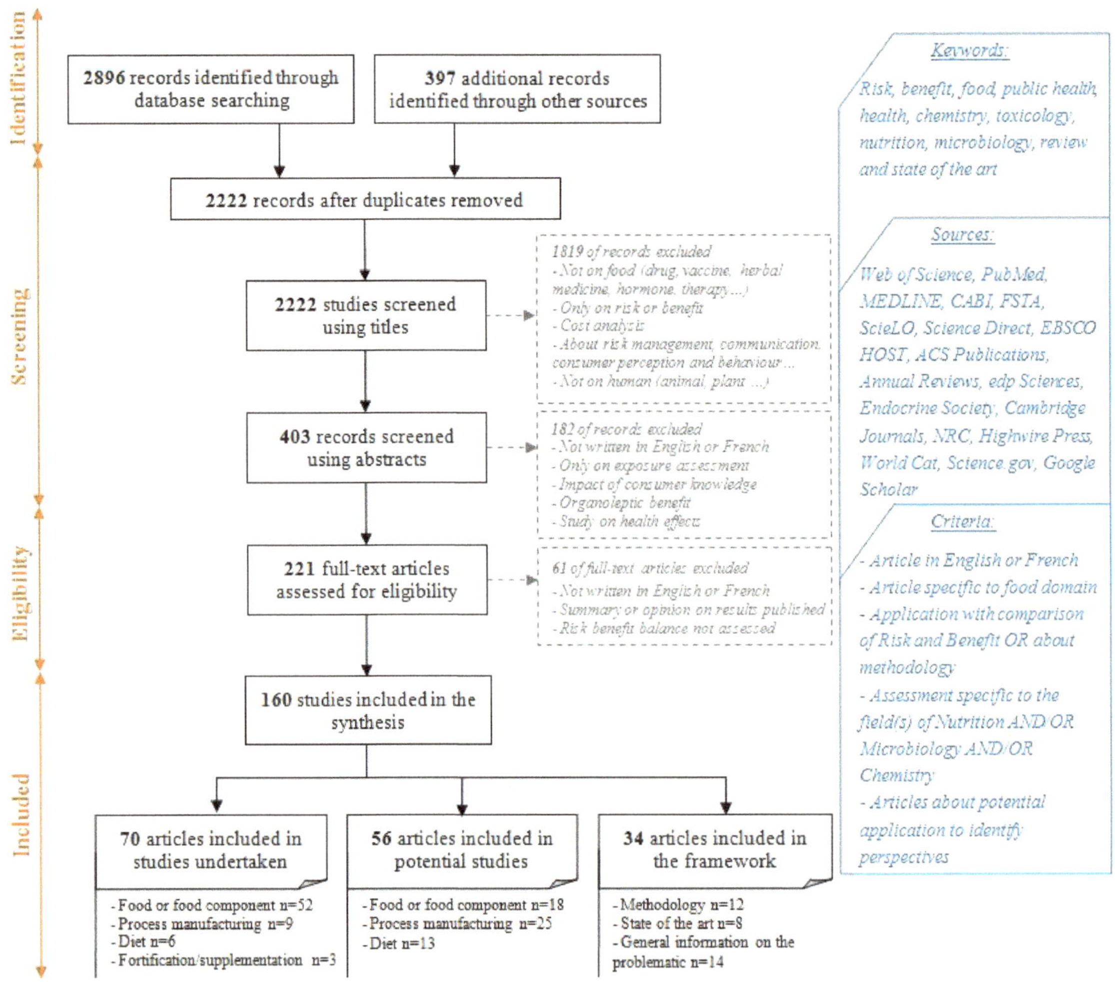

Fig. 2. Over view of the applied data search process and the results based on PRISMA's four-phase diagram [10]

Moreover, the three available studies [11-13] that integrated these three disciplines compared chemical and nutritional risks-benefits using safety reference values and gave recommendations on hygiene practices, which cannot be assimilated into a proper quantitative nutrition-chemical-microbial RBA. More generally, microbial risk is not often assessed in RBA and rarely in a quantitative way. Recently, Berjia et al. [14] carried out a comparison of nutritional benefits and microbiological risks associated with fish consumption.

3.1.3 Comparison criteria used in RBA

Different criteria are used to compare risks and benefits:

1 *Comparison of risks and benefits under constraints, based upon safety reference values.* This is a comparison of scenarios of consumer exposure. For each scenario of consumption, consumers are exposed to different risks and benefits related to the field of chemistry and/or nutrition and/or microbiology. The aim of this comparison is to set a threshold in accordance with safety levels set by food safety agencies. Regarding the risks identified, this threshold is set below the maximum levels of tolerable exposure (i.e. Acceptable Daily Intake, Tolerable Daily Intake, Upper Limit) and in agreement with nutritional intake recommendations (Recommended Daily Allowance, Estimated Average Requirement). Above this threshold, consumers could be exposed to a risk. Then, benefits are maximized, if possible, with respect to this threshold. This comparison can be considered semi-

quantitative because the RBA output is not expressed in a quantitative way (even if the assessment in chemical, microbiological or nutritional field might be quantitative). In addition, the process is likely to be iterative: RBA conclusions will be revised as often as the safety levels are reviewed. A comparison under constraints has been performed 46 times among the 70 studies Fig. 5.

2 *Comparison of risks and benefits based upon health endpoint*. For example, risk can be expressed as the probability of increasing the prevalence of coronary heart disease and benefit as the probability of decreasing this prevalence. It might also be expressed using the intellectual quotient (IQ) endpoint. A comparison based upon health endpoint has been performed 15 times among the 70 studies Fig. 5. Only articles that compared health endpoints one by one were included in this group.

3 *Comparison of risks and benefits based upon a composite metric* like the Disability Adjusted Life Years (DALY). This aims to compare quantitatively the impact of different diseases all together, contrary to the last group. It provides a comprehensive assessment of the consequences of a disease by integrating the quality of life lost(w) after the disease onset, the duration of the disease (Years of Life with Disability, YLD) and Years of Life Lost (YLL) [15]. At an individual scale, the DALY metric is calculated as indicated in Eq. 1, and is illustrated in Fig. 6 by the case of a person who has fallen sick and died after a period of life with a disability.

$$DALY = w.\ YLD + YLL \quad (1)$$

The use of the DALY metric as a comparison criterion requires many data, which are unfortunately not always available. However, to avoid this problem, epidemiological data can be used to inform the probabilities of falling ill, dying and recovering, as was done by Hoekstra et al. [16] and Berjia et al. [14]. A comparison using a composite metric has been performed 9 times among the 70 studies Fig. 5.

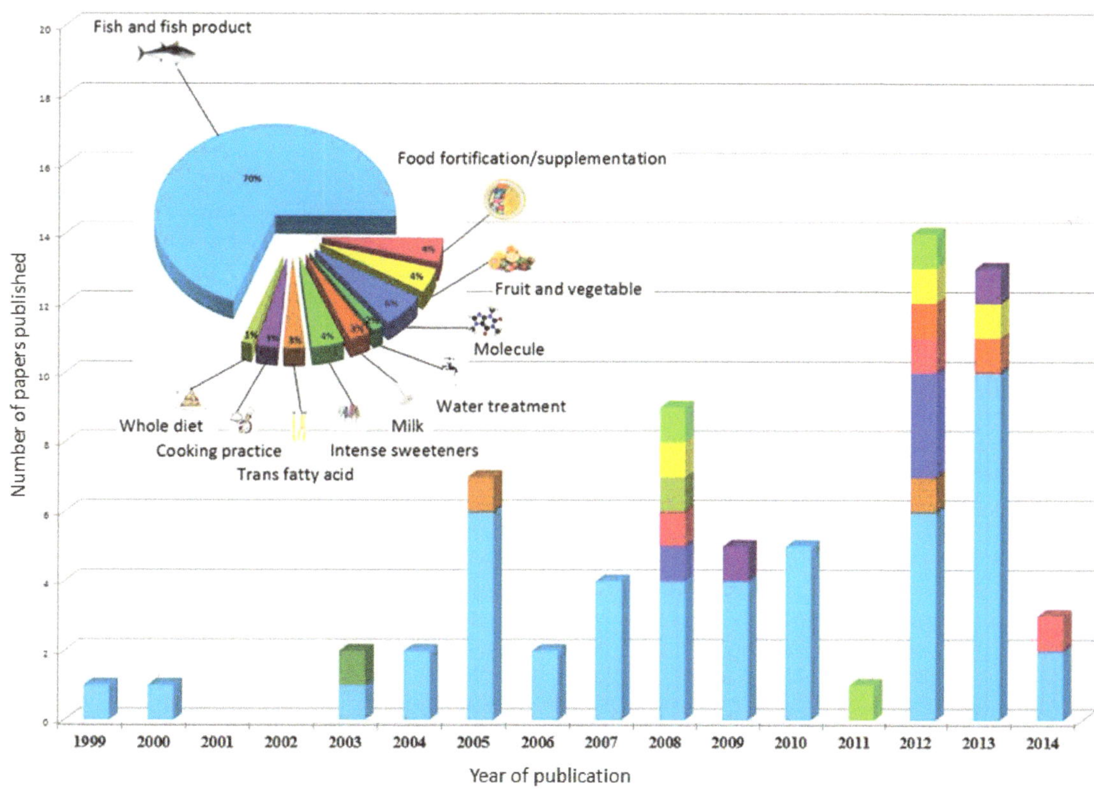

Fig. 3. Classification of the 70 studies performed by year and food category

3.1.4 Main RBA applications

Studies undertaken in recent years have resulted in progress in scientific knowledge in RBA. They have also enabled the food authorities to make recommendations on food consumption, such as the EFSA on fish consumption [18]. More generally, RBA research has led to promising applications, which can be schematically split into two categories: those leading to recommendations by food safety authorities and those leading to process and formulation design by manufacturers.

The applications are listed below. However, it is important to keep in mind that the conclusions presented here are extremely summarized, and can by no means be considered definitive statements concerning recommendations. More information on each study is provided in Table S1 (Appendix). However, for a comprehensive view of the study, please refer to the original paper.

- ***Applications leading potentially to recommendations***
  - ***Impact of a specific food on health***

The most investigated case study is fish. Fish contains docosahexaenoic (DHA) and eicosapentaenoic fatty acids (EPA) recognized for their health benefits but it is also contaminated by pollutants such as methyl mercury and dioxins, sources of adverse effects now clearly demonstrated. Fish composition is also dependent on fish species, fish feeding and place of production, which considerably influence its chemical contamination and fatty acid content. In addition, health effects vary greatly according to the subpopulation affected, which is a major issue in RBA. This topic has been investigated for 15 years and is still in progress because of its complexity. Overall, each study tackles the same subject (fish consumption) but brings new information by studying particular conditions (assimilated into co-variables in the analysis) affecting the risk-benefit assessment.

The overall recommendation is to consume two fish dishes per week, including one with fatty fish [11,18-20], while alternating fish species, production type (farmed or wild) and production location. The recommendation varies from strictly two portions per week of fatty fish, including ¼ of lean fish [21,22], to two to three servings per week [12,23]. Some studies also give specific recommendations according to the subpopulation at risk, such as women of childbearing age and children [24].

Other studies have compared the impact of risks and benefits on specific health endpoints and have given ranges of recommendations to minimize the risk of stroke [25], coronary heart disease (CHD) [26,27], and IQ change in the newborn [28,29] or stroke and fetal development disturbance [30].

In addition, as highlighted by Cardoso et al. [31], the risk-benefit balance of fish consumption varies between countries. RBAs have been carried out at a country level in Norway [32], the Netherlands [14,16], Poland [33,34], France [35], China [36-40], the USA [41,42] and Bermuda [43]. In addition, the type of fish species could change the risk-benefit balance [44-48]. Likewise, the type of farming may have an impact [49,50]. As a result of these two factors (population and fish species), some specific populations could be negatively impacted by fish consumption. For instance, the Portuguese population, which consumes about 57kg of fish per year, should favor certain fish species to limit the potential risk due to high intake [51-53]. Likewise, the Inuit population should limit its fish consumption [54]. Conversely, the Kahnawake community south of the St Laurence river, also high fish consumers, is not exposed to risk [55].

The complexity of the assessment of fish intake is increased by the fact that fish consumption by pregnantor lactating women or women of childbearing age could impact the newborns' neurodevelopment and thus increase or decrease their IQ [56-58].

Finally, a few quantitative RBAs regarding fish consumption have been performed, providing figures that enable RBA recommendations to be deciphered. For example, in the US adult population, the current fish consumption enables to gain 5000 healthy years per year per 100000 people, calculation based on the Washington state [59]; also in US, a 50% increase in fish consumption could save 120000 years annually of perfect health for people [60]. More specifically, based on a French study on 1011 people, it was concluded that a weekly consumption of 1104 g of fish could save between 97 and 285 healthy years annually [35]. This example demonstrates that a quantitative comparison of risks and benefits is more transparent and objective than a comparison under constraints.

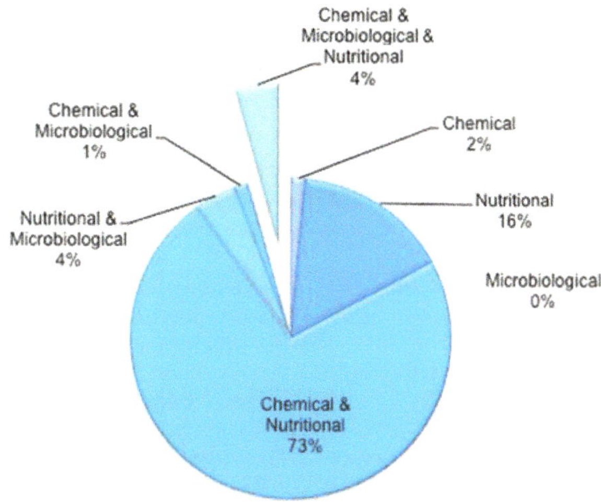

Fig. 4. RBA studies performed classified by scientific fields, based on 70 studies

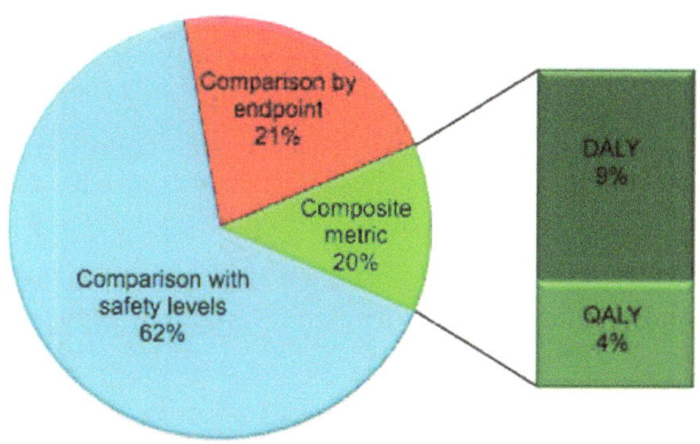

Fig. 5. RBA studies performed classified by type of comparison, based on 70 studies

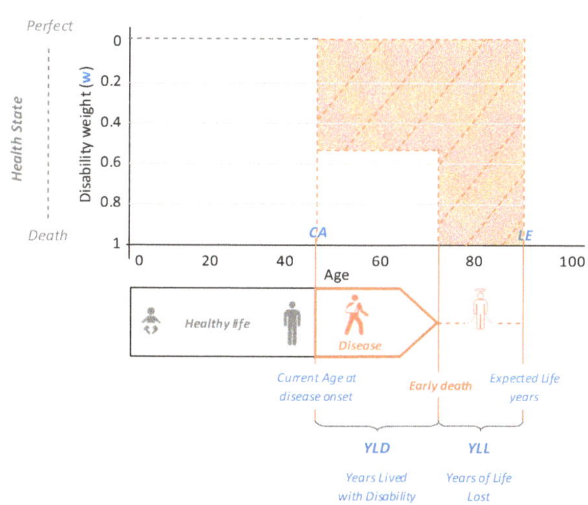

Fig. 6. Illustration of DALY adapted from Tijhuis et al. [17] with the case of a person who falls sick and dies after a period of disability

Recommendations concerning other food categories have also been given. Although not based upon a quantitative comparison, it has been pointed out that the intake of fruits and vegetables [61,62] and soy proteins [48] should be increased since these food categories do not expose consumers to risk. In contrast, the intake of trans fatty acids should be limited [63,64].

- ***Impact of a particular type of diet on health***

The type of diet has also been studied through RBA to assess its overall impact on health.

Replacement of sugar by intense sweeteners has been judged healthy because it prevents over weight and caries [64,65] although risks can outweigh benefits for children who are high consumers of soft drinks with a potential risk of exceeding the acceptable limit of intense sweetener intake [66].

Software has been developed to assess individually risk-benefit related to diet. Some programs are specific to a product, e.g. fish consumption [67,68], while others include a wide range of foods [69,70].

- ***Applications leading potentially to process and formulation design***

- ***Impact of manufacturing process on health***

The manufacturing process is identified as a source of risks and benefits because it could introduce risk and/or benefit or modify the risk-benefit balance.

Water treatment decreases microbial contamination but introduces chemical risk at the same time. The balance has been quantitatively assessed by Havelaar et al. [71]who demonstrated that the benefit outweighed the risk. Milk treatment is also beneficial because it decreases microbial risk in spite of biochemical reactions [72].

RBA may be used as a tool to optimize the process line by assessing the impact of different production parameters on the risk-benefit balance. Rigaux et al. [73] has optimized the thermal sterilization of vegetables to maximize vitamin concentration without exposing consumers to microbial risk. Likewise, the thermal process of cookies might be optimized to enhance their antioxidant activity while limiting the formation of harmful compounds [74]. The type of thermal process also has an influence on food composition and thus on the risk-benefit balance. For instance, a comparison of fish cooking processes demonstrated that grilling is healthier than boiling or roasting [75]. More generally, to optimize the thermal process, it is necessary to analyze altogether the potential loss of nutritional properties, the possible formation of hazardous molecules such as acrylamide [72,76] and benzo(a) pyrene [64], and the efficiency of microbial inactivation.

- ***Impact of food formulation on health***

The positive impact of bread supplementation with folic acid on public health has been quantitatively assessed. In the Netherlands, a small supplementation of 70 µg per 100 g of bread could save 7000 healthy years annually [64] and a higher supplementation (i.e. 140 µg per 100 g of bread) could save 11812 healthy years annually [77].

It has been reported that margarine supplemented with plant sterol could save eight healthy years per 1000 people [78].

3.1.5 Studies on the positive and negative health effects associated with food consumption

Besides RBA studies, there were also 56 studies on the positive and negative health effects, which could potentially be used in RBA. A list of the main subjects of interest is provided below.

First, some foods or food components have been identified as ambivalent, i.e. food for which it is not straight forward to assess whether the risk is higher than the benefit or *vice versa.* Among them, it is worth mentioning: coffee [79-84], tea [85,86], alcohol [87-90], broccoli [91], meat [92,93], chocolate [94], phytoestrogen [95,96], is oflavone [97] and nitrite/nitrate [98,99].

Other issues related to food agricultural practices and food manufacturing practices have been pointed out [100]: organic food production [101], use of pesticides [102,103], use of genetically modified organisms [104-108], the thermal process [109], irradiation of food [110-112], use of artificial sweeteners [113-115], use of antimicrobials [116], red meat cooking practices [117], food fortification[118], the occurrence of the Maillard reaction [119], milk treatment [120,121], etc.

Finally, RBA related to diets, such as the Mediterranean diet [122,123], a raw diet [124], vegetarianism [125,126], and baby food infant formulae or breastfeeding [127-133], could be of interest.

3.2 Methodology of Risk-benefit Assessment

Risk-benefit assessment (RBA) is an emerging discipline and its framework is still in progress. However, important works have been carried out by European scientists to develop the RBA approach.

The search identified 34 documents related to the RBA framework. Twelve of them dealt with the methodology step by step. Among them, four papers were published by safety agencies, the EFSA [2,4] and RIVM [5,77], four others by the European projects BRAFO [6,134] and QALIBRA [8,135] and four by scientific researchers [136-139]. The European BEPRARIBEAN project [7] also contributed to developing this framework through six' states of the art' in risk-benefit analysis [17,140-144], concluded in Tijhuis et al. [145]. Fourteen other papers added information about the framework. The International Life Sciences Institute organized a session about the risk-benefit balance of food at the North America Annual Meeting in 2013; a presentation was made about the risk-benefit analysis of food [146], another about risk and benefit for chemical contaminants [147] and a third dealt with the risk-benefit assessment of nutrient intake[148]. Two other European projects, BENERIS [9,149] and Plantlibra [150], addressed this issue, two theses [56,151] were published, and other scientific researchers published articles [15,152-156] on more specific points of the framework.

The first work on RBA methodology was carried out by the EFSA in 2006 [4] followed in 2010 by their recommendations on risk-benefit analysis methodology [2]. In parallel, the RIVM published a decision tree [5]. Then the BRAFO working group suggested an integrative approach [6], applied its methodology to case studies [48,64,72] and published a consensus document [134].

Other works have contributed to the RBA framework development. For example, the QALIBRA project has provided online software [8] which enables a quantitative comparison of risk and benefit to be made based on DALY (Eq. 1) and Quality Adjusted Life Years (QALY). Within the BENERIS project, an information and exchange web-platform has been created [9]. The BEPRARIBEAN project has enabled good practices to be established in risk-benefit analysis [7,145] within various scientific fields: Medicine [140], Environmental Health [141], Food Microbiology [142], Economics and Marketing-Finance [143], Consumer Perception [144], and Food and Nutrition [17].

The RBA methodology is based on the risk assessment framework [3] universally applied in the fields of microbiology and chemistry, but a risk-benefit comparison step is added. The RBA framework is described below in detail and summarized in Fig. 7.

First, according to the papers investigated, there is a consensus to start the RBA by a preliminary step consisting of "0. Problem definition" [2,5,6,8], in order to define the case study (a food, a food compound or a diet), the (sub) population targeted, and different scenarios of consumer exposure to be assessed (reference and alternative scenarios).

Then, RBA mirrors a traditional risk assessment [2,4], which includes four steps: hazard identification, exposure assessment, hazard characterization and risk characterization [3]. However, the terminologies used need to be adapted to integrate the benefit assessment. In fact, in a risk assessment, the term "hazard" is used to define a biological, chemical or physical agent able to cause an adverse health effect [157]. The risk is thus "a function of the probability of an adverse health effect and the severity of that effect, consequential to a hazard(s) in food" [1]. The term "benefit" is unanimously used to mirror the risk but we found different terms used to mirror the term "hazard": "positive effect" [6,64,72,134], "benefit" [4,77,78], "positive health effect" [2,158], "beneficial effect" [17], etc. Nevertheless, in the field of nutrition, the same agent could be a source of risk and benefit depending on the consumer exposure [136]. In this context, we propose to use a more general term to encompass the term hazard and its counterpart on the benefit side. We have named this term "Health Effect Contributing Factor" (HECF) and we define it as an agent able to cause an adverse or a positive health effect in the case of exposure. We chose this term because an HECF could be positive and negative, thus applicable in the nutrition field. In addition, as a positive or beneficial (health) effect is the consequence of a benefit and not its source, as a hazard is for a risk, the use of the term HECF can skirt this problem. In the same way, we have grouped together the terms risk and benefit under the expression "Health Impact" (HI), which we define as a function of the probability of an adverse or positive health effect and the severity of that effect, resulting from exposure to an HECF. A positive HI is a benefit and a negative HI is a risk. In this conceptual framework, a decrease in risk is considered a benefit and a decrease in benefit is considered a risk Fig. 7.

After defining the problem (step 0), risk and benefit are assessed in parallel in each field (nutrition, chemistry and microbiology) following the risk assessment steps. If we introduce the terminologies proposed above, we can name the next four steps as follows:

"1. Identification of HECF",
"2. Exposure assessment",
"3. Characterization of HECF" and
"4. Health impact characterization".

At any step, even if the assessment is qualitative or semi-quantitative, EFSA and BRAFO [2,6] advise stopping the assessment if risk outweighs benefit or *vice versa*. Yet, Berjia et al. [14] illustrated in a cold-smoked salmon study that a quantitative comparison of risk and benefit could reverse the risk-benefit balance. However, due to a lack of data, a quantitative comparison is often not feasible. For these reasons, we suggest an alternative after step 4 Fig. 7. If the consumer is not exposed to both risk and benefit, there is no interest in performing a risk-benefit comparison, and the assessment is only performed from the risk side or from the benefit side. If the data available are too scarce to carry out a quantitative comparison, a comparison with a composite metric is not feasible but a comparison under constraints could be undertaken.

When the appropriate data are available and the risk-benefit comparison is of interest, a quantitative RBA can be performed. The assessment is extended to step"5. Harmonization of HI in the same metric" and then to step "6. Assessment of different scenarios of consumer exposure".

To move harmonization forward (step 5), there are still scientific bottlenecks. Indeed, risk assessment differs in each field because each has its own characteristics while the risk-benefit comparison aims to integrate all the results in the same metric. Performing a quantitative RBA is thus difficult due to the lack of a common unit to express the risk. Chemical risk assessment often expresses the risk as the probability of exceeding a threshold, or a safety reference value; microbiological risk assessment output is the probability of getting sick or dying from a disease; nutritional health assessment integrates two elements: deficiency or excess of a food component, and homeostasis (internal regulation to maintain a compound ata relatively constant concentration).

Finally, assessors report their conclusions to the decision-making managers who select the best scenario. At this stage, it is important to keep in mind that the best scenario is not necessarily the one corresponding to the best benefit-risk balance as the managers have to take other considerations into account, such as economic factors or food availability.

Two recent studies clearly illustrate how a quantitative RBA in the fields of microbiology/nutrition and chemistry/nutrition could be applied from step 0 to step 6 Fig. 7. They both carried out a full quantitative comparison of risks and benefits using DALY as a comparison criterion.

Berjia et al. [14] were the first scientists to perform a quantitative RBA in the fields of microbiology and nutrition. They balanced the risk of listeriosis due to cold-smoked salmon consumption with the health benefit due to omega-3 intake. They concluded that a change in the consumption of smoked salmon from the reference scenario (women 23 g/day and men 20 g/day) to the alternative scenario (40 g/day for adults) could save 9343 DALYs in the Danish population (5.57 million inhabitants), if the product was consumed before four weeks of storage. The sensitivity analysis highlighted that the net impact on health depends on the storage time of the product before consumption: from five weeks onwards, the net health impact is reversed and the overall effect is negative because of the increasing risk of listeriosis.

The second example of RBA was performed in the disciplines of chemistry and nutrition, which are currently those most explored. Hoekstra et al. [16] balanced the risk and benefit of fish consumption in Denmark. The net public health impact resulting from a change in the consumption of fish from 100 g/day to 200g/day could save 2.7 DALYs per 1000 people.

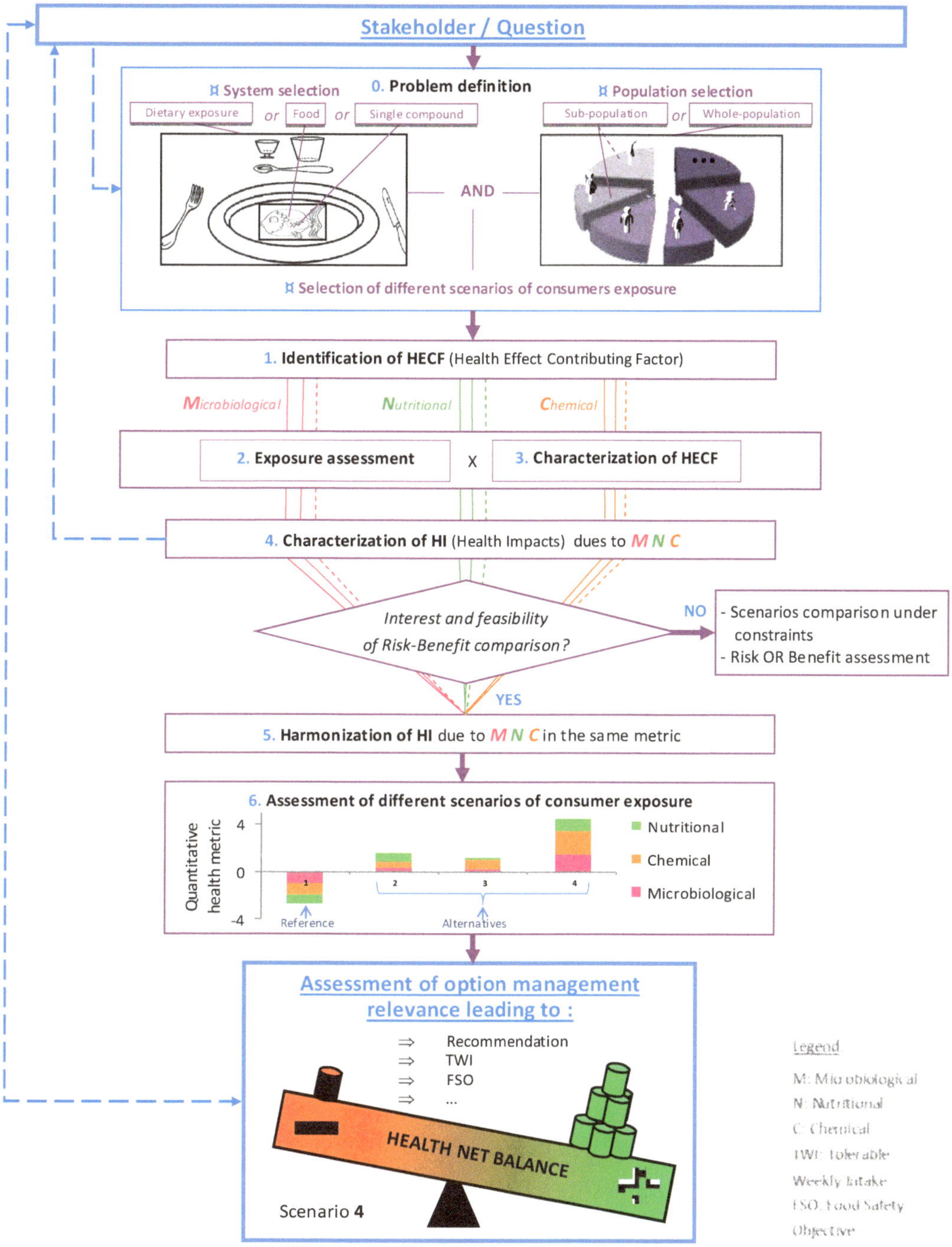

Fig. 7. Summary of risk-benefit assessment methodological framework based on different literature sources

4. CONCLUSION

The risk-benefit assessment discipline emerged at the beginning of the 21st century. RBA studies are intended to address various issues concerning the food supply-chain "from farm to fork". Although the first and most popular studies were related to fish consumption (48 of the 70 studies analyzed in this review), research has now diversified into a wider range of food categories such as fruits, vegetables and soy protein. The majority of RBA studies aimed to compare chemical risk with nutritional benefit (51 out of 70). The number of RBAs integrating components of nutrition, chemistry and microbiology was relatively low (3 out of 70); moreover, they were not fully quantitative but limited to a comparison under constraints (i.e. comparison of consumer exposure to reference safety levels).

Although the methodology is still in progress, these studies followed the same overall methodology based on the universal risk assessment framework [3] as advised by the EFSA [2]. Risks and benefits are first assessed independently and then compared with each other. This comparison can be made under constraints (46 out of 70 studies), based on health endpoints (15 out of 70) or using a composite metric such as DALY (9 out of 70). This lattermetric is a practical tool to compare the effect of different diseases on health, integrating their severity and duration. To generalize further the use of a composite metric as a comparison criterion, the harmonization of scientific approaches needs to be enhanced; in particular, output risk (or benefit) assessment has to be expressed in a common unit.

To conclude, RBA is currently recognized as a scientific discipline with a wide range of applications. It is becoming a tool used in public health management, for instance in food recommendations on fish consumption [11,18,28,32]. It might be used in the future by food manufacturers as an aid in process and formulation design [72,77].

COMPETING INTERESTS

Authors have declared that no competing interests exist.

REFERENCES

1. IPCS. IPCS Harmonization Project - IPCS risk assessment terminology. WHO, Geneva. 2004;1-122.
2. EFSA. Guidance on human health risk-benefit assessment of foods. Scientific opinion. EFSA Journal. 2010;8(1673):1-40.
3. WHO. Risk assessment. Accessed 12 February 2014.
   Available: http://www.who.int/foodsafety/micro/riskassessment/en/index.html.
4. EFSA. EFSA Science Colloquium 6. Risk benefit analysis of foods: methods and approaches. Scientific Colloquium Series of the European Food Safety Authority No.6; 2006.
5. Fransen H, de Jong N, Hendriksen M, Mengelers M, Castenmiller J, Hoekstra J, et al. A tiered approach for risk-benefit assessment of foods. Risk Anal. 2010;30(5):808-816.
6. Hoekstra J, Hart A, Boobis A, Claupein E, Cockburn A, Hunt A, et al. BRAFO tiered approach for benefit-risk assessment of foods. Food Chem Toxicol. 2012;50:684-698.
7. Verhagen H, Tijhuis MJ, Gunnlaugsdóttir H, Kalogeras N, Leino O, Luteijn JM, et al. State of the art in benefit–risk analysis: Introduction. Food Chem Toxicol. 2012;50(1):2-4.
8. Hart A, Hoekstra J, Owen H, Kennedy M, Zeilmaker MJ, de Jong N, et al. Final qalibra framework for risk-benefit assessment (2010). Accessed 26 February 2014.
   Available:http://www.qalibra.eu/showFile.cfm?filename=D29c_final_framework_paper.pdf.
9. BENERIS. Beneris. Accessed 26 February 2014; 2011.
   Available: http://en.opasnet.org/w/Beneris.
10. Moher D, Liberati A, Tetzlaff J, Altman DG. Preferred reporting items for systematic reviews and meta-analyses: The PRISMA Statement. Plos Med. 2009;6(7).
11. ANSES. Avis de l'Agence nationale de sécurité sanitaire de l'alimentation, de l'environnement et du travail relatif aux recommandations sur les bénéfices et les risques liés à la consommation de produits de la pêche dans le cadre de l'actualisation des repères nutritionnels du PNNS. In: 2012-SA-0202, A.S.n., (Ed.), Maisons-Alfort, France; 2013. French.
12. Nesheim MC, Yaktine AL. Seafood choices: Balancing benefits and risks. Institute of Medicine, National Academies: Washington, DC; 2007.
13. VKM. Benefit and risk assessment of breastmilk for infant health in Norway - Opinion of the Steering Committee of the Norwegian Scientific Committee for Food Safety, Norwegian Scientific Committee for Food Safety (VKM). Norway; 2013.
14. Berjia FL, Andersen R, Hoekstra J, Poulsen M, Nauta M. Risk-benefit assessment of cold-smoked salmon:

Microbial risk versus nutritional benefit. Eur J Food Res Rev. 2012;2(2):49-68.
15. Gold MR, Stevenson D, Fryback DG. HALYs and QALYs and DALYs, oh my: Similarities and differences in summary measures of population health. Annu Rev Publ Health. 2002;23:115-134.
16. Hoekstra J, Hart A, Owen H, Zeilmaker M, Bokkers B, Thorgilsson B, et al. Fish, contaminants and human health: Quantifying and weighing benefits and risks. Food Chem Toxicol. 2013;54:18-29.
17. Tijhuis MJ, de Jong N, Pohjola MV, Gunnlaugsdottir H, Hendriksen M, Hoekstra J, et al. State of the art in benefit-risk analysis: Food and nutrition. Food Chem Toxicol. 2012;50(1):5-25.
18. EFSA. Opinion of the Scientific Panel on contaminants in the food chain on a request from the European Parliament related to the safety assessment of wild and farmed fish Question N° EFSA-Q-2004-22 Adopted on 22 June 2005. EFSA Journal. 2005;236:1-118.
19. AFSSA. Opinion of the French Agency for Food Safety on benefits / risks associated with fish consumption. In: 2008-SA-0123, A.S.n., (Ed.), Maisons-Alfort, France; 2008. French.
20. SACN/COT. Scientific Advisory Committee on Nutrition/Committee of Toxicity. Advice on fish consumption: benefits & risks. London:TSO ; 2004.
21. Sirot V. An analytical approach to risk / benefit of the consumption of fish and seafood, Maisons Alfort, Paris; 2010.French.
22. Sirot V, Leblanc JC, Margaritis I. A risk-benefit analysis approach to seafood intake to determine optimal consumption. Br J Nutr. 2012;107(12):1812-1822.
23. Becker W, Darnerud PO, Petersson-Grawé K. A risk-benefit analysis based on the occurrence of Dioxin/PCB, Methyl Mercury, n-3 Fatty Acids and Vitamin D in Fish, National Food Administration, Sweden; 2007.
24. Balshaw S, Edwards JW, Daughtry BJ, Ross KE. Risk-benefit analysis of fish consumption: Fatty acid and mercury composition of farmed southern bluefin tuna, Thunnus maccoyii. Food Chem. 2012;131(3):977-984.
25. Bouzan C, Cohen JT, Connor WE, Kris-Etherton PM, Gray GM, Konig A, et al. A quantitative analysis of fish consumption and stroke risk. Am J Prev Med. 2005;29(4):347-352.
26. König A, Bouzan C, Cohen JT, Connor WE, Kris-Etherton PM, Gray GM, et al. A Quantitative Analysis of Fish Consumption and Coronary Heart Disease Mortality. Am J Prev Med. 2005;29(4):335-346.
27. Mozaffarian D, Rimm EB. Fish intake, contaminants, and human health - evaluating the risks and the benefits. Jama-J Am Med Assoc. 2006;296(15):1885-1899.
28. FAO/WHO. Joint FAO/WHO Expert Consultation on the Risks and Benefits of Fish Consumption, Roma; 2010.
29. Ginsberg GL, Toal BF. Quantitative approach for incorporating methylmercury risks and Omega-3 fatty acid benefits in developing species-specific fish consumption advice. Environ Health Perspect. 2009;117(2):267-275.
30. FDA. Draft report of quantitative risk and benefit assessment of consumption of commercial fish, focusing on fetal neurodevelopmental effects (Measured by Verbal Development in Children) and on coronary heart disease and stroke in the general population. Draft report (2009). Accessed 4 March 2014.
Available:http://www.fda.gov/food/foodborneillnesscontaminants/metals/ucm088794.htm.
31. Cardoso C, Bandarra N, Lourenco H, Afonso C, Nunes M. Methylmercury risks and EPA plus DHA benefits associated with seafood consumption in Europe. Risk Anal. 2010;30(5):827-840.
32. VKM. Fish and seafood consumption in Norway-Benefits and risks. Norwegian scientific committee for food safety, March 2006. English summary (2006). Accessed 13 March 2014. Available: http://www.vkm.no/dav/83cf7bd765.pdf.
33. Usydus Z, Szlinder-Richert J, Polak-Juszczak L, Kanderska J, Adamczyk M, Malesa-Ciecwierz M, et al. Food of marine origin: Between benefits and potential risks. Part 1. Canned fish on the Polish market. Food Chem. 2008;111(3):556-563.
34. Usydus Z, Szlinder-Richert J, Polak-Juszczak L, Komar K, Adamczyk M, Malesa-Ciecwierz M, et al. Fish products available in Polish market – Assessment of the nutritive value and human exposure to

dioxins and other contaminants. Chemosphere. 2009;74(11):1420-1428.
35. Guevel MR, Sirot V, Volatier JL, Leblanc JC. A risk-benefit analysis of French high fish consumption: A QALY approach. Risk Anal. 2008;28(1):37-48.
36. Chen MYY, Wong WWK, Chung SWC, Tran CH, Chan BTP, Ho YY, et al. Quantitative risk-benefit analysis of fish consumption for women of child-bearing age in Hong Kong. Food Addit Contam. 2014;31(1):48-53.
37. Du ZY, Zhang J, Wang CR, Li LX, Man QQ, Lundebye AK, et al. Risk-benefit evaluation of fish from Chinese markets: Nutrients and contaminants in 24 fish species from five big cities and related assessment for human health. Sci Total Environ. 2012;416:187-199.
38. Gao YX, Zhang H, Yu X, He JL, Shang X, Li X, et al. Risk and benefit assessment of potential neurodevelopmental effect resulting from consumption of marine fish from a coastal archipelago in China. J Agric Food Chem; 2014.
39. Zhang DP, Zhang XY, Yu YX, Li JL, Yu ZQ, Wu MH, et al. Tissue-specific distribution of fatty acids, polychlorinated biphenyls and polybrominated diphenyl ethers in fish from Taihu Lake, China, and the benefit-risk assessment of their co-ingestion. Food Chem Toxicol. 2012;50(8):2837-2844.
40. Zhang DP, Zhang XY, Yu YX, Li JL, Yu ZQ, Wang DQ, et al. Intakes of omega-3 polyunsaturated fatty acids, polybrominated diphenyl ethers and polychlorinated biphenyls via consumption of fish from Taihu Lake, China: A risk–benefit assessment. Food Chem. 2012;132(2):975-981.
41. Sidhu KS. Health benefits and potential risks related to consumption of fish or fish oil. Regul Toxicol Pharmacol. 2003;38(3):336-344.
42. Gochfeld M, Burger J. Good fish/bad fish: A composite benefit-risk by dose curve. Neurotoxicology. 2005;26(4):511-520.
43. Dewailly E, Rouja P, Dallaire R, Pereg D, Tucker T, Ward J, et al. Balancing the risks and the benefits of local fish consumption in Bermuda. Food Addit Contam. 2008;25(11):1328-1338.
44. Cardoso C, Afonso C, Lourenço H, Nunes ML. Seafood consumption health concerns: The assessment of methylmercury, selenium, and eicosapentaenoic + docosahexaenoic fatty acids intake. Food Control. 2013;34(2):581-588.
45. Dewailly E, Ayotte P, Lucas M, Blanchet C. Risk and benefits from consuming salmon and trout: A Canadian perspective. Food Chem Toxicol. 2007;45(8):1343-1348.
46. Gladyshev MI, Sushchik NN, Anishchenko OV, Makhutova ON, Kalachova GS, Gribovskaya IV. Benefit-risk ratio of food fish intake as the source of essential fatty acids vs. heavy metals: A case study of Siberian grayling from the Yenisei River. Food Chem. 2009;115(2):545-550.
47. Loring PA, Duffy LK, Murray MS. A risk-benefit analysis of wild fish consumption for various species in Alaska reveals shortcomings in data and monitoring needs. Sci Total Environ. 2010;408(20):4532-4541.
48. Watzl B, Gelencser E, Hoekstra J, Kulling S, Lydeking-Olsen E, Rowland I, et al. Application of the BRAFO-tiered approach for benefit-risk assessment to case studies on natural foods. Food Chem Toxicol. 2012;50:699-709.
49. Foran JA, Good DH, Carpenter DO, Hamilton MC, Knuth BA, Schwager SJ. Quantitative analysis of the benefits and risks of consuming farmed and wild salmon. J Nutr. 2005;135(11):2639-2643.
50. Hites RA, Foran JA, Carpenter DO, Hamilton MC, Knuth BA, Schwager SJ. Global assessment of organic contaminants in farmed salmon. Sci. 2004;303(5655):226-229.
51. Afonso C, Cardoso C, Lourenço HM, Anacleto P, Bandarra NM, Carvalho ML, et al. Evaluation of hazards and benefits associated with the consumption of six fish species from the Portuguese coast. J Food Compos Anal. 2013;32(1):59-67.
52. Afonso C, Lourenço HM, Cardoso C, Bandarra NM, Carvalho ML, Castro M, et al. From fish chemical characterisation to the benefit-risk assessment – Part A. Food Chem. 2013;137(1–4):99-107.
53. Maulvault AL, Cardoso C, Nunes ML, Marques A. Risk-benefit assessment of cooked seafood: Black scabbard fish (*Aphanopus carbo*) and edible crab (*Cancer pagurus*) as case studies. Food Control. 2013;32(2):518-524.
54. Laird BD, Goncharov AB, Egeland GM, Chan HM. Dietary advice on inuit

traditional food use needs to balance benefits and risks of Mercury, Selenium, and n3 Fatty Acids. J Nutr. 2013;143(6):923-930.

55. Chan HM, Trifonopoulos M, Ing A, Receveur O, Johnson E. Consumption of freshwater fish in Kahnawake: Risks and Benefits. Environ Res. 1999;80(2):213-222.
56. Gradowska PL. Food benefit-risk assessment with bayesian belief networks and multivariable exposure-response. Thesis. Netherlands: Delft University of Technology. 2013;154.
57. Leino O, Karjalainen AK, Tuomisto JT. Effects of docosahexaenoic acid and methylmercury on child's brain development due to consumption of fish by Finnish mother during pregnancy: A probabilistic modeling approach. Food Chem Toxicol. 2013;54(0):50-58.
58. Zeilmaker MJ, Hoekstra J, Eijkeren JCHv, Jong Nd, Hart A, Kennedy M, et al. Fish consumption during child bearing age: A quantitative risk-benefit analysis on neurodevelopment. Food Chem Toxicol. 2013;54:30-34.
59. Ponce RA, Bartell SM, Wong EY, LaFlamme D, Carrington C, Lee RC, et al. Use of quality-adjusted life year weights with dose-response models for public health decisions: A case study of the risks and benefits of fish consumption. Risk Anal. 2000;20(4):529-542.
60. Cohen JT, Bellinger DC, Connor WE, Kris-Etherton PM, Lawrence RS, Savitz DA, et al. A quantitative risk–benefit analysis of changes in population fish consumption. Am J Prev Med. 2005;29(4):325-334.
61. EFSA. Nitrate in vegetables - scientific opinion of the panel on contaminants in the food chain (Question No EFSA-Q-2006-071) Adopted on 10 April 2008. EFSA Journal. 2008;689:1-79.
62. Reiss R, Johnston J, Tucker K, DeSesso JM, Keen CL. Estimation of cancer risks and benefits associated with a potential increased consumption of fruits and vegetables. Food Chem Toxicol. 2012;50(12):4421-4427.
63. AFSSA. Risks and health benefits of foods brought by trans fatty acids - Recommendations; 2005. French.
64. Verhagen H, Andersen R, Antoine J-M, Finglas P, Hoekstra J, Kardinaal A, et al. Application of the BRAFO tiered approach for benefit-risk assessment to case studies on dietary interventions. Food Chem Toxicol. 2012;50:710-723.
65. Hendriksen MA, Tijhuis MJ, Fransen HP, Verhagen H, Hoekstra J. Impact of substituting added sugar in carbonated soft drinks by intense sweeteners in young adults in the Netherlands: Example of a benefit-risk approach. Eur J Nutr. 2011;50(1):41-51.
66. Husoy T, Mangschou B, Fotland TO, Kolset SO, Jakobsen HN, Tommerberg I, et al. Reducing added sugar intake in Norway by replacing sugar sweetened beverages with beverages containing intense sweeteners - A risk benefit assessment. Food Chem Toxicol. 2008;46(9):3099-3105.
67. Domingo JL, Bocio A, Falco G, Llobet JM. Benefits and risks of fish consumption Part I. A quantitative analysis of the intake of omega-3 fatty acids and chemical contaminants. Toxicology. 2007;230(2-3):219-226.
68. Domingo JL, Bocio A, Marti-Cid R, Llobet JM. Benefits and risks of fish consumption Part II. RIBEPEIX, a computer program to optimize the balance between the intake of omega-3 fatty acids and chemical contaminants. Toxicology. 2007;230(2-3):227-233.
69. Marti-Cid R, Bocio A, Llobet JM, Domingo JL. Balancing health benefits and chemical risks associated to dietary habits: RIBEFOOD, a new Internet resource. Toxicology. 2008;244(2-3):242-248.
70. TECNATOX. RIBEFOOD (2013). Accessed 6 March 2014. Available: http://www.tecnatox.cat/ribefood-2013.
71. Havelaar AH, De Hollander AEM, Teunis PFM, Evers EG, Van Kranen HJ, Versteegh JFM, et al. Balancing the risks and benefits of drinking water disinfection: Disability adjusted life-years on the scale. Environ Health Perspect. 2003;108(4):315-321.
72. Schütte K, Boeing H, Hart A, Heeschen W, Reimerdes EH, Santare D, et al. Application of the BRAFO tiered approach for benefit-risk assessment to case studies on heat processing contaminants. Food Chem Toxicol. 2012;50:724-735.
73. Rigaux C. Monte Carlo methods and second-order Bayesian inference for assessing microbiological risks and

benefits in the nutrionnels transforamtion vegetables. PhD in Biostatistics. Paris, France: Agro Paris Tech. 2013 ;205. French.
74. Morales FJ, Martin S, Acar OC, Arribas-Lorenzo G, Goekmen V. Antioxidant activity of cookies and its relationship with heat-processing contaminants: A risk/benefit approach. Eur Food Res Technol. 2009;228(3):345-354.
75. Costa S, Afonso C, Bandarra NM, Gueifão S, Castanheira I, Carvalho ML, et al. The emerging farmed fish species meagre (*Argyrosomus regius*): How culinary treatment affects nutrients and contaminants concentration and associated benefit-risk balance. Food Chem Toxicol. 2013;60(0):277-285.
76. Seal CJ, Mul Ad, Eisenbrand G, Haverkort AJ, Franke K, Lalljie SPD, et al. Risk-benefit considerations of mitigation measures on acrylamide content of foods - a case study on potatoes, cereals and coffee. Br J Nutr. 2008;99(Suppl. 2):46.
77. Hoekstra J, Verkaik-Kloosterman J, Rompelberg C, van Kranen H, Zeilmaker M, Verhagen H, et al. Integrated risk-benefit analyses: Method development with folic acid as example. Food Chem Toxicol. 2008;46(3):893-909.
78. Hoekstra J, Fransen HP, van Eijkeren JCH, Verkaik-Kloosterman J, de Jong N, Owen H, et al. Benefit-risk assessment of plant sterols in margarine: A QALIBRA case study. Food Chem Toxicol. 2013;54:35-42.
79. Butt MS, Sultan MT. Coffee and its Consumption: Benefits and Risks. Crit Rev Food Sci Nutr. 2011;51(4):363-373.
80. Ranheim T, Halvorsen B. Coffee consumption and human health - beneficial or detrimental? - Mechanisms for effects of coffee consumption on different risk factors for cardiovascular disease and type 2 diabetes mellitus. Mol Nutr Food Res. 2005;49(3):274-284.
81. Richling E, Habermeyer M. A Risk-benefit analysis Is coffee healthy? Chem Unserer Zeit. 2014;48(1):12-20.
82. Ruxton C. Health aspects of caffeine: Benefits and risks. Nursing Standard. 2009;24(9):41-50.
83. Ruxton CHS. The impact of caffeine on mood, cognitive function, performance and hydration: a review of benefits and risks. Nutr Bulletin. 2008;33(1):15-25.
84. Taylor SR, Demmig-Adams B. To sip or not to sip: The potential health risks and benefits of coffee drinking. Nutr Food Sci. 2007;37(6):406-418.
85. Gramza-Michalowska A. Caffeine in tea camellia sinensis-content, absorption, benefits and risks of consumption. J Nutr Health Aging. 2014;18(2):143-149.
86. Schwalfenberg G, Genuis SJ, Rodushkin I. The benefits and risks of consuming brewed tea: beware of toxic element contamination. J Toxicol. 2013;2013:370460-Article ID 370460.
87. Ellison RC. Balancing the risks and benefits of moderate drinking. Ann N Y Acad Sci. 2002;957(1):1-6.
88. Foster RK, Marriott HE. Alcohol consumption in the new millennium – weighing up the risks and benefits for our health. Nutr Bulletin. 2006;31(4):286-331.
89. Mukamal KJ, Rimm EB. Alcohol consumption: Risks and benefits. Curr Atheroscler Rep. 2008;10(6):536-543.
90. Thakker KD. An overview of health risks and benefits of alcohol consumption. Alcohol Clin Exp Res. 1998;22:285-298.
91. Latte KP, Appel K-E, Lampen A. Health benefits and possible risks of broccoli - An overview. Food Chem Toxicol. 2011;49(12):3287-3309.
92. Biesalski HK. Meat as a component of a healthy diet – are there any risks or benefits if meat is avoided in the diet? Meat Sci. 2005;70(3):509-524.
93. McAfee AJ, McSorley EM, Cuskelly GJ, Moss BW, Wallace JMW, Bonham MP, et al. Red meat consumption: An overview of the risks and benefits. Meat Sci. 2010;84(1):1-13.
94. Watson RR, Preedy VR, Zibadi S. Chocolate in health and nutrition (2013). Accessed 12 February 2014.Available: http://dx.doi.org/10.1007/978-1-61779-803-0.
95. AFSSA. Safety and benefits of phytoestrogens provided by food - Recommendations;2005.French.
96. Wagner JD, Anthony MS, Cline JM. Soy phytoestrogens: Research on benefits and risks. Clin Obstet Gynecol. 2001;44(4):843-852.
97. Andres S, Abraham K, Appel KE, Lampen A. Risks and benefits of dietary isoflavones for cancer. Crit Rev Toxicol. 2011;41(6):463-506.

98. Milkowski A, Garg HK, Coughlin JR, Bryan NS. Nutritional epidemiology in the context of nitric oxide biology: A risk–benefit evaluation for dietary nitrite and nitrate. Nitric Oxide. 2010;22(2):110-119.
99. Tang Y, Jiang H, Bryan NS. Nitrite and nitrate: Cardiovascular risk-benefit and metabolic effect. Curr Opin Lipidol. 2011;22(1):11-15.
100. van Boekel M, Fogliano V, Pellegrini N, Stanton C, Scholz G, Lalljie S, et al. A review on the beneficial aspects of food processing. Mol Nutr Food Res. 2010;54(9):1215-1247.
101. AFSSA. Risk assessment and nutritional and health benefits of foods derived from organic farming, Maisons-Alfort, France; 2003.French.
102. Harman JWUSGAO. Food safety: Difficulties in assessing pesticide risks and benefits: Statement of John W. Harman. U.S. General Accounting Office: Washington, D.C.; 1992.
103. Seiber JN, Ragsdale NN. Examining risks and benefits associated with pesticide use: An overview, pesticides: Managing risks and optimizing benefits. American Chemical Society. 1999;1-6.
104. AFSSA. GMOs and food: Can we identify and evaluate the health benefits? 2004. French.
105. Amin L, Hamdan F, Hashim R, Samani MC, Anuar N, Zainol ZA, et al. Risks and benefits of genetically modified foods. Afr J Biotechnol. 2011;10(58):12481-12485. French.
106. Arnst B. Current and potential benefits and risks for the genetic modification of foods for humans. Proceedings-Nutrition Society of Australia. 2000;24:12-13.
107. Kramkowska M, Grzelak T, Czyzewska K. Benefits and risks associated with genetically modified food products. Ann Agr Env Med. 2013;20(3):413-419.
108. Purchase IUB. Genetically modified foods for human health and nutrition the scientific basis for benefit-risk assessment: Special issue. Elsevier: Amsterdam; 2003.
109. Deutsche F. Thermal processing of food. Potential health benefits and risks; 2007.
110. Acheson DWK. To irradiate or not to irradiate: What are the risk-benefit arguments in relation to food safety? Clin Infect Dis. 2001;33(3):375-375.
111. Ekanem E, Muhammad S, Thompson C. Potential risks and benefits of food irradiation: a review of the literature. J Food Distribution Res. 2005;36(1):229-229.
112. United States GAO. Food irradiation: Available research indicates that benefits outweigh risks: Report to congressional requesters. The Office: Washington, D.C. (P.O. Box 37050, Washington, D.C. 20013); 2000.
113. Bukhamseen F, Novotny L. Artificial sweeteners and sugar substitutes - some properties and potential health benefits and risks. Res J Pharm Biol Chem Sci. 2014;5(1):638-641.
114. Gardner C. Non-nutritive sweeteners: Evidence for benefit vs. risk. Curr Opin Lipidol. 2014;25(1):80-84.
115. Tombek A. Update on sweeteners: Is there anything new about the benefits and risks? Ernahrungs Umschau. 2010;57(4):196-200.
116. Ilg Y, Kreyenschmidt J. Review: Benefits and risks of the use of antimicrobial components in the food chain. Arch Lebensmittelhyg. 2012;63(2):28-34.
117. Berjia FL, Poulsen M, Nauta M. Burden of diseases estimates associated to different red meat cooking practices. Food Chem Toxicol. 2014;66(0):237-244.
118. Brzozowska A. Food fortification and diet supplementation - benefits and risk. Zywnosc. 2001;8(4):16-28.
119. Somoza V. Five years of research on health risks and benefits of Maillard reaction products: An update. Mol Nutr Food Res. 2005;49(7):663-672.
120. Claeys WL, Cardoen S, Daube G, Block Jd, Dewettinck K, Dierick K, et al. Raw or heated cow milk consumption: Review of risks and benefits. Food Control. 2013;31(1):251-262.
121. Neaves P. Unpasteurised milk: Do the risks outweigh the benefits? Food Sci Technol. 2000;14(1):38-40.
122. Anonymous. Brief critical reviews - Mediterranean diet and coronary heart disease: Are Antioxidants Critical? Nutr rev. 1999;57(8):253.
123. Grosso G, Mistretta A, Frigiola A, Gruttadauria S, Biondi A, Basile F, et al. Mediterranean diet and cardiovascular risk factors: a systematic review. Crit Rev Food Sci Nutr. 2014;54(5):593-610.

124. Cunningham E. What is a raw foods diet and are there any risks or benefits associated with it? J Am Diet Assoc. 2004;104(10):1623-1623.
125. Dagnelie PC. Nutrition and health--potential health benefits and risks of vegetarianism and limited consumption of meat in the Netherlands. Ned Tijdschr Geneeskd. 2003;147(27):1308-1313.
126. Sabate J. The public health risk-to-benefit ratio of vegetarian diets: changing paradigms; 2001.
127. de Zegher F, Sebastiani G, Diaz M, Dolores Gomez-Roig M, Lopez-Bermejo A, Ibanez L. Breast-feeding vs formula-feeding for Infants born small-for-gestational-age: Divergent effects on fat mass and on circulating IGF-I and high-molecular-weight adiponectin in late infancy. J Clin Endocr Metab. 2013;98(3):1242-1247.
128. Fewtrell MS, Kennedy K, Murgatroyd PR, Williams JE, Chomtho S, Lucas A. Breast-feeding and formula feeding in healthy term infants and bone health at age 10 years. Br J Nutr. 2013;110(6):1061-1067.
129. Frank JW, Newman J. Breast-feeding in a polluted world: uncertain risks, clear benefits. Can Med Assoc J. 1993;149(1):33-37.
130. Goldman AS, Hopkinson JM, Rassin DK. Benefits and risks of breastfeeding. Adv Pediatr. 2007;54:275-304.
131. Harris GS, Highland HJ. Birthright denied: The risks and benefits of breastfeeding. J Nurse Midwifery. 1979;24(6):41-41.
132. Mead MN. Contaminants in human milk: Weighing the risks against the benefits of breastfeeding. Environ Health Perspect. 2008;116(10):427-434.
133. Serreau R, Rigourd V, Amirouche A, Cheung K, Fauroux V. Allaitement maternel ou pas ? Balance benefices/risques. Rev Med perinat. 2011;3(4):206. French.
134. Boobis A, Chiodini A, Hoekstra J, Lagiou P, Przyrembel H, Schlatter J, et al. Critical appraisal of the assessment of benefits and risks for foods, 'BRAFO Consensus Working Group'. Food Chem Toxicol. 2013;55:659-675.
135. Hart A, Hoekstra J, Owen H, Kennedy M, Zeilmaker MJ, de Jong N, et al. Qalibra: A general model for food risk-benefit assessment that quantifies variability and uncertainty. Food Chem Toxicol. 2013;54:4-17.
136. Renwick AG, Flynn A, Fletcher RJ, Muller DJ, Tuijtelaars S, Verhagen H. Risk-benefit analysis of micronutrients. Food Chem Toxicol. 2004;42(12):1903-1922.
137. Palou A, Pico C, Keijer J. Integration of risk and benefit analysis-the window of benefit as a new tool? Crit Rev Food Sci Nutr. 2009;49(7):670-680.
138. Renwick AG, Dragsted LO, Fletcher RJ, Flynn A, Scott JM, Tuijtelaars S, et al. Minimising the population risk of micronutrient deficiency and over-consumption: A new approach using selenium as an example. Eur J Nutr. 2008;47(1):17-25.
139. van der Voet H, de Mul A, van Klaveren JD. A probabilistic model for simultaneous exposure to multiple compounds from food and its use for risk-benefit assessment. Food Chem Toxicol. 2007;45(8):1496-1506.
140. Luteijn JM, White BC, Gunnlaugsdottir H, Holm F, Kalogeras N, Leino O, et al. State of the art in benefit-risk analysis: Medicines. Food Chem Toxicol. 2012;50(1):26-32.
141. Pohjola MV, Leino O, Kollanus V, Tuomisto JT, Gunnlaugsdottir H, Holm F, et al. State of the art in benefit-risk analysis: Environmental health. Food Chem Toxicol. 2012;50(1):40-55.
142. Magnússon SH, Gunnlaugsdóttir H, van Loveren H, Holm F, Kalogeras N, Leino O, et al. State of the art in benefit–risk analysis: Food microbiology. Food and Chemical Toxicology. 2012;50(1):33-39.
143. Kalogeras N, Odekerken-Schroder G, Pennings JM, Gunnlaugsdottir H, Holm F, Leino O, et al. State of the art in benefit-risk analysis: economics and marketing-finance. Food Chem Toxicol. 2012;50(1):56-66.
144. Ueland O, Gunnlaugsdottir H, Holm F, Kalogeras N, Leino O, Luteijn JM, et al. State of the art in benefit-risk analysis: Consumer perception. Food Chem Toxicol. 2012;50(1):67-76.
145. Tijhuis MJ, Pohjola MV, Gunnlaugsdottir H, Kalogeras N, Leino O, Luteijn JM, et al. Looking beyond borders: Integrating best practices in benefit-risk analysis into the field of food and nutrition. Food Chem Toxicol. 2012;50(1):77-93.

146. ILSI. North America Annual Meeting Richard Williams Risk benefit analysis of food (2013). Accessed 20 January 2014. Available: http://www.ilsi.org/Pages/ViewItemDetails.aspx?WebID=eb9dea86-6210-452e-a97a-62327e4490ec&ListID=06bbbf30-51e5-4053-add9-4d053cd120e4&ItemID=370.
147. ILSI. North America Annual Meeting Robert Buchanan Risk Benefit for Chemical Contaminants (2013). Accessed 20 January 2014. Available: http://www.ilsi.org/Pages/ViewItemDetails.aspx?ListID={06bbbf30-51e5-4053-add9-4d053cd120e4}&ItemID=369&URL=http://www.ilsi.org/Publications/Video64.pdf
148. ILSI. North America Annual Meeting Alicia Carriquiry A Risk Benefit Approach to Assess Nutrient Intake - Do We Need a New DRI (2013). Accessed 20 January 2014. Available: https://www.youtube.com/watch?v=Vyt5yqob0UE&feature=youtu.be.
149. Tuomisto JT. Integrated benefit and risk analysis for assessing food safety and health benefit; Results from QALIBRA and BENERIS Foreword. Food Chem Toxicol. 2013;54:1-2.
150. Larranaga-Guetaria A. PlantLIBRA: PLANT food supplements, levels of Intake, Benefit and Risk Assessment The Regulatory Framework for plant food supplements in the EU. Agro Food Industry Hi-Tech. 2012;23(5):20-22.
151. Berjia FL, Method development in risk-benefit assessment and burden of disease estimation of food. Thesis. Denmark: National Food Institute, Technical University of Denmark. 2013;172.
152. Pascal G. Safety impactthe risk/benefits of functional foods. Eur J Nutr European Journal of Nutrition. 2009;48(S1):33-39.
153. Verkerk RHJ. The paradox of overlapping micronutrient risks and benefits obligates risk/benefit analysis. Toxicology. 2010;278(1):27-38.
154. Peleg M, Normand MD, Corradini MG. A method to estimate a person or group health risks and benefits from additive and multiplicative factors. Trends Food Sci Tech. 2012;28(1):44-51.
155. Sand S, Becker W, Darnerud PO. Aspects of risk-benefit assessment of food consumption directions for the future. Nordic Council of Ministers, Copenhagen; 2008.
156. Burlingame B, Pineiro M. The essential balance: Risks and benefits in food safety and quality (vol 20, pg 139 2007). J Food Compos Anal. 2007;20(8):739-739.
157. FAO/WHO. Exposure assessment of microbiological hazards in food. Guidelines. Microbiological Risk Assessment. Series, No. 7; 2008.
158. Hellberg RS, DeWitt CAM, Morrissey MT. Risk-benefit analysis of seafood consumption: A review. Compr. Rev. Food. Sci. Food Saf. 2012;11(5):490-517.

APPENDIX

Table S1. Summary of the main results of risk-benefit assessment studies

Comparison based on*	Scientific domain**	Main results	Reference (first author, year)
• *Food component(s) is/are a source of risk(s) and benefit(s)*			
		Fish	
- Safety levels	N/C	Fish consumption is high in Portugal (≈57kg/year). Assessment of the three most consumed species demonstrated that its consumption should be limited to one serving/week of silver scabbard fish or three servings/week of hake or ray.	Afonso, 2013 [52]
	N/C	A daily consumption of 160 g of fish muscle (6 species studied) does not expose consumers to risk and contributes to nutritional benefit. Consumption of liver should be avoided and a weekly consumption of *L. whiffiagonis* is recommended.	Afonso, 2013 [51]
	N/C	Consumption of two portions of fish per week is recommended including one portion with a high content of EPA and DHA, but with changes in species and points of production (subgroup specifications are given).	AFSSA, 2008 [19]
	N/C/M	The ANSES agency recommends that the general population consume 200 g/week of fish (including 100 g of fish with a high content of EPA and DHA). Specific recommendations are given for the sensitive subpopulation. It also advises specific hygiene measures.	ANSES, 2013 [11]
	N/C	A list of intake recommendations is given for different subpopulations (infants, healthy adults, CHD patients and hyperglyceridemia patients) depending on fish and fish species to achieve the recommended weekly intake (RWI) without exceeding the tolerable weekly intake (TWI).	Balshaw, 2012 [24]
	N/C	Consuming fish two to three times a week decreases cardiovascular diseases, the risk of osteoporosis and fractures. Fish with up to 1 mg/kg methyl mercury should be limited to one serving per month. Pregnant or lactating women may consume one of the three weekly portions with a high omega 3 content.	Becker, 2007 [23]
	N/C	The assessment of fish consumption ineight European countries highlighted that the probability of being exposed to risk and benefit depends on the fish species. Countries with a low fish intake could be subject to small risk and benefit (Italy and the United Kingdom) or low risk but high benefit (Germany and the Netherlands) while high consumers are exposed to both (France, Spain, Portugal and Iceland).	Cardoso, 2010 [31]
	N/C	The Portuguese population exceeds the provisional tolerable weekly intake (PTWI) of methyl mercury regarding the consumption of hake, ray and silver scabbardfish without achieving the relative daily allowance (RDA) and the relative daily intake	Cardoso, 2013 [44]

		(RDI) of Selenium, EPA and DHA. They advise limiting the consumption of these three fish species to less than one meal/week.	
	N/C	The Kahnawake community of south of the St Laurence river is not exposed to chemical risk due to fish consumption and fishing.	Chan, 1999 [55]
	N/C	Salmon and trout sold in Quebec can be regularly eaten to take advantage of nutritional benefit without exposing consumers to chemical risk (e.g. farmed Atlantic salmon can be consumed in one serving/day).	Dewailly, 2007 [45]
	N/C	43 fish species from Bermuda were analyzed and recommendations are given by subgroup. For example, women of childbearing age should not consume predatory fish while other subgroups should limit their consumption to one portion per week or month.	Dewailly, 2008 [43]
	N/C	In China, a consumption of 80 to 100 g/day of marine oily fish from the Chinese market is associated with potential nutritional benefit without exposing consumers to chemical risk.	Du, 2012 [37]
	N/C	No difference between wild and farmed fish has been identified. The advantage of farmed fish is that the contaminant level can be controlled and decreased by modification of fish feeding. A consumption of one to two portions/week is advised with restrictions for sensitive groups.	EFSA, 2005 [18]
	N/C	A daily consumption of Siberian grayling from Yenisei River provides the RDI of EPA but could exceed reference doses (RfD) of chromium. Concentration may vary according to month.	Gladyshev, 2009 [46]
	N	A curve of the balance of net benefit-harm is created with estimated thresholds. However, more data are required to estimate thresholds and asymptotes using this curve.	Gochfeld, 2005 [42]
	N/C	Wild salmon have significantly fewer chemical contaminants than farmed salmon and a higher EPA content. Farmed salmon from Europe contains a higher level of chemical contaminants than those from South and North America and a similar EPA content.	Hites, 2004 [50]
	N/C	In Canada, 35% of the Inuit population is exposed to chemical risk due to consuming fish contaminated by methyl mercury. To decrease this risk and keep the benefit, the consumption of ringed seal liver could be replaced by ringed seal meat, ringed seal blubber, beluga mukluk or Arctic char, for example.	Laird, 2013 [54]
	N/C	The Portuguese adult consumption of black scabbardfish should be limited to 90 g grilled meat and 120 g of fried meat. Edible crab brown meat should not exceed 27 g boiled meat per week and its consumption should be avoided by children and lactating or pregnant women.	Maulvault, 2013 [53]
	N/C/M	A consumption of 270 g to 340 g/week of fish is advised. Children under 12 years	Nesheim, 2007 [12]

		old and pregnant and lactating women should limit tuna consumption to150 g/week and avoid predatory fish. Other subgroups can consume more fish but they should change fish (and seafood) species. There is additional benefit by including seafood high in EPA and DHA. Microbial risk could be limited by hygiene practices during handling and cooking.	
	N/C	Consumption of two portions/week, including one oily fish, decreases CVD risk and improves fetal development. Pregnant and lactating women should select certain fish.	SACN/COT, 2004 [20]
	N/C	Fish consumption has been assessed as safe in the State of Michigan. A list of the top 11 fish was established to increase benefits.	Sidhu, 2003 [41]
	N/C	Consumption of 181-213 g/week of certain fatty fish species and 26-72 g/week of lean fish or shellfish provides a good risk-benefit balance.	Sirot, 2010 [21]
	N/C	A consumption of 200 g/week of selected fatty fish and 50 g/week of lean fish maximizes benefit and minimizes risk.	Sirot, 2012 [22]
	N/C	Consumption of canned fish from the Polish market presents higher benefit than risk. Limitation depends on fish species.	Usydus, 2008 [33]
	N/C	Fish products from the Polish market vary greatly in terms of potential beneficial and adverse health effects; recommendation of quantity depends on species.	Usydus, 2009 [34]
	N/C	Consuming fish from Taihu Lake to achieve RDA of EPA and DHA does not expose consumers to chemical risk (PCBs and PBDEs).	Zhang, 2012 [40]
	N/C	The risk-benefit ratio has been assessed for four fish species from Taihu Lake in China and for three muscles (dorsal, ventral and tail) and three viscera (heart, liver and kidney). The current Chinese fish consumption does not present a risk, except for ventral and tail consumption of top mouth cutler that should be avoided.	Zhang, 2012 [39]
	N/C	It is recommended that the Norwegian population increase their fish consumption to achieve two meals of fatty fish per week.	VKM, 2006 [32]
	N/C	Consumption of 200 g/week of farmed salmon decreases CHD incidence and increases contaminant intake but still below the PTWI.	Watzl, 2012 [48]
- Endpoint	N/C	Fish consumption (from one to twelve servings per week) decreases the relative risk (RR) of stroke compared with the scenario of no consumption.	Bouzan, 2005 [25]
	N/C	In HongKong, moderate fish consumption by pregnant women is a source of benefit for the IQ of their children with a gain of 0.79 to 5.7 points if they vary the species.	Chen, 2014 [36]
	N/C	A consumption by pregnant women of one to seven servings/week of fish (depending on fish species) decreases CHD and increases the future newborn IQ. Details are given for each subgroup and as a function of fish species.	FAO/WHO, 2010 [28]
	N/C	CurrentUS fish consumption prevents 30000 deaths per year from CHD and 20000 deaths per year from stroke. Women of childbearing age should increase their fish	FDA, 2009 [30]

		consumption to 340 g/week to improve fetal neurodevelopment.	
	N/C	RDI of EPA and DHA could not be achieved through farmed or wild salmon consumption without exposing consumers to carcinogenic risk. Intake recommendations are given depending on fish market location.	Foran, 2005 [49]
	N/C	The IQ gained by children during their mother's pregnancy is positive with a consumption of 175 g/week and 450 g/week of 30 fish species from Zhoushan in China; optimal weekly consumption is given for every species. Consumption of *Scoliodon sorrakowah* is not recommended.	Gao,2014 [38]
	N/C	Risk and benefit due to fish consumption are assessed to optimize newborn visual recognition memory (VRM) and limit CHD. A table of intake recommendations depending on species is provided.	Ginsberg, 2009 [29]
	N/C	To ensure their child's IQ is more than 100 points, Finnish pregnant women should reduce their consumption of vendace by 13%, white fish by 18%, perch by 31%, and pike by 90% and increase their intake of Atlantic salmon by 2% and Baltic herring by 4%.	Gradowska, 2013 [56]
	N/C	A small increase in fish consumption decreases CHD mortality risk by 17% and non-fatal heart disease risk by 27%.	König, 2005 [26]
	N/C	Current fish consumption by Finnish pregnant women generates compensation in effects on infant's IQ. Fatty fish consumption creates a gain in IQ and lean fish consumption an adverse IQ effect.	Leino, 2013 [57]
	N/C	Salmon consumption presents more health benefit than risk. However, the risk-benefit balance of Arctic grayling, pike, sablefish and halibut cannot be assessed because data depend on regions and studies.	Loring, 2010 [47]
	N/C	Consumption of one to two servings/week reduces CHD risk by 36% and total mortality rate by 17%. Women of childbearing age, pregnant or lactating should consume two servings/week with species restrictions.	Mozaffarian, 2006 [27]
	N/C	Women's fish intake during pregnancy causes a decrease in newborn IQ for most species consumed. Risk clearly outweighs benefit (until 11 IQ points lost with swordfish), and only a few species slightly improve the IQ (+1 point for mackerel).	Zeilmaker, 2013 [58]
- DALY/QALY	N/M	Consumption of 40 g/day of cold-smoked salmon by the Danish population could improve population health with a potential gain of 10000 healthy years annually if the product is consumed before 4 weeks of storage.	Berjia, 2012 [14]
	N/C	In US, an increase of 50% in fish consumption by the adult population, except women of childbearing age, could save 120000 healthy years annually.	Cohen, 2005 [60]
	N/C	In France, a higher fish intake (1104 g/week) than the current consumption (334 g/week) could save between 97 and 285 healthy years based on the French study CALIPSO on 1011 people.	Guevel, 2008 [35]

	N/C	The Dutch population could improve their health with a consumption of 200 g of fish/week. Onaverage, 2.7 healthy years per 1000 people could be gained every year compared to the current consumption.	Hoekstra, 2013 [16]
	N/C	In Washington state, adult consumption of fish has net beneficial effects on health with a gain of approximately 5000 healthy years saved per year per 100000 people but the net health balance is negative for women of childbearing age.	Ponce, 2000 [59]
		Fruits and vegetables	
- Safety levels	N	Overall, consumption of 400 g of vegetable per day is a source of beneficial effects and does not expose consumers to a relevant risk due to nitrate intake.	EFSA, 2008 [61]
- Endpoint	N/C	An increase of one serving of vegetable and one of fruit per day could prevent 20000 cancer cases and create 10 cases due to pesticide consumption.	Reiss, 2012 [62]
		Soy protein	
- Safety levels	N	With a consumption of 25 g/day of soy protein, beneficial effects clearly outweigh the potential risk: reduction of CVD, breast and prostate cancer risk.	Watzl, 2012 [48]
		Trans fatty acids	
- Safety levels	N	The substitution of 5% of the energy intake from saturated fatty acids by 5% from carbohydrates brings beneficial and adverse health effects related to the same disease (CVD).	Verhagen, 2012 [64]
	N	A consumption of more than 2% of trans fatty acids within the total energy food intake improves CHD risk. A suggestion of an UL of 1% of trans fatty acids within the total energy food intake and a mention of %trans fatty acids of total fatty acids on food labeling is made.	AFSSA, 2005 [63]
• The manufacturing process is a source of risk(s) and benefit(s)			
		Milk treatment	
- Safety levels	N/M	Microbial benefit (reduction of microorganisms) from heat treatment outweighs potential risk due to the reduction of lysine and the inactivation of bioactive molecules.	Schütte, 2012 [72]
		Water treatment	
- DALY/QALY	C/M	Water treatment by ozonation decreases Cryptosporidium parvum infection but introduces chemical risk due to bromate. The overall health effect is a gain of one healthy year per million people annually.	Havelaar, 2003 [71]
		Vegetable transformation	
- Safety levels	N/M	The green bean process could be optimized to achieve the RDA without exceeding a microbial threshold of *G. stearothermophilus* by reducing waiting times and blanching duration and by increasing the sterilizing value or by decreasing the pH of the end product.	Rigaux, 2013 [73]
		Cookie process	

- Safety levels	N/C	Heat processing of cookies produces harmful compounds and modifies antioxidant activity depending on time, temperature, sugar and leavening agents. The risk-benefit ratio on compound quantity is lower at low temperature and small duration but the impact on health is not quantified.	Morales, 2009 [74]
		Fish culinary treatment	
- Safety levels	N/C	The comparison of three fish cooking practices (boiling, grilling and roasting) has demonstrated that grilling fish is the best fish treatment tooptimize nutritional benefit and limit chemical risk with a limitation of two meals/week.	Costa, 2013 [75]
		Acrylamide formation	
- Safety levels	N/C	The use of sodium bicarbonate to bake products should reduce acrylamide concentration but it could cause a nutritional loss and generate other unknown molecules.	Seal, 2008 [76]
	N/C	Reduction of acrylamide in potato and cereal-based products through measures applied in production is desirable.	Schütte, 2012 [72]
		Benzo(a)pyrene formation	
- Safety levels	C	The use of artificial smoked flavor or industrial smoking control is beneficial to reduce the risk of benzo(a)pyrene.	Schütte, 2012 [72]
• Diet is a source of risk(s) and benefit(s)			
		Breastfeeding	
- Safety levels	N/C/M	Benefit associated with breastfeeding outweighs risks due to contaminants and contributes to an efficient neurodevelopment, the creation of defense against infection and the reduction of obesity risk.	VKM, 2013 [13]
		Replacement of sugar by intense sweetener	
- Safety levels	N	Substitution of sugar by intense sweeteners in beverages decreases sugar consumption (too high for adolescents) but acesulfame K intake becomes close to the acceptable daily intake (ADI) and benzoic acid ADI could be exceeded.	Husoy, 2008 [66]
	N	The substitution of sugars by low calorie sweeteners in beverages is associated with benefit: it limits caries risk, prevents overweight and chronic disease risk.	Verhagen, 2012 [64]
	N	For young adults in the Netherlands, the substitution of 100% sugar by intense sweeteners in beverages is beneficial in caries prevention and body mass reduction and does not expose this population to potential risk.	Hendriksen, 2011 [65]
		Individual assessment of risk and benefit exposure	
- Safety levels	N/C	RIBEPEIX is software to assess risk-benefit associated with individual fish consumption according to chemical and nutritional safety reference values.	Domingo, 2007 [67,68]
	N/C	RIBEFOOD is an application available online to assess individual overall diet according to safety reference values. The software guides consumers to find food substitution to improve their risk-benefit balance.	Marti-Cid, 2008 [69]

• RBA is used in food formulation			
		Margarine fortification	
- DALY/QALY	N	Margarine fortification with plant sterol in the Netherlands should save 8 healthy years annually per 1000 people.	Hoekstra, 2013 [78]
		Bread supplementation	
- DALY/QALY	N	In the Netherlands, bread fortified with 140 µg/100 g folic acid should result in 11812 healthy years saved annually.	Hoekstra, 2008 [77]
	N	In the Netherlands, a small bread fortification of 70 µg/100 g folic acid should result in 7000 healthy years saved every year with a loss of 53 healthy years.	Verhagen, 2012 [64]

** Risk-benefit comparisons are sorted into three groups, 'safety levels', 'endpoint' and 'DALY/QALY' which are explained in section 3.1.3; ** N: Nutrition, C: Chemistry, M: Microbiology*

4

Symposium Report: Effective and Safe Micronutrient Interventions, Weighing the Risks against the Benefits

Ingrid Bielderman[1*], Maaike J. Bruins[2], Reina Engle-Stone[3], Noel W. Solomons[4], Hans Verhagen[5,6], Lindsay H. Allen[7], Lynnette M. Neufeld[8], Klaus Kraemer[1] and Jane Badham[1]

[1]Sight and Life, Basel, Switzerland.
[2]DSM Biotechnology Center, Delft, The Netherlands.
[3]Department of Nutrition, University of California, Davis, CA, USA.
[4]Center for Studies of Sensory Impairment, Aging and Metabolism (CeSSIAM), Guatemala City, Guatemala.
[5]National Institute for Public Health and the Environment (RIVM), Bilthoven, The Netherlands.
[6]University of Ulster, Northern Ireland Centre for Food and Health (NICHE), Coleraine, Northern Ireland, UK.
[7]US Department of Agriculture (USDA), ARS Western Human Nutrition Research Center, Davis, CA, USA.
[8]Global Alliance for Improved Nutrition (GAIN), Geneva, Switzerland.

Authors' contributions

This work was carried out in collaboration between all authors. Authors IB and MJB wrote the first draft of the manuscript. All authors have contributed to, read, and approved the final manuscript.

ABSTRACT

Micronutrient fortification of staple foods can be an effective strategy to combat micronutrient malnutrition. When planning on fortification, challenges faced include the collection of essential information on population food and nutrient intake patterns, as well as the use of this information in a method to select appropriate fortification levels. A symposium was organized aimed at discussing the existing approaches to set effective and safe micronutrient fortification levels and to outline the challenges and needs in this area. Two different approaches to establish effective and safe fortification levels for food fortification were presented. In the first approach, the Estimated Average Requirement (EAR) and Tolerable Upper Intake Level (UL) are used as cut-points in the micronutrient intake distribution to evaluate and simulate effective and safe micronutrient intakes.

Corresponding author: Email: ingridbielderman@gmail.com

This was exemplified by challenges encountered in Guatemala and Cameroon towards unequal vitamin A intake distribution and the impact of the food vehicle choice. Secondly, the risk-benefit approach was presented as an approach in which risks and benefits of micronutrient intakes can be quantified and balanced in order to optimize fortification benefits with the least risks and to allow decision making. This was illustrated by a case on folic acid fortification in The Netherlands.

Irrespective of the approach, food and nutrient intake data are required to identify potential vehicles for fortification, quantify the nutrient gap to be addressed, and set the appropriate level of fortification based on consumption pattern. Such information is rarely available to the quality and extent ideal to set fortification levels and requires regular updating, as exemplified in the case of sugar fortification in Guatemala. While the EAR cut-point method can be used to determine the proportion of the population meeting their required and safe nutrient intakes and set goals, risk-benefit assessment may offer an answer to commonly-asked questions as to whether, and at which levels, the benefits of increasing micronutrient intakes outweigh the risks.

Keywords: Public health; fortification; risk-benefit assessment; deficiency; toxicity; micronutrients.

ACRONYMS AND ABBREVIATIONS

ADI	:	*Acceptable Daily Intake*
BRAFO	:	*Benefit Risk Analysis of Foods*
CeSSIAM	:	*Center for Studies of Sensory Impairment, Aging and Metabolism*
DALY	:	*Disability-Adjusted Life Year*
EAR	:	*Estimated Average Requirement*
EFSA	:	*European Food Safety Authority*
FAO	:	*Food and Agriculture Organization of the United Nations*
GAIN	:	*Global Alliance for Improved Nutrition*
HCES	:	*Household Consumption and Expenditure Surveys*
IMAPP	:	*Intake Monitoring, Assessment and Planning Program*
INCAP	:	*Institute of Nutrition of Central America and Panama*
JECFA	:	*Joint FAO/WHO Expert Committee on Food Additives*
IOM	:	*Institute of Medicine of the National Academies*
IU	:	*International Units*
LOAEL	:	*Lowest Observed Adverse Effect Level*
NCI	:	*US National Cancer Institute*
NOAEL	:	*No Observed Adverse Effect Level*
QALY	:	*Quality-Adjusted Life Year*
RAE	:	*Retinol Activity Equivalents*
RDA	:	*Recommended Daily Allowance*
RE	:	*Retinol Equivalents*
RIVM	:	*National Institute for Public Health and the Environment*
RNI	:	*Recommended Nutrient Intake*
UL	:	*Tolerable Upper Intake Level*
USDA	:	*US Department of Agriculture*
WHO	:	*World Health Organization*
WFP	:	*World Food Programme*

1. INTRODUCTION

When the level of micronutrient intake is insufficient to meet the requirement, this may result in typical signs and symptoms, which have been fully characterized for each of the vitamins and minerals [1]. For some micronutrients, excessive intake is also a risk factor [2]. The adverse effects related to long-term excessive intake of micronutrients are less-well understood. Few data on human case reports or case-control studies are generally available to establish safe intake limits or sometimes insufficient to establish a safe limit.

Because of the obvious clinical benefit of overcoming micronutrient inadequacies, most fortification programs have been focusing on micronutrient inadequacies whereas overexposure received little attention, as there is not yet a solid understanding of the risk that comes with excessive intakes. For most micronutrients there is little concern about excess intake from the amounts typically used in fortification programs. However, the growing number of foods being fortified and simultaneous micronutrient initiatives has risen concern about the safe upper intakes of micronutrients in some developing countries. When a micronutrient inadequacy is widespread, the benefits of addressing the inadequacy are likely to outweigh the risks of excessive intakes. Yet, it is critical to understand whether overconsumption of micronutrients with an upper safe level of intake is actually a true concern. Related to this, it is important to understand how to set micronutrient levels in food that are adequate and safe for the population.

The conference organizers identified a need to understand the risks related to inadequate and excessive intake of micronutrients with fortification. How to set effective yet safe levels of micronutrients in fortified foods is considered a major challenge when planning on food fortification, particularly in the case of micronutrients with a relatively narrow margin of safe intake between the actual intake and the UL, as for folic acid and vitamin A [3].

Sight and Life, the Global Alliance for Improved Nutrition (GAIN), and the Dutch National Institute for Public Health and the Environment (RIVM) organized a symposium titled “Effective and safe micronutrient interventions: weighing the risks against the benefits”. The symposium took place on the 3rd of June 2014 on the second day of the Micronutrient Forum Global Conference, held in Addis Ababa, Ethiopia, from June 2-6, 2014. The symposium aimed to discuss existing approaches available to set micronutrient levels in fortified foods, and discuss their values, advantages, disadvantages, and shortcomings, including data needs. Practical examples were added to the program to provide insight into the benefits and challenges experienced in food fortification using these different approaches. This symposium report summarizes the presentations and the questions and discussion raised by the audience.

This report summarizes the five invited presentations and discussions that took place during the symposium. Prof Dr Lindsay H Allen (US Department of Agriculture (USDA), ARS Western Human Nutrition Research Center, Davis, CA, USA) presented on public health risks and consequences of too low and too high micronutrient intakes in developing countries; Dr Reina Engle-Stone (University of California, Davis, CA, USA) presented on risks of vitamin A inadequacy and chronic or intermittent excess of vitamin A intake in Cameroon; Dr Noel Solomons (Center for Studies of Sensory Impairment, Aging and Metabolism, Guatemala) presented on vitamin A intake distribution among the Guatemalan population: contemporary perplexing perspectives; Prof Dr Hans Verhagen (National Institute for Public Health and the Environment (RIVM), Bilthoven, The Netherlands), presented on integrated risk-benefit assessment in food and nutrition; and Dr Maaike Bruins (DSM Biotechnology Center, Delft, The Netherlands) presented on the challenge of setting micronutrient fortification levels.

2. PUBLIC HEALTH RISKS AND CONSEQUENCES OF TOO LOW AND TOO HIGH MICRONUTRIENT INTAKES IN DEVELOPING COUNTRIES

Prof Dr Lindsay H Allen (US Department of Agriculture (USDA), ARS Western Human Nutrition Research Center, Davis, CA, USA).

Micronutrient deficiencies or excess intakes of micronutrients can exert a large impact on public health. The prevalence of inadequate and excess intakes can be assessed by making use of dietary intake data and comparing intakes to the recommended values. In addition, biochemical markers of micronutrient deficiency or excess may also be used to evaluate severity of a public health problem or the effectiveness of a public health intervention. Clinical and biochemical markers of deficiencies are generally available for most micronutrients, but are rarely available to detect excessive micronutrient intakes (Table 1). Clinical signs and symptoms of deficiency usually manifest only when the deficiency is severe, and they are often quite non-specific and therefore may be less useful. Biochemical markers of nutrient exposure/status usually change at an earlier stage of deficiency and are more nutrient-specific.

The goal of increasing micronutrient intake by fortification is to provide approximately 97.5% of individuals in a population group with an intake of each micronutrient that meets their Estimated Average Requirement (EAR) while no more than approximately 2.5% of individuals in that population exceed their Tolerable Upper Intake Level (UL) (Fig. 1) [4].

In other words, the probability of both nutrient inadequacy and excess must be acceptably low. In this regard, it is important to note that the UL is an upper safe level of intake within a margin of safety and not a "toxic" level. The UL is derived by dividing the No Observed Adverse Effect Level (NOAEL) or the Lowest Observed Adverse Effect Level (LOAEL) by an uncertainty factor, to account for the uncertainties associated with extrapolating from the observed data. The size of the uncertainty factor depends on features such as the severity of the adverse effect and level of uncertainty about the data, providing a margin of safety.

Biochemical and clinical data collected from a population do not provide quantitative estimates of how much of each nutrient should be added as a fortificant and which food(s) to use as vehicle for fortification. Collection of food intake data is critical before implementing micronutrient interventions [5]. It provides information on micronutrient intakes, shows prevalence of inadequate and excess intakes within a population, reveals the main food sources consumed and can therefore be used to identify the most suitable food vehicles for fortification and predict prevalence of inadequate and excess intakes after fortification (Fig. 2).

Well-collected food intake data provide important information on the amount of micronutrients that need to be added to the diet through fortification or dietary diversification, or possibly through supplementation in particular high-risk subgroups of the population. When planning on fortification, the gap between usual intake and the EAR needs to be identified for all micronutrients in vulnerable population groups. In addition, information on the usual distribution of intake of foods that could be fortified is needed. To this end, intake data need to be collected by food surveys over two days on approximately 100 people per group. Then the prevalence of inadequate intake of each micronutrient needs to be estimated, as well as the local intake of potential food vehicles.

Subsequently, the effect of fortification with different levels of micronutrients can be simulated, including scenarios that result in inadequate and excessive intakes. The software planning program IMAPP (Intake Monitoring, Assessment and Planning Program) can be used to simulate different fortification scenarios [6]. It requires entering nutrient intake data, but it corrects for bioavailability and day-to-day variation in intake and estimates prevalence of intakes below the EAR. The software also simulates the proportion of a population group with intakes below the EAR and above the UL assuming different levels of fortification of food vehicles under consideration.

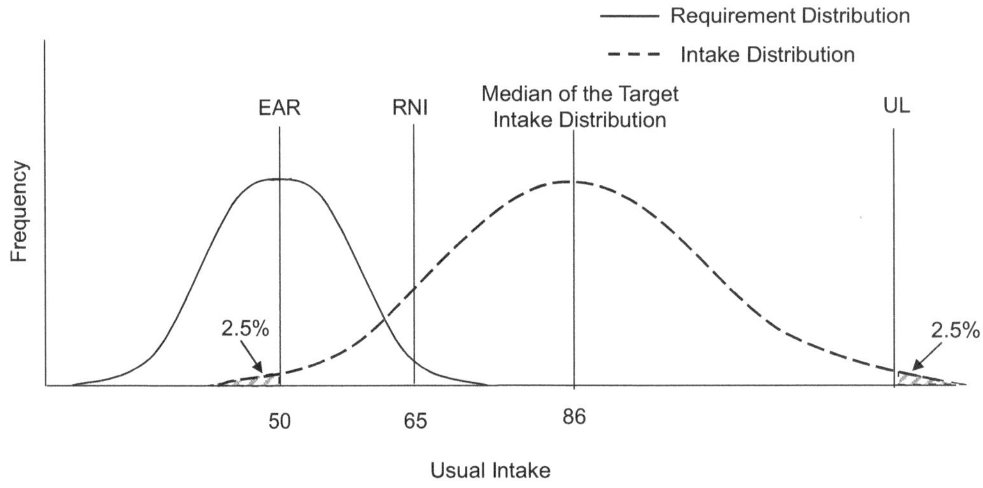

Fig. 1. The principle of the EAR cut-point method

For each nutrient, plan so that <2.5% of a population group has intakes <EAR and <2.5% exceed the UL. Modified from [4]

Table 1. Micronutrients of concern – indicators of too little and/or too much

Nutrient	Clinical symptom of severe deficiency	Biochemical marker of deficiency	Clinical symptom of excess intake	Biochemical marker of excess intake
Iron	√	√	√	No
Vitamin A	√	√	√	√?
Iodine	√	√	√	√
Zinc	√	(√)	No	No
Folate	√	√	No	√?
Vitamin B12	√	√	No	No
Thiamin	√	√	√	No
Riboflavin	√	√	No	No
Vitamin B6	√	√	√	No
Vitamin D	√	√	√	√

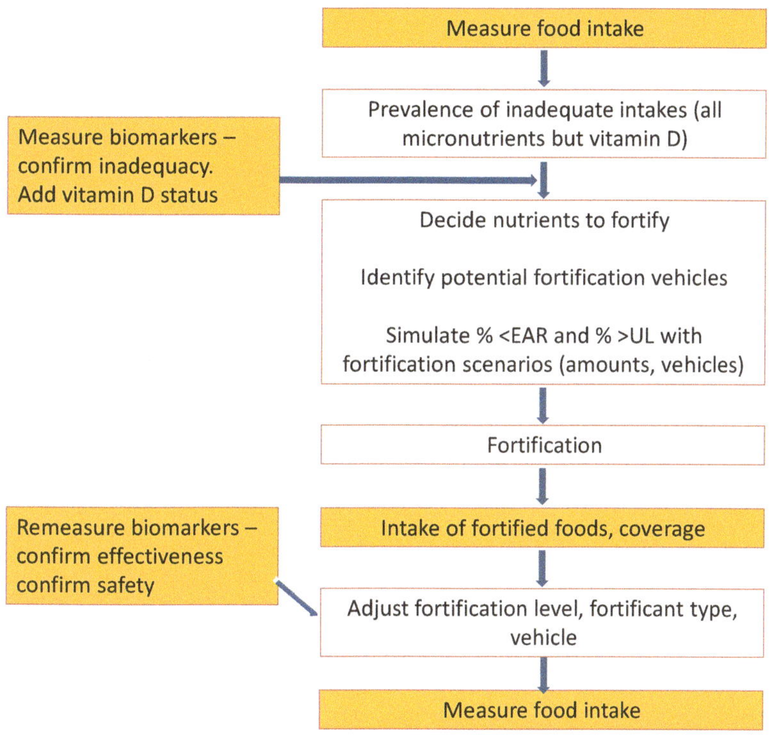

Fig. 2. Key steps in food fortification programming

Food intake data are critical in defining the prevalence of inadequate micronutrient intakes in a population. Next to food intake data, biochemical data on micronutrient status may assist in confirming and complementing food intake data pre- and post-fortification

Despite their usefulness and importance, nationally representative surveys of individual dietary intakes are rarely performed in developing countries as it is relatively time-consuming to collect all the data and to calculate nutrient intakes by converting food consumption data to nutrient intakes using a food composition database. The use of dietary intake data also has some limitations; the collected data can be subject to recall bias and bias in recording "good/bad" foods [7]. Another limitation of food intake data is that recording of a single intake day is not usually representative of a person's usual intake due to day-to-day variation. However, the IMAPP program corrects for day-to-day variation of intake, either by estimating this variation from two days of intake data per individual, or offering an estimate established in other populations. Another limitation of relating intake data to the EAR and UL is the uncertainty around these reference values. For some micronutrients, the EAR, and especially the UL, is based on a limited number of data, or the data were considered insufficient to even establish an EAR or a UL.

Dietary intake data may not be available, in which case the World Health Organization (WHO) and Food and Agriculture Organization of the United Nations (FAO) method allows one to estimate the usual *per capita* intake of fortification vehicle(s) from household intake data or national food balance sheets [4]. However, the disadvantage of this approach is that prevalence of intakes below the EAR or above the UL is not estimated and vulnerable members of households or populations are not identified.

Biochemical data on micronutrient status may assist in confirming and complementing intake data concerning whether a micronutrient deficiency is a national problem (Fig. 2, Table 1) [8]. For vitamin D, no intake data can be used, thus for this nutrient status assessment relies on measuring plasma/serum 25-hydroxyvitamin D. Changes in biochemical indicators of status can also be useful in monitoring and evaluating the nutritional impact of a fortification program, provided that a suitable biochemical marker of status is available (Fig. 2). In addition, alongside food intake data, biochemical markers of excessive intake, if available, may assist in monitoring the safety of a micronutrient intervention program (Fig. 2). However, collecting biochemical indicators in a field-setting can be invasive, expensive, and laborious and for several micronutrients, no good biochemical markers are available. Biochemical markers such as plasma zinc, calcium, iron and retinol perform relatively poorly due to homeostatic regulation of blood levels and/or potential influence by infection or inflammation, not reflecting true micronutrient status.

Prevalence estimates of specific micronutrient inadequacies based on intake data and biochemical markers may not agree. For example, results may show a high prevalence of inadequate folate intakes accompanied by lower frequency of low plasma folate concentrations [8]. In this case, errors in food composition data might be responsible and should be checked. Moreover, the criteria used to define "inadequacy" in terms of EAR cut-off value versus biomarker cut-off value often differ, particularly in at-risk groups such as infants and young children and pregnant and breastfeeding women. This may consequently result in different prevalence rates of inadequacy depending on the method used.

For most micronutrients, there is little concern about excess intake from the amounts typically used in fortification programs. However, concerns about exceeding the UL exist for some micronutrients due to the narrow range between the EAR and the UL, particularly when several micronutrient programs (supplementation, staple food fortification, home fortification with micronutrient powders or pastes) are implemented. Table 2 shows the particular cases of concern, which include overconsumption of iron from multiple oral supplements, of folic acid from oral supplements and food fortification, and of vitamin A from oral supplements and (multiple) fortified foods and micronutrient powder. Micronutrient imbalance may also be of concern; both folate and vitamin B12 are important determinants of fetal growth and development. During pregnancy, the increased requirement for folate is met with iron and folic acid supplements and often folic acid fortified flour as well, while vitamin B12 deficiency usually remains untreated [9].

The availability of valid and reliable data determines our ability to develop evidence-based micronutrient intervention programs and policies to achieve nutrition goals. Use of both intake data and biomarkers reflecting nutrient exposure, status, and functional effects is critical for assessing the problem and monitoring/evaluating the efficacy of a program. Current food consumption assessment tools may be subject to substantial measurement errors. Therefore, future efforts should invest in improved methods for collecting and analyzing food intake data. Prevalence of deficiency estimated from dietary intake data and biochemical markers should be compared; if the biochemical deficiency/ inadequacy does not confirm the intake data, then a dietary deficiency is unlikely to exist. Research is needed to enable better comparison and interpretation of differences in biochemical markers compared to dietary indicators. Also investment in new technologies may lead to discovery of novel and suitable biochemical markers. We need consensus on the relative strengths or weaknesses and applicability of various biomarkers of micronutrient status under specific conditions. Many biomarkers have been identified; yet formal consensus is limited regarding the best biomarkers for particular nutrients and applications. It has been recognized that a process is needed to harmonize decision making about which nutrition biomarkers are best suited for a given use, under specific conditions and settings, and whether and when to use infection and inflammation biomarkers to correct biochemical values. A number of concerted actions/programs/networks, such as EURRECA (EURopean micronutrients RECommendations Aligned) and BOND (Biomarkers of Nutrition for Development), have been initiated aiming to provide further consensus on biomarker use and evidence-informed guidelines.

3. RISKS OF INADEQUATE VITAMIN A INTAKE AND CHRONIC OR INTERMITTENT EXCESS OF VITAMIN A IN CAMEROON

Dr Reina Engle-Stone, University of California, Davis, CA, USA.

In Cameroon, dietary intake data have been used with success to predict the effects of large-scale micronutrient interventions on inadequate and excessive intakes [10]. In addition, a kinetic model of hepatic retinol concentrations was used to model the effect of multiple vitamin A interventions (fortification and supplementation) on liver vitamin A stores of young children.

A national micronutrient survey conducted in Cameroon in 2009 was used as the data source to simulate nutrient intakes prior to and after large-scale food fortification [11]. Dietary intake data were collected by 24-hour dietary recall interviews (with duplicates in a subset) from 1002 households with at least one woman 15-49 years of age and at least one child 12-59 months of age. Usual intake distributions of vitamin A were estimated according to the method developed at the US National Cancer Institute (NCI) [12]. Inadequate and excessive vitamin A intakes were estimated as the prevalence of vitamin A (pro-vitamin A and retinol) intakes below the EAR and the prevalence of retinol intakes above the UL.

The effects of different vitamin A intervention scenarios on vitamin A intakes and the prevalence of intakes below the EAR were simulated by adjusting individual nutrient intake values to reflect the new intervention scenario and then recalculating the distribution of usual intakes by the NCI method. Analyses were conducted for women and non-breastfed children in three different regions of Cameroon. The following scenarios were simulated: 1) the effect of increasing the vitamin A content of refined oil (which is currently fortified in Cameroon); 2) the effect of fortifying a second food vehicle with vitamin A, i.e., wheat flour, sugar and bouillon cubes. The prevalence of intakes below the EAR for vitamin A in the Cameroonian population suggests a need for interventions to increase vitamin A intakes. The regional patterns of low intake are consistent with the data for low plasma retinol-binding protein concentrations that were measured among children and women as a biomarker of vitamin A deficiency [13].

Simulation of vitamin A intakes suggested that the current oil fortification program would increase vitamin A intakes, but complementary interventions are needed to meet the gap between the EAR and usual intake of vitamin A, especially in the north of the country. The simulations also indicated that fortification of additional food vehicles would not only increase dietary vitamin A, but may also increase excessive intakes of preformed vitamin A in subgroups of children, depending on the food vehicle and level of fortification. The simulations can be used to determine the optimal fortification level in different food vehicles that would result in the lowest prevalence of the population with intakes below the EAR and the lowest prevalence of intakes above the UL. Of the vehicles examined, bouillon cubes was considered the best selection for a second fortification vehicle with respect to balancing inadequate and excessive intakes because bouillon cube consumption was homogenous throughout the country.

It has to be noted that dietary intake modeling has some limitations, such as error in dietary assessments and uncertainty around cutoffs for inadequate and excessive intake, and the fact that physiological risks of intakes between the UL and the LOAEL are unknown. Because, young children in some countries are exposed to both fortification and periodic high-dose vitamin A supplements, any risk of these overlapping interventions should be investigated. The potential risk of excessive vitamin A intake via supplements is, however, difficult to estimate because high-dose periodic supplementation is difficult to translate into an equivalent daily intake value. For that purpose, an Excel-based kinetic model of liver vitamin A was developed at the University of California, Davis. The model can be used to estimate the potential risk of inadequacy and toxicity (based on liver stores) of specific scenarios meant to improve the vitamin A status of deficient populations [14]. The model was used to assess the risk of intermittent excessive liver vitamin A through high-dose vitamin A supplements on top of multiple programs providing daily vitamin A at different fortification levels in Cameroon. The model's built-in parameters for children 1-2 years of age included: Average liver weight; 2.2% loss of hepatic stores per day [15] and retention of dietary vitamin A (50%). Hepatic retinol concentrations were then modeled for different daily intakes of vitamin A (derived from the dietary simulations), with or without high-dose supplementation. A sensitivity analysis was also performed by estimating the hepatic retinol concentrations, assuming 50% or 70% retention of vitamin A.

Fig. 3 shows an example of a sensitivity analysis for the effect of multiple vitamin A interventions on liver retinol concentrations over time for a young child. Although the upper safe limit of liver vitamin A stores is unknown, a physiologic range of liver vitamin A concentration is between 20 and 300 µg/g liver [16]. Even though some of the scenarios with multiple fortification vehicles resulted in retinol intakes greater than the UL, simulations using the liver model suggested that these intakes would not be associated with hepatic retinol concentrations greater than 300 µg/g liver, even following administration of a high-dose vitamin A supplement (although in some sensitivity analyses liver vitamin A concentrations transiently surpassed this threshold).

Assumed is a total daily vitamin A intake of 1700 µg RAE/d, 50% (blue line) or 70% (purple line) retention of dietary vitamin A, and administration of a high-dose vitamin A capsule (200,000 IU) at 12 months of age. The daily vitamin A intake represents the 95th percentile of predicted total vitamin A intakes of urban, non-breastfeeding Cameroonian children 12-59 months of age under a modeled program scenario including fortified oil (12 mg RAE/kg) and fortified wheat flour (5.9 mg RAE/kg). Hepatic retinol concentrations are shown in relation to the cutoffs (red lines) for inadequate and excessive liver vitamin A concentrations of 20 and 300 µg retinol/g liver, respectively [16].

In summary, dietary intake data are useful to predict the effects of micronutrient interventions on risk of inadequate and excessive intakes, as shown in this case study of large-scale food fortification in Cameroon. Based on the dietary intake simulations, there is evidence that young Cameroonian children (1-4 years of age) may benefit from large-scale vitamin A fortification, but some subgroups may be at risk of exceeding their UL, when multiple food vehicles and/or high fortification levels are in place. The biological risks of intakes between the UL and LOAEL are unknown. In this respect, more research is needed to identify the risks of intakes at the high end of the intake spectrum. A kinetic model of liver vitamin A concentrations may assist in exploring the contribution of high-dose vitamin A supplements to liver stores in children, in combination with dietary intake data. The data from Cameroon suggested that there is a low risk of excessive liver stores under the scenarios examined. For both the intake simulation model and the kinetic model, the quality of the results depends on the quality of the data used. Moreover, empirical validation of the models is still needed. Using these models, the risk of inadequate or excessive intake will depend on whether the targeted population subgroup is reached, stressing the importance of targeted programming. Whether the target group is reached through fortification strongly depends on the choice of the food vehicle and its intake distribution among the population. Therefore, it is important to collect data on intake distribution and not just data on average amount consumed. Besides collecting dietary intake data and program coverage data, it is essential to monitor the actual nutrient content of fortified foods to ensure that the program has the intended impact.

4. VITAMIN A INTAKE DISTRIBUTION AMONG THE GUATEMALAN POPULATION: CONTEMPORARY PERPLEXING PERSPECTIVES

Dr Noel Solomons, Center for Studies of Sensory Impairment, Aging and Metabolism, Guatemala.

A national survey of Guatemala in 1965-1967 found that 27% of children under-five were deficient in vitamin A based on having serum retinol levels below 20 µg/dL. Starting at the end of 1975, sugar was fortified with vitamin A at 15 mg/kg to address public vitamin A deficiency owing to strong advocacy by Guillermo Arroyave of the Institute of Nutrition of Central America and Panama (INCAP). Mandatory sugar fortification took place from 1975-1977. From October 1975 through November 1977, surveys were performed in children and lactating mothers showing that only 9% of children suffered low levels of vitamin A in their blood [17]. Immediately thereafter, fortification stopped due to a number of reasons, and vitamin A deficiency reappeared. By 1987, a general study found a situation similar to that of 1965. Voluntary fortification recommenced in 1987, and since that time until today, all sugar was to be fortified at the refinery level of about 15 mg/kg and household level has been constant at about 10 mg/kg. This fortification level of 10-20 mg/kg sugar was recommended such that it would add an average intake range of 400-800 µg retinol equivalents (µg RE/d). This rationale was based on an estimated average sugar consumption by pregnant women at that time of 40 g/d and aimed to provide them 600 (µg RE/d).

Nowadays, many foods are fortified with vitamin A in Guatemala. Many processed foods are fortified with vitamin A such as breakfast cereals and protein-rich *atoles* (e.g., Incaparina), or are made with vitamin A-fortified sugar (e.g. sweet bread, soft drinks). In addition, children could potentially receive vitamin A through special formulated products such as vitamin A-fortified corn soy blend distributed by the government, micronutrient powders distributed by the World Food Programme (WFP), or fortified lipid spread. Moreover, when enrolled in a vitamin A supplementation program, children can receive six-monthly oral supplements of 200,000 International Units (IU) of retinyl palmitate.

The concern for adverse health effects does not only reside at the lower end of the vitamin A intake exposure or status spectrum. The upper end of the status or intake exposure continuum also can produce adverse consequences.

In order to be able to evaluate and manage potential inadequate and excessive intakes of vitamin A, several (inter)national committees, for example the Joint FAO/WHO Expert Committee on Food Additives (JECFA), the European Food Safety Authority (EFSA) and United States Institute of Medicine's Food and Nutrition Board (IOM) used by both Canada and United States, have set nutrient intake values such as the EAR, the Reference Nutrient Intake (RNI) or Recommended Daily Allowance (RDA) (Table 3) and the UL (Table 4).

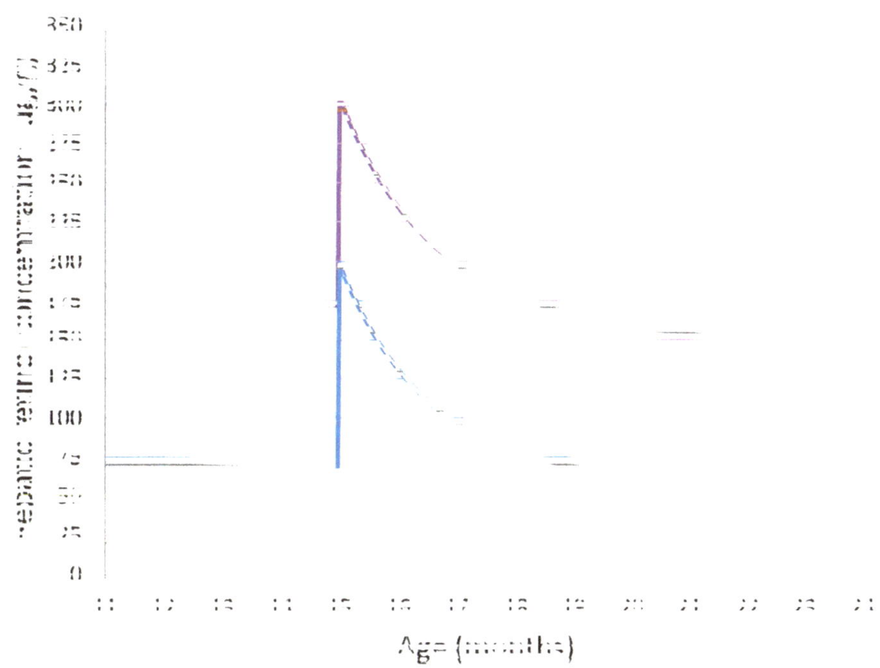

Fig. 3. Predicted changes in hepatic retinol concentration over time among young children

Table 2. Risks of excessive micronutrient intakes involved in supplementation and/or fortification

Nutrient	EAR/UL	Risk	Fortification or supplement
Iron	6/45	Inflammation, gastrointestinal discomfort	Supplement
Vit A	625/3000	Liver toxicity, 2X ↑ bone turnover markers, teratogenicity	High-dose supplement + fortification
Iodine	95/1100	Elevated TSH	Excess universal salt iodization, fortification
Zinc	9/40	↓ Copper status	Supplement
Folate	320/1000	↓ B12 status	Fortification, supplement, especially if both
Vit B12	2/–	None	
Thiamin/Vit B1	1.1/–	None	
Vit B6	1.1/100	Neurological damage	Supplement

- In a first step, information needs to be collected on the quantity of the food vehicle and micronutrients consumed in order to quantify the nutrient gaps to be addressed. Biochemical data on micronutrient status can be supportive in confirming whether a micronutrient deficiency is a national problem.
- In a second step the appropriate level of fortification based is determined based on consumption pattern micronutrient intake gap to be addressed. To that end, intakes above the EAR and below the UL are often used to estimate the proportion of the population meeting their required and safe intake, respectively.
- The same process is applied to monitor whether a fortification program is effective and safe, requiring regular updating of the data.
- The value and limitations of dietary intake data, nutrient reference values, and biochemical markers are discussed.

Dietary adequacy is the habitual consumption from all sources of the recommended amount of total vitamin (preformed vitamin A and provitamin A carotenoids) by the individual (RNI, RDA) or by the population group (EAR). The EAR of vitamin A for adults of 600 µg/d was derived from the requirements to maintain liver stores to meet vitamin A needs of a healthy population for 4 months, estimated to be approximately 20 µg/g liver [16]. The RNI or RDA is assumed to be the EAR plus twice the standard deviation of 20%, i.e., calculated by multiplying by 1.4 [20]. Dietary excess of vitamin A is the habitual consumption of dietary preformed vitamin A in excess of the UL in relation to the norms for different segments of the population by age and sex. Because little is known about the risks of excessive intakes, the UL is established by applying an uncertainty factor to the level where no or the lowest adverse effects were observed (NOAEL and LOAEL) from case reports or cohort studies. The UL of vitamin A relates to preformed vitamin A and not provitamin A from carotenoids. Provided it is from dietary sources, there is no UL to provitamin A carotenoid intake, with the only consequence being yellowing of the skin (carotenodermia).

However, there are some caveats related to the ULs for preformed vitamin A. The UL of 3000 µg/d for adults as established by the EFSA is based on a risk of congenital birth defects forming in the embryos during the first trimester of pregnancy [21]. This UL has been extrapolated to other age and gender groups based on relative body size (metabolic weight). However, the risk of congenital birth defects in pregnant women is not relevant to other life stages or gender groups. The IOM and WHO/FAO extrapolated the UL of 3000 µg/d only to all women of childbearing age. The UL established by IOM and WHO/FAO for other gender and life stage groups, is based on the risk of hepatotoxicity of which lowest effects were observed at 14,000 µg RAE/d with an uncertainty factor of 5 and a correction factor for body weight. Since the evidence of adverse effect risks in adults for bone demineralization and skeletal fracture risk for intakes of preformed vitamin A intakes exceeding 1500 µg/d was non-conclusive for adults and absent for children, the EFSA, WHO/FAO and IOM did not take it into consideration.

The EAR reflects the average requirement of vitamin A in a population group, i.e., the intake that meets the requirement of half of this group, whereas the other half is at risk of inadequacy (Fig. 4). The UL reflects an intake level of preformed vitamin A for a population group that is likely still to be safe in this group.

- Cameroon is among the countries that have used dietary intake with success to predict the effects of large-scale micronutrient interventions on the population proportions having inadequate vitamin A intakes and exceeding safe vitamin A intakes using the EAR and UL as cut-point.
- Children (1-4 years of age) may benefit from large-scale vitamin A fortification, but some subgroups may be at risk of exceeding their UL, when multiple food vehicles and/or high fortification levels would be implemented.
- This illustrates that for micronutrients with an UL close to the usual intake, shifting the intake distribution between the EAR and the UL is a challenge.
- Based on a kinetic model of liver vitamin A concentrations predicted that there is a low risk of excessive liver stores under the different scenarios of food fortification and high-dose supplementation examined.

Table 3. Total daily recommended vitamin A intakes expressed as Recommended Nutrient Intake (RNI) by WHO/FAO and recalculated* Estimated Average Requirement (EAR)

Population group	RNI (µg RE**/d)	EAR (µg RE/d)
0-6 mo infants	375	--
6-12 mo infants	400	--
Toddlers (1-3 y)	400	286
Preschool (4-6 y)	450	321
Childhood (7-9 y)	500	357
Adolescent male (10-18 y)	600	428
Adolescent female (10-18 y)	600	428
Adult male (>18 y)	600	428
Adult female (>18 y)	500	357
Pregnancy	800	571
Lactation	850	607

** EAR values are adjusted from WHO/FAO [4] by recalculating with 1.4 factor*
*** RE: Retinol Equivalent*

Table 4. Tolerable Upper Intake Levels (UL) of preformed vitamin A established by The IOM (USA, Canada) [18] and the WHO/FAO [4], and the EFSA (Europe) [19]

Population group	UL (RAE*/d) IOM WHO/FAO	Population group	UL (RAE/d) EFSA
6-12 mo	600 [a & #]	6-12 mo	Not determined
1-3 y	600 [a #]	1-3 y	800 [b $]
4-8 y	900 [a #]	4-6 y	1000 [b $]
9-13 y	1700 [a #]	7-10 y	1500 [b $]
14-18 y male	2800 [a #]	11-14 y	2000 [b $]
		15-17 y	2600 [b $]
>18 y male	3000 [a]	>18 y	3000 [b]
14-18 y female	2800 [b #]		
>18 y female	3000 [b]		

** RAE: Retinol Activity Equivalent*
[a] Derived from a LOAEL for hepatotoxicity of 14000 µg RAE/d for adults and an uncertainty factor of 10
[b] Derived from a NOAEL for teratogenicity of 3000 µg RAE/d for adult women
[&] WHO/FAO has no UL established for the age 6-12 mo
[#] Corrected for body weight
[$] Corrected for metabolic weight

A public health problem of endemic hypovitaminosis A is declared based on evaluation of a representative sample (national, sub-national) of serum retinol levels from children aged 6-59 months [22]. Action to address a public health problem of endemic hypovitaminosis A is justified when more than 20% of the surveyed sample has a retinol concentration of <20 µg/dL (<0.70 µmol/L). The Guatemalan nutritional survey of 2008-2009, encountered only 3% of retinol samples below 20 µg/dL, with 1.8% remaining after an adjustment for inflammation [23]. Guatemala could therefore formally classified as a nation without a hypovitaminosis A problem, which can be attributed to the program of sugar fortification.

CeSSIAM has contributed what could be considered confirmatory data. This comes from a regional, convenience-sample survey conducted on the Pacific coastal plain; there was not a single sample among 582 serum specimens with a low retinol level below 20 µg/dL; meanwhile,13 of 268 samples from children (4.8%) and 1 of 314 (0.3%) of adult women had a marginal retinol concentration below 30 µg/dL. However, this survey may not necessarily tell us that vitamin A intake is adequate in the whole population. There may be geographically uneven intake distributions in the country and there may still be existing pockets of low vitamin A consumption with associated low status for any segment(s) of the Guatemalan population. Meanwhile, with all concurrent vitamin A initiatives available, there could be a summation of sources of preformed vitamin A in other pockets that poses risk of consumption in excess of the UL on an occasional or habitual basis. A survey in the Western Highlands of Guatemala showed that pockets of persistent sub-adequacy indeed still exist. The Western Highlands of Guatemala showed that 25% of urban pregnant and lactating women had inadequate vitamin A intakes below their (WHO/FAO) RNI of 800 and 850 µg RE/d, respectively (Fig. 5). Contrary to rural areas where inadequate intake prevalence was much higher; 84% of the women did not meet their RNI. This result should ideally be confirmed by the modified retinol dose response as a biomarker of vitamin A inadequacy.

Vitamin A intake data among women of reproductive age from fortified sugar alone has been collected. Fig. 6 shows that in urban populations, 70% of women consumed more than their RNI of 500 µg RE/d of vitamin A from sugar (light-blue) whereas in rural populations, their consumption was less than their RNI of µg 500 µg RE/d (dark-blue) (Fig. 6). Whether these women with intakes below their RNI were really at risk of inadequacy should be confirmed by the modified retinol dose response as a biomarker of vitamin A status.

The survey data pointed out that five of the 40 women in the urban region consumed preformed vitamin A in excess of their UL of >3000 µg on the day prior to interview. This is the UL for vitamin A, edging toward the NOAEL of 4500 µg/d [18].

With all the vitamin A initiatives, it is not unlikely that pockets of habitual excess exposure also exist for children. Hypothetically, a 4 year-old Guatemalan child could over consume with all of the sources of vitamin A superimposed in the diet and from the health care system. This would result in a maximal intake of 3847 µg RAE/d of vitamin A (Table 5).

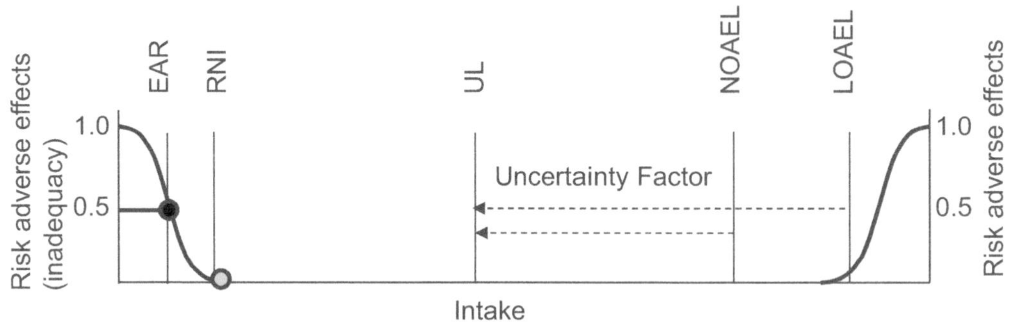

Fig. 4. Relationship between dietary nutrient reference values and risk of inadequacy or excessive intakes

Adjusted from [2]. EAR: Estimated Average Requirement, RNI

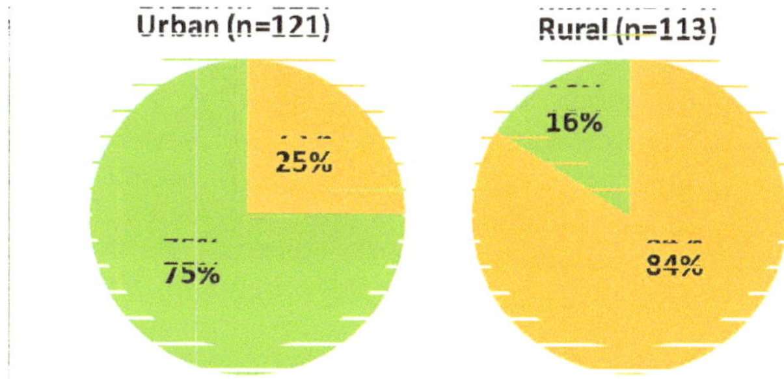

Fig. 5. Adequate and inadequate vitamin A intakes among pregnant and lactating women in urban and rural Western Highlands of Guatemala

Proportions with intakes above and below the RNI specific for lactating and pregnant women (WHO/FAO) are indicated in green and yellow, respectively

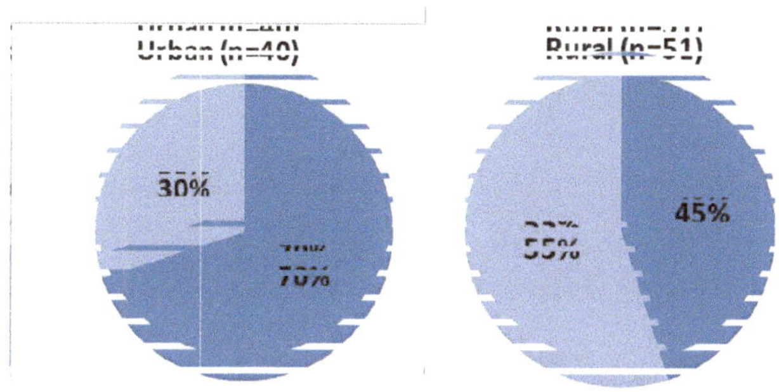

Fig. 6. Vitamin A intakes from fortified sugar among women in urban and rural Western Highlands of Guatemala

Proportions with intakes above and below the RNI specific for lactating and pregnant women (WHO/FAO) are indicated in light blue and dark-blue, respectively

Table 5. Hypothetical high intake scenario of a 4-year old child over-exposed to vitamin A

Source	Form	µg RAE/d vitamin A provided (% of total)
Basic Guatemalan family diet, containing predominantly herbs, fruit, carrots and squashes	Provitamin A carotenes	928 (24%)
Sugar	Preformed vitamin A	1014 (24%)
Other fortified foods	Preformed vitamin A	1636 (43%)
Oral supplement	Preformed vitamin A	267 (7%)*
Total		3847

** Prorated per day over 183 days*

Whether a child with intakes above the UL is at risk of excessive intakes is ideally not only based on food intake data, but also confirmed by elevation of one or another of the still not strongly validated putative indices for excessive vitamin A status. When retinol concentrations in the liver reach about 300 µg/g, circulating levels of retinol may rise in association with increasing circulating fasting retinyl esters [16]. Early studies proposed the use of fasting plasma retinyl ester concentrations greater than 10% of total serum vitamin A concentration (exceeding the mobilization and transport capacity of Retinol-Binding Protein 4) as a biomarker reflecting excessive intakes [24,25]. However, this notion has been challenged by a study in 2001 with 6547 adults showing no relationship between high fasting retinyl ester concentrations and liver dysfunction [26]. More recently, retinoic acid metabolites like 9-cis retinoic acid have been studied as a function of vitamin A intake. Some of these metabolites mediate the functions of vitamin A required for growth and development but in high amounts may mediate some of the adverse effects [27]. Despite the large amount of new understanding gained about vitamin A metabolism, the debate continues concerning the use of suitable vitamin A exposure biomarkers.

Since sugar fortification became a national program in Guatemala, vitamin A deficiency is officially no longer a public health problem. This could be considered a historical success story of addressing a nutritional problem in a developing country. However, preformed vitamin A is now consumed in Guatemala from many sources. Therefore, this may also be considered a potential adverse scenario of uncontrolled and uncoordinated summation of sources of preformed vitamin A.

Hypovitaminosis A still exists in pockets of the country, particularly in rural areas. The question now is, what is the way forward for Guatemala? INCAP has proposed a modification in the fortification based on two considerations. The first is the change in the fortification process for the nation. In the early years, all sugar was fortified at the time of its initial harvesting and milling, in the range of 10-20 mg/kg, taking into consideration a loss over the ensuing year. Indeed, sugar collected in households had a concentration ~10 mg/kg, as analyzed by the surveillance system at INCAP. The new procedure involves so-called *just-in-time* fortification, in which vitamin A is constantly being added to sugar just before it goes from the warehouse to market. No longer is post-fortification loss a consideration. The INCAP-recommended range, moreover, has moved down to 5-9 mg/kg, for a median of 7 mg/kg. Given the increase in median sugar consumption of adult women, to an estimated 75 g per day, the 7 mg/kg would provide 525 RE, close to their RNI of 500 RE. As of the moment, the Government of Guatemala is studying the proposed change.

If implemented, this would reduce preformed vitamin A intake by only 350-400 μg/d in the case of our example of the overexposed preschool child (Table 5), but would reduce the percentage of rural reproductive-age women with sugar-based intakes meeting their RNI by approximately 20%. Especially when the intake distribution is right-tailed, exposing a population to interventions with additional vitamin A will reduce the proportion with intakes below the EAR but may unavoidably result in a another (smaller) proportion exceeding their UL. Not allowing the UL to be exceeded may leave a proportion of the population with intakes below the EAR unaddressed (Fig. 7). More understanding is needed in Guatemala on how adequate intakes for the majority at all ages can be promoted without causing excess intakes, and on how specific rural populations can be targeted for outreach in their moments of increased requirements. In summary, the complex situation requires a multifaceted approach based on the total diet and all available sources of vitamin A.

5. INTEGRATED RISK-BENEFIT ASSESSMENT IN FOOD AND NUTRITION

Prof Dr Hans Verhagen, National Institute for Public Health and the Environment (RIVM), The Netherlands. Risk-taking is normal in everyday life if there are associated (perceived) benefits. Risk-benefit assessment compares the risk of a situation to its related benefits and addresses the acceptability of the risk. Risk-benefit assessment in food and nutrition is relatively new. It weighs the beneficial against the adverse effects that a food (or food component) may have, in order to facilitate better informed policy decisions regarding public health issues. It is rooted in the recognition that good food and nutrition can improve health and that some risk may be acceptable if benefits are expected to outweigh such risk.

In the field of food and nutrition, risks are not accepted. In contrast, risks associated with drugs or medicines are generally accepted. Concepts of balancing risks and benefits is common in evaluation of medicines [28]. Moreover, for economics and marketing finance, financial risk management decisions are even a necessity [29]. In the field of food and nutrition, the health burden related to unsafe food is a lot smaller than the health burden related to eating unhealthily or eating too much. For example, the benefits related to eating fruits and vegetables and reduction of cardiovascular disease, cancer, etc, by far outweigh the risks of pesticide contamination [30]. Agricultural pesticides are monitored regularly by government authorities, and are below acceptable daily intake (ADI) levels. Yet residues remain a consumer issue and are mainly a perceived risk, unlike a real risk.

Traditionally, the prevalence of population intakes below the EAR is taken to reflect the risk of inadequacy and intakes above the UL are taken to reflect the risk of excess intakes.

Alternatively, more sophisticated risk-benefit assessments can be used to estimate health benefits and risks of increasing micronutrient intakes without making use of nutrient reference values. Risk-benefit models make use of a step-wise approach as support to decision making related to changes in risks and benefits of food or nutrient intakes. In each subsequent step, additional data on occurrence and consumption will allow the assessment to be refined.

Risk-benefit approaches differ from traditional cut-point methods. Cut-point methods use prevalence of population nutrient intakes below or above nutrient reference values as an estimate of "risk". Risk-benefit approaches are rather based on intake dose-response effects rather than nutrient reference values as are cut-off points. Risk-benefit approaches use incidence data of adverse effects opposed to prevalence data exceeding a nutrient reference value. Incidence takes into account the number of people affected. Moreover, other dimensions of the magnitude of the risk, such as duration and severity of the adverse health effect, can be integrated into a common health metric such as the Disability-Adjusted Life Year (DALY) or Quality-Adjusted Life Years (QALY). Moreover, by making use of the DALY or QALY as a common metric of "health risk", deterministic or probabilistic risk-benefit evaluations can be performed. Thus, in risk-benefit assessments, benefits and risks are assessed in one approach and may conditionally be expressed into one metric. This allows the comparison of adverse and beneficial effects to be qualitative as well as quantitative.

For example, by using the Benefit Risk Analysis of Foods (BRAFO) risk-benefit approach, the risks and benefits of increasing folic acid intakes via bread fortification in the Dutch population were assessed [31,32]. In a first step (Table 6), an inventory was made of the established adverse health effects in the literature associated with increasing folic acid intake. Subsequently the change in incidence rates of these adverse health effects was estimated depending on the increasing folic acid level used to fortify flour for making bread (Table 7). Based on these estimates of changing incidence rates, the benefits may still not clearly outweigh the risks (Table 7). A next step can be performed in which also other dimensions of risk magnitude such as duration and severity of the adverse health effects are integrated into a common health metric such as the DALY. Table 8 shows the net health benefit of bread fortification with folic acid at the level of 70 µg/100 g of flour. The health loss resulting from masked vitamin B12-deficiency in terms of DALYs appeared to be negligible compared to the health gain resulting from prevented neural tube defects.

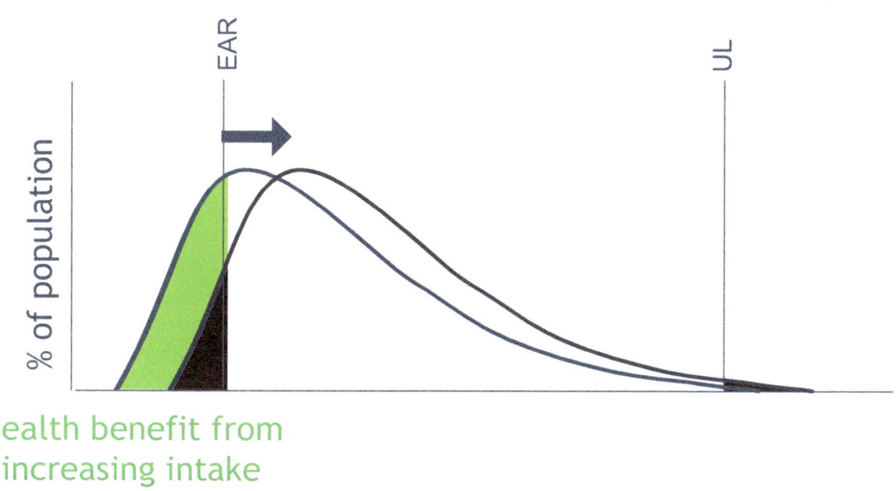

Fig. 7. Shifting vitamin A intakes in a population may result in intakes above the Tolerable Upper Intake Level (UL) while leaving a proportion with intakes below the Estimated Average Requirements (EAR)

- More than 25 years ago, after the vitamin A sugar fortification program was suspended in Guatemala, surveys pointed out that vitamin A deficiency was again a public health problem. Since fortification recommenced in Guatemala in 1987, first voluntary followed by a mandated program, remarkable progress has been made in reducing vitamin A deficiency.
- Nowadays, all sugar for table use and processing, as well as many commercial foods are fortified with vitamin A. Nevertheless, pockets of persistent vitamin A sub-adequacy may still exist in Guatemala as shown by a recent survey while it is not unlikely that pockets of habitual excess exposure also exist.
- The changes in sugar consumption, number of fortified foods, and fortification levels over the years illustrate the need to re-assess the situation on a regular basis.
- Since vitamin A deficiency in Guatemala and vitamin A post-fortification loss are no longer considered problems, lower sugar fortification levels are under consideration. Yet, this will not provide the full solution to the complex problem to address inadequate vitamin A intakes while avoiding excessive intakes.

Table 6. Hazard–benefit identification: Established health effects of folate and/or folic acid

Benefits
Reduced incidence of neural tube defects
Reduced incidence of megaloblastic anaemia
Reduced incidence of stroke
A reduced incidence in colorectal cancer
Risks
Increased incidence of neurological damage due to masking of vitamin B12 deficiency
Increase in the incidence of colorectal cancer
Accelerated progression of colorectal cancer

Table 7. Benefits and risks of bread fortification with folic acid at 70 µg/100 g flour (in The Netherlands): Results as relative or absolute change in incidence rates with folic acid fortification

Benefits	**Δ Incidence relative (absolute)**
Reduced incidence of neural tube defects	- 37% (-83)
Risks	
Increased incidence of neurological damage due to masking of vitamin B12 deficiency	+1% (53)

Table 8. Benefits and risks of flour fortification with folic acid at 70 µg/100 g flour: results as change in DALYs*

Benefits	**Δ DALYs**
Reduced incidence of neural tube defects	- 5474
Risks	
Increased incidence of neurological damage due to masking of vitamin B12 deficiency	+ 53

**DALY: Disability Adjusted Life Year*

Nutrients may confer both beneficial and adverse health effects. Measures directed at safety may lead to suboptimal or inadequate intake levels of micronutrients from a benefit perspective. However, not allowing benefits to occur in order to guarantee 100% safety is a risk-management decision much the same as accepting some risk in order to achieve more benefits. A risk-benefit assessment can help risk managers to make more informed and balanced risk-benefit decisions.

One step further than risk-benefit considerations are cost-benefit considerations in which the costs of the respective options are compared one with another. A review on the health-related economic impact of nutrition interventions published in 2012 found that of the 56 published cost-benefit studies, 24 studies related to micronutrient interventions, 25 to healthy diets, and 7 to functional foods. The majority of food and nutrition interventions were highly cost-effective [33], suggesting that investing in adequate nutrition can significantly contribute positively to public health and society.

In conclusion, risk-benefit assessment can be a valuable and transparent approach to provide the best possible science-based answer to complicated questions with a large, and potentially optimal, impact on public health. Risk-benefit approaches assist in calculating the risks versus the benefits for public health of food or nutrient interventions. When using common health metrics, the costs of the intervention versus the benefits for public health can be calculated. Ultimately, public health can profit from risk-benefit and cost-benefit assessments. Risk-benefit approaches can help us to understand the magnitude of the risks and assist in making evidence-informed policy decisions in nutrition for the betterment of public health (and reducing their costs).

6. THE CHALLENGE OF SETTING MICRONUTRIENT FORTIFICATION LEVELS

Dr Maaike Bruins, DSM Biotechnology Center, Delft, The Netherlands.

Micronutrient deficiencies impose a major disease burden on the affected persons and the societies in which they live. Micronutrient intervention programs (as part of a broader approach) may be implemented to increase the micronutrients of concern. In addition to the conventional approaches of micronutrient supplementation and large-scale food fortification, novel products have been developed for fortifying foods or complementary foods in the household using fat-based spreads and micronutrient powders.

When setting micronutrient levels for foods or food supplements, the intake levels need to be effective in mitigating the adverse health effects from inadequacy while at the same time needing to be safe. A legal framework provides the means to regulate minimum and maximum limits for micronutrients in foods or food supplements. However, there is currently no broad scientific consensus on the general principles or approaches for the assessment of benefits and risks to human health of micronutrient deficiency or overexposure and the two are usually conducted independently.

Two main approaches exist in assessing risks of inadequate and excessive intakes that can be used in setting effective micronutrient levels for foods or food supplements. The usual approach in nutrition planning makes use of nutrient reference values as cut-points to shift the population intake distribution. The planning goal for a population (life stage and gender) group is to achieve an acceptably low prevalence of intakes below the EAR, reflecting inadequate intakes and acceptably low prevalence of intakes above the UL, reflecting too high intakes [4,20]. This method requires an estimate of the nutrient intake distribution within the main population groups, rather than just mean intakes.

The EAR is the nutrient intake level that is estimated to meet the requirements of half of a population group, but is inadequate for the other half [20]. The EAR is based on adverse effect markers of inadequacy reported in literature. Evidence on adverse effects due to excessive nutrient intake is often scarce, and usually large uncertainty exists around the level of intake at which adverse effects start. The UL is the level of intake that is likely to pose no adverse effects from too high intakes as it is set a factor lower (depending on the uncertainty and severity) than the highest level at which no adverse effect is observed (NOAEL) or the lowest level at which a relevant adverse effect was observed (LOAEL) [20]. For some micronutrients, a high uncertainty results in a low UL close to the high end intake levels, particularly for children [20,34]. This may pose a dilemma as to whether fortification levels should be set relatively low to ensure a safety margin at the high end of the intake spectrum while leaving a proportion of the population at risk of inadequate intakes, or *vice versa*.

Another approach for deciding on raising nutrient intake levels via foods or food supplements is to apply a risk-benefit assessment [32,35,36]. The risk-benefit approach takes into consideration the differences in adverse health consequences at both ends of the nutrient intake spectrum. The adverse health consequences of inadequate and excessive nutrient intakes are balanced against each other. Decisions are made in steps; at each step, the benefits are weighed against the risks of the nutrient intervention; and a advanced step is invoked if uncertainty still exists [32,35,36].

The relevant adverse health effects related to inadequate and excessive intakes of the nutrient are inventoried in the different population groups. Positive and negative effects before and after the nutrient intervention are expressed as a semi-quantitative or quantitative measure. Modelling of the intake-response relationship of the adverse effects is an important aspect of risk-benefit assessments. Different dimensions of risks can be integrated into a common metric such as the DALY or QALY. These aspects include incidence of the adverse effect (number of individuals yearly affected), the age of onset and time lived with the adverse effect (temporarily, permanent), and the severity or disability of the adverse effect. The risk-benefit approach allows assumptions for the assessments to be made, as long as these assumptions are explicitly stated.

Different fortification levels can convey different types of benefits and risks as illustrated by a risk-benefit assessment case for folic acid fortified flour in The Netherlands [31]. When expressed as incidence rates, at 70 µg folic acid per 100 g of flour, neural tube defects, megaloblastic anemia, and colorectal cancer would be prevented by 83, 2425 and 405 cases/year, respectively, whereas undiagnosed vitamin B12 deficiencies would increase by 53 cases/year (Table 9). In terms of people yearly affected, prevented cases of megaloblastic anemia may appear the biggest benefit; however, when taking into account factors as severity and time lived with the adverse effect, by expressing in DALYs, it becomes clear that preventing neural tube defects has a far bigger positive health impact (5474 DALYs prevented) than reducing megaloblastic anemia (24 DALYs prevented). The largest public health impact was estimated to be achieved with levels between 140 and 240 µg folic acid per 100 g of flour.

Targeted micronutrient intervention programs can be a strategy if the target population at highest risk of micronutrient deficiency is well-defined, usually young children and women of childbearing age [4]. Mass fortification is usually more cost-effective than other strategies, particularly if the micronutrient deficiency is widespread among the general population [4]. The success of mass fortification largely depends on the choice of the food vehicle; the food should be affordable to the target population and be consumed regularly and in constant amounts by the whole population [4].

Usual intake distributions of most micronutrients tend to be skewed. For example, a large proportion of the population may consume a small amount of the micronutrient, while a small proportion of the population may consume a large amount. In micronutrient programming, the challenge is to reach the populations at risk of inadequacy while avoiding intakes in those segments that do not need it. I.e. the ideal food carrier has a normal and narrow intake distribution within the population with little excessive low and high intakes. Preferably, the quantity of micronutrient required to address a deficiency is divided over multiple instead of one food [37]; lower micronutrient levels in different foods reduces the chance of excessive intakes while increases the likelihood that a food is consumed by the group most in need (Fig. 8).

In conclusion, methods to assess micronutrient levels in foods and food supplements taking into account risks of inadequacy versus risks of excess differ in approach and concept. Table 10 summarizes the differences in the two main approaches. Consumption data of micronutrients from the usual diet and other sources are required for both approaches but unfortunately are rarely collected. Estimating the additional levels of micronutrients to foods or supplements that are effective in compensating micronutrient deficiencies and meanwhile safe, can be challenging when using the UL as the UL does not give any indication of the magnitude of any risk associated with intakes that exceed the UL [2]. Gaining understanding of the benefits and risks associated with increasing micronutrient intakes, and how serious a risk is, may assist in making decisions. Implementation of micronutrient interventions inherently requires policy decisions on the acceptability of risks of too low or too high intakes.

7. AUDIENCE DISCUSSION

The presentations were followed by a general discussion between the audience and the presenters, who took place in the discussion panel. The specific remarks and questions from the audience and the presenter's answers are summarized.

- The principle of risk-benefit assessment is that one must decide what risks to accept in order to get the benefits.
- Decisions are made in a step-wise approach, continuing as uncertainty exists, each step being more refined.
- Risk-benefit approaches are based on the concept that adverse health effects dose-dependently change in response to shifting (micronutrient) intakes.
- In risk-benefit approaches, different dimensions of risks are considered including incidence, duration and severity, in order to quantify and balance the benefits against the risks.
- The use of quantitative measures allows to assess the cost-effectiveness of a food or nutrient intervention.

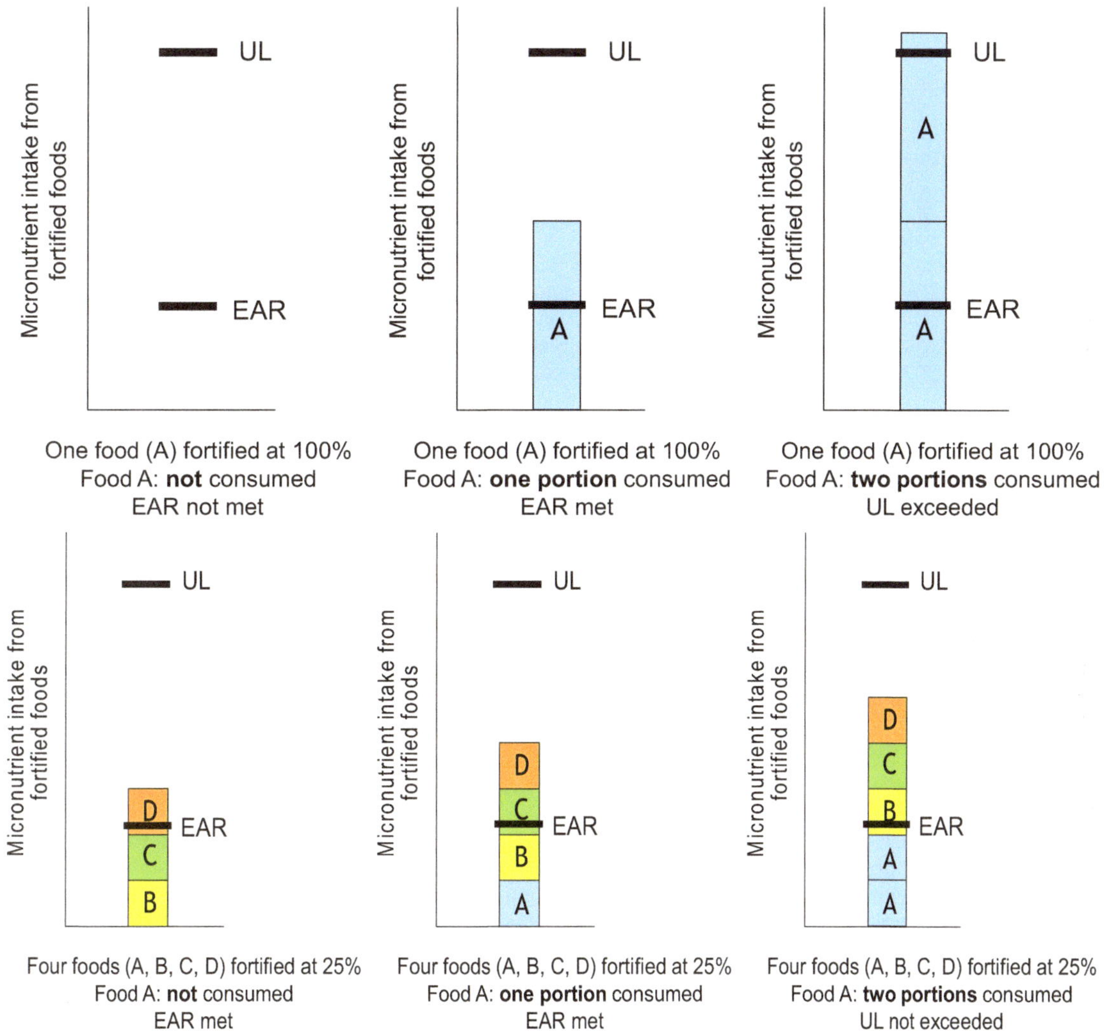

Fig. 8. Fortification of one food or multiple food vehicles at lower level: likelihood of exceeding the Tolerable Upper Intake Level (UL) and meeting the Estimated Average Requirement (EAR)
Intake if one food at 100% (upper panels) or four foods at 25% (lower panels) would be fortified. When fortifying only food A at 100%, and food A is not consumed, the EAR would not be met while consumption of two portions of food A would exceed the UL. When fortifying four foods at 25%, and food A is not consumed, the EAR would still be met while consumption if two portions of food A would not exceed the UL. This figure is published in: [38]

Table 9. Changes in DALYs* (prevented: Minus, increased: Plus) when fortifying flour with folic acid at 70, 140, 280, 420 μg/100 g flour

		Δ DALY			
		70 μg	140 μg	280 μg	420 μg
Neural tube defects	Absolute Δ in incidence	-83	-118	-150	-166
	Change in DALYs	-5474	-7710	-9812	-10855
Megaloblastic anemia	Absolute Δ in incidence	-2425	-3197	-4012	-4432
	Change in DALYs	-24	-32	-40	-44
Undiagnosed vitamin B12 deficiencies	Absolute Δ in incidence	53	76	121	166
	Change in DALYs	53	76	120	165
Colorectal cancer	Absolute Δ in incidence	-405	-749	-445	1954
	Change in DALYs	-2217	-4146	-167	21740
Total	Total change in DALYs	-7662	-11812	-9899	11006

** DALY: Disability Adjusted Life Year*

Table 10. Methods and approaches to set micronutrient levels to foods and supplements

	Method using nutrient reference cut-points	Risk-benefit approach
Risks	% population intake below EAR* or above UL**	Health impact Semi-quantitative Quantitative
Input	Intake distribution	Intake distribution Intake-adverse effect incidence curves Severity, duration
Simulation	Intake distribution	Intake distribution Relationship intake vs. adverse effect incidence Health impact/disease burden (increased or prevented)

**EAR: Estimated Average Requirement*
***UL: Tolerable Upper Intake Level*

Intervention 1: One of the messages that was not made clear from the presentations is that often, the vulnerable group that is used to compare intakes against the EAR is different from the groups that are at most risk of excessive intakes. For example: we may not need to be concerned with pregnant women not reaching their EAR for nutrients but the group that may be at most risk of excess will be adolescent boys who eat a lot of unhealthy foods. That wasn't made explicit, but it is important. If we do these calculations even though pregnant women may be our target group and we want to get 200 samples of dietary intake for the women, we also need information on diet for the other groups who we think are at risk of excess.

Answer Prof Allen to intervention 1: Typically the group that runs the highest risk of exceeding the upper limit is adult men, because their overall food intake is higher. It turns out that adult men are, indeed, the issue.

Intervention 2: We should make sure that we choose the right food vehicle for fortification with the right levels and the right nutrients in the food and I wanted to comment on the work that has been ongoing in many different regions on the harmonization of fortification levels of staple foods for many different regions. I wanted to hear your thoughts on that sort of process. How to balance between having harmonized standards for country versus individual levels?

Answer Prof Allen to intervention 2: There is value in at least setting a range of harmonized values of reference nutrient intakes to give people some guidance on what is unlikely to ever be useful and what would be too much. Each country and population is so different. The presentations by Dr Engle-Stone and Dr Solomons exemplify how, in a single population, you have these major groups and not tiny little groups that have similar food patterns. I feel strongly you should collect all the information and act upon it. Harmonized reference intake values are designed for populations rather than for individuals; for most individuals in that group they may apply whereas for some individuals may not.

- At intakes equal to the EAR half of the population is at risk of inadequacy. Intakes equal to the UL are unlikely to cause adverse effects as they comprise a margin of uncertainty.
- The usual method to set fortification levels is to minimize inadequate intakes and optimize safe intakes by minimizing the proportion of the population with intakes below the EAR and above the UL, respectively.
- When doubt exists about the benefits versus the risks of increasing micronutrient intakes, a risk-benefit approach may assist in decision making by balancing benefits against risks
- It is important to consider different dimensions of (reduced) risks involved with increasing micronutrient intakes, i.e. (reduced) incidence, duration and severity, when balancing benefits against risks.
- Dividing the quantity of micronutrient -required to fill the micronutrient gap- over multiple instead of one food (i.e. fortify multiple carriers at a lower content) is a suitable strategy to reach groups at risk of inadequacy while avoiding excessive intakes in other groups.

Intervention 3: I think we are handicapped with the approach to figure out what risks and benefits are. Simply, we often do not know the distribution in a population, this may be bimodal in Guatemala, which is one thing. The other thing is if you go out and simulate then you consider everything at an average level. Moreover, there is a lot of intra-individual variation, heterogeneity; I feel we do not exactly know what we are doing. I do also not have an answer, I have that observation. Can we find a way out? I think it is impossible but as Prof Allen was saying we need to collect all the data.

Answer Dr Solomons to intervention 3: I give you praise for the term "heterogeneity". I am interested in heterogeneity in response to nutrient exposures, I think it is the underlying background that we are only now beginning to be able to identify with techniques such as 'omics' where there are explanations for the different responders; but we do not have the money or the technology to identify them, so we treat them as a mixture with a distribution that is smooth. But you have tremendous heterogeneity in genetic, ethnic, life experience features, etc. I am an opponent of "one size fits all". Targeting and screening are important parts of public health that have been neglected. This arises more as a philosophical issue of a misguided notion of equity with which public health began. When it comes to fortification and supplementation, if you do not respect heterogeneity it will come back to bite you with instances of excess.

Answer Prof Allen to intervention 3: We dealt a little bit with heterogeneity of requirements but what we haven't dealt with is the heterogeneity in response to excessive levels. I work in a center that can measure exquisitely small changes in response using 'omic' methods to almost any one single food intervention, or one single nutrient intervention. In that context, I would like to say that I utterly reject using the LOAEL as a criterion for balancing risk-benefit; we just do not know what we are doing. There is no reason to feed anybody anything that leads to excess exposure in 1% of the population, there is chronic exposure when dealing with fortification. Gene expression changes in response to excess nutrient intakes every minute. Maybe I am exaggerating, but that is the way I see it. A body that is having to deal with this all the time and we should never get on that excessive end because we do not know what is happening.

Intervention 4: Addresses question to Prof Allen; In your presentation, you said something about fortification of flour and you wondered why people do not add vitamin B12 because you think it should be there. You have done work in Kenya and found high rates of deficiency in schoolchildren; meanwhile, we have just finished work in Kenya in pregnant women and there was no evidence of deficiency. So I wonder with fortification policy and guidelines, you are talking about national level? We have a lot of diversity in terms of regions and target groups, e.g., children, women, tribes, dietary patterns. When you say that we should just add vitamin B12 because we can, how would you comment on your bases for that?

Answer Prof Allen to intervention 4: This is a complicated question. Which cut-point method did you find deficiency with? There is a fair amount of debate about that. I actually mentioned folic acid – not vitamin B12 - because there is a substantial amount of data. We have some data too, that folic acid is not masking diagnosis of vitamin B12 deficiency: it is rather exacerbating vitamin B12 deficiency, it makes status worse and we have seen this in a number of populations. In Bangladesh and India, they often give 2000 µg/d of folic acid along with supplements in pregnancy and sometimes fortification as well but no vitamin B12. Safety of vitamin B12 has no tolerable upper intake level. At least people should get their requirements for these nutrients. Therefore, if you know that intake is low and there is a high prevalence of deficiency, it should be given. If you do not have deficiency, then do not give it as a fortificant or supplement. I am not saying always give it by any means.

Answer Dr Solomons to intervention 4: In Guatemala, because there is no other preparation that is cheap and available, we give children 5000 µg of folic acid with iron every week as part of our national norm. I do not know about their genomics and epigenetics: but because it is cheap and available, they give it. There is no real reason; children are not that fast growing that they are going to run out of their red blood cell folate. In my opinion, what we fortify - folic acid the chemical, totally oxidized form - is different from food-folate forms. So what we fortify with, is folic acid and is 100% oxidized. Preformed vitamin A is as incomparable to carotene provitamin A as oxidized folic acid is to the folate naturally found in food.

Answer Prof Kraemer to intervention 4: Folic acid is the totally oxidized form and usually not occurring in the body, so depending on the dose, it is reduced and methylated when taken up into the body. The dilemma is that the evidence on neural tube defect reduction is coming from folic acid and not from other forms. So it is still used and it might be unethical to use other forms without evidence that other folate forms reduce neural tube defects. I agree, the ideal form in the diet is 5-methyltetrahydrofolic acid, but oxidized folic acid is used for technical reasons primarily.

Intervention 5: It is a really important discussion. You are making this analysis targeting with step one: perfect food intake data, perfect data on neural tube defects, perfect data on compliance in your country. Some countries do not have any of those things. So how can we make a model to go back one more step? How much money should we spend trying to get the data to make the right decision?

Answer Prof Verhagen to intervention 5: If you look at the original paper by Hoekstra from 2008 [28] on folic acid fortification in The Netherlands, it already identifies that there are a lot of things that we do not know. We have to make assumptions and deal with uncertainties even in The Netherlands, yet, in research what we always say that more research is necessary otherwise we wouldn't be here. What I think is that we should work on the basis of what we have, in real risk-benefit it is the challenge to come to a quantitative basis including all the limitations and uncertainties. Step away from qualitative considerations because that is what we have done in the past with folic acid fortifications. We have prevention of neural tube defects and masking of vitamin B12 deficiency, which are good and bad. So we did not undertake any action based on qualitative findings. Make it quantitative, even with all the uncertainties, so you can indicate that potentially it is good for public health, and you should not ignore the limitations but realize there are limitations and work on that basis. This is considered in The Netherlands, because we did perform a study for the Dutch government, and they were aware of the data; they knew that the fortification of flour with folic acid would be good for public health, but they did not do it for political reasons. We would expose everybody in the population to get benefit, but only a subset of females can get pregnant, not males. There is also another way of getting folic acid in women that can get pregnant, that is - by supplementation. That was the national policy decision following calculations. In summary, for risk-benefit considerations -go quantitative- even with all the limitations, otherwise you will not do a proper job.

Intervention 6: I want to hear your opinion on the use of proxies for dietary intake and add a comment. I chaired the group in 2008 that had to deal with consumption for the guidelines. We did not use availability from FAO food balance sheets because that is useless insofar as it does not have a distribution. In 2008, we decided what dietary intake data was available in the world. Available to us in 2008 were 75 national household consumption and expenditure surveys (HCES). We used those to establish ranges of intake of flour with which to set recommended levels later on. We used the food fortification formulator [34], because Dr Omar Dary was part of the working group, in order to check on potential excess. We had to do a national representative survey, a 24-hr recall with 2 measurements in order to move forward. This implies another question for Dr Solomons, I am also not for a one fits all. We should do a screening. But are we going to have all the tools to do so and individualize public health interventions? Why do we need national guidelines? They are not recipes? I have not used the risk-benefit method and I like it a lot. Without using DALYs, once you put them quantitative, the whole conclusion changes and we should apply that for public health reasons.

Answer Prof Allen to intervention 6: These proxy countrywide data, how would they handle the description of the differences across Cameroon for example? I do not know a country in which that situation doesn't exist, and that bothers me. These proxies give no textural context: not by poverty strata, not by region, not only by whatever. I would rather have a proxy of percentage energy intake from animal source foods than I would something from a survey like the HCES. I think there probably could be other proxies. I have done dietary intake data surveys, and the fact is that they are not so bad. On the Disability-Adjusted Life Year issue, again I think we do not have the data. The neural tube defects is an easy problem, whereas others are more difficult and you have a lot of different nutrients to be concerned with.

Answer Dr Solomons to intervention 6: The Hippocratic dictum "First do no harm" applies not only to the individual in the clinical setting of medicine, but to population sectors in public health as well. We do not have to screen every single person, but rather find the regions in the country where there are gaps - gaps in any way, e.g. by geographic, ethnic, and economic characteristics. The dosage or coverage needs to be compensated - up or down - whatever the intervention is that is affecting their lives.

Intervention 7: I was interested to hear about the idea of multiple fortified foods at low levels. I was confronted the whole time with the question of having enough and not too much. About 10 years ago, the European Commission was considering legislating maximum safe levels of fortification in Europe. In order to get ready for that, the ILSI Europe model authored by Prof Albert Flynn et al. [35] was used. If you have got reasonable intake data, one may simply calculate the 95th percentile of intake of a micronutrient from food, and add the 95th percentile of the micronutrient from a supplement until you have a certain level of intake. In all cases below the UL, take the difference between this intake value and the UL and divide it into 20 slices of 100 kcal and make some corrections for percentage of food that is fortified. After that, you come up with a maximum safe fortification level per 100 kcal. It doesn't go into risk assessment beyond using the UL. I can imagine this could be useful in a context with good intake data. In the end the European Commission decided not to legislate. Is this a reasonable model? Maybe a little bit less complicated than doing the risk benefit assessment?

Answer Prof Verhagen to intervention 7: I know this Flynn model. This model provides for safe levels of fortification by 100 kcal and why that works best. Risk-benefit of micronutrients is something different insofar as it also takes into account the risks of deficiency. What you are rightly doing is comparing intakes with an UL, and passing beyond the UL is still safe. If you want to do real risk-benefit calculations, you do not need to compare with an UL: ignore the UL and go for lowest adverse effect level. Then you can really compare. The UL is just a safe limit. If you go for risk benefit, ignore the UL, it is a risk management issue, but only at the LOAEL does an adverse effect start that is where you focus on. That constitutes a completely different comparison.

Intervention 8: When I hear about excess intake, the evaluation point has just been pointed out as the UL. If we were to apply this to developing countries the UL is inappropriate. In the sense that it is very much based on Western ideas of how to measure excessive intakes. In developing countries, the issue is not so much excessive intakes in people alone, but also risk of infectious diseases. The risk of excessive folic acid intakes in a person with malaria may be different from that in a healthy person. Folic acid supplementation may reduce the efficacy of anti-malaria drugs. The anti-malaria drug Cotrimoxazol is given as primary drug intervention for malaria and prophylaxis in HIV-infected adults, but folic acid supplementation may impair its effectiveness and increase the risk of malaria. For neural tube defects, folic acid may be of benefit; in Africa, the number of neural tube defects is more than 6000 per year [36], but for malaria-treated individuals folic acid may not be effective and malaria is more prevalent than neural tube defects. Some sort of risk-benefit analysis could be applied to assess the risks and benefits of folic acid fortification in developing countries.

8. SUMMARY

Several key messages emerged from the symposium. Before considering a micronutrient fortification program, it is essential to collect food and nutrient intake data in the different population groups, supported by ancillary information such as biochemical data on nutritional status, and possibly clinical data. This helps to gain insight into the groups to be targeted, target intake levels and the most suitable food vehicles for fortification. In a second step, the data are used to simulate the effect of different food vehicles and micronutrient food contents. The EAR and UL can be used to estimate the effect of different fortification scenarios on the proportion of the population with adequate and safe intakes, or a more elaborated risk-benefit approach can be used to model the benefits and possible risks. Repeated food intake data collection after implementation allows for subsequent monitoring and evaluation of the programs' efficacy and safety.

At intakes equal to the EAR, 50% of the population is at risk of inadequacy and intake equal to the UL is unlikely to represent a risk of adverse effects. Shifting micronutrient intakes such that the EAR is met in the target population without exceeding the UL in any of the population groups is a challenge in mass fortification. In order to decide on fortification strategies, and acceptability of risks, understanding of the potential risks below the EAR and above the UL and their magnitude in terms of health impact is essential.

In Cameroon dietary intake data and biochemical markers have successfully been used to inventory the public health problem of vitamin A deficiency. Simulation of the current oil fortification in Cameroon suggested that vitamin A intakes would increase, however, insufficiently to meet the requirements. Fortification of additional food vehicles would help to meet the dietary gap, however, depending on the food vehicle and level of fortification, also increase intakes in subgroups of children in excess of their UL. When modelling the effect of different fortified foods and high-dose supplements on the contribution of vitamin A to liver stores in children, minimal risks of excessive liver stores were shown. In Guatemala, most of the sugar and many foods are nowadays fortified at levels that deliver effective vitamin A amounts for preschoolers. This has been important in reducing prevalence rates of vitamin A deficiency and related adverse health effects. Still, intakes below the EAR associated with low status exist for vitamin A in segments of the Guatemalan population. Meanwhile, it cannot be excluded that the expanding number of vitamin A-fortified foods puts some preschoolers at risk of exceeding their UL. This illustrates the difficulty of shifting the intake distribution such that both intakes below the EAR and above the UL are minimized.

The choice of the food vehicle and its distribution of intake in the population is important as illustrated for vitamin A fortification in Guatemala and Cameroon; the food should be consumed by the whole population, yet in constant amounts. However, fortifying one food at the target level may not adequately reach the target population and fortifying multiple foods at high levels may overexpose some groups. Dividing the target level over multiple foods can be a strategy to more equally distribute intake of the target micronutrient(s).

A risk-benefit assessment provides insight into the possible risks and benefits associated with increasing food or nutrient intakes. It can simulate how health benefits and health risks change as result of increasing micronutrient intakes; the benefits of reducing a micronutrient inadequacy versus the risks of excessive consumption. The approach has been used to evaluate the largest net health gain with different food fortification scenarios as exemplified by folic acid-fortified flour in The Netherlands. The risks and benefits are quantitatively expressed by considering the number of people affected, severity, and duration of the health effect. By expressing the benefits and risks quantitatively, they can be balanced, allowing decision making on the most appropriate fortification level.

Given the potential of long-term health benefits of successful fortification programs, it is essential to ensure proper investment in collection of food intake data and preferably biochemical and clinical markers to evaluate and monitor whether intakes are adequate and safe. Population intake distributions in relation to the EAR and UL provide rough estimates of adequate and safe intakes, while risk-benefit approaches can provide more in-depth assessment of quantitative health benefits and risks at the two ends of the intake spectrum. Risk-benefit modelling offers an important answer to the most commonly-asked question as to whether the benefits of increasing micronutrient intakes outweigh the risks. More data collection and advanced modeling will lead to smaller uncertainties around the benefits and risks. Lessons can be learned from experiences in different countries implementing micronutrient programs, that knowledge can be applied to provide guidance to public health policy and program design.

9. CONCLUSION

In conclusion, risk-benefit assessment can be a valuable and transparent approach to provide the best possible science-based answer to complicated questions with a large, and potentially optimal, impact on public health. Risk-benefit approaches assist in calculating the risks versus the benefits for public health of food or nutrient interventions. When using common health metrics, the costs of the intervention versus the benefits for public health can be calculated. Ultimately, public health can profit from risk-benefit and cost-benefit assessments. Risk-benefit approaches can help us to understand the magnitude of the risks and assist in making evidence-informed policy decisions in nutrition for the betterment of public health (and reducing their costs).

ACKNOWLEDGEMENTS

We acknowledge the contributions of Kenneth H Brown and Marjorie J Haskell (UC Davis) and Martin Nankap and Alex O Ndjebayi (Helen Keller International) to the Cameroon vitamin A analyses and the assistance of Marieke Vossenaar (CeSSIAM) in the data-analysis and presentation preparation for the Guatemalan setting.

COMPETING INTERESTS

Maaike J Bruins is an employee of DSM. Ingrid Bielderman, Klaus Kraemer, and Jane Badham are employees of Sight and Life, a humanitarian nutrition think tank of DSM.

REFERENCES

1. Shenkin A. Micronutrients in health and disease. Postgrad Med J. 2006;82:559-67.
2. Renwick AG, Flynn A, Fletcher RJ, Muller DJ, Tuijtelaars S, Verhagen H. Risk-benefit analysis of micronutrients. Food Chem Toxicol. 2004;42:1903-22.
3. European Commission. Scientific committee on food. Report of the Scientific Committee on Food on the Revision of Essential Requirements of Infant Formulae and Follow-on Formula. 2003;1-208.
4. World Health Organization, Food and Agricultural Organization of the United Nations. Guidelines on food fortification with micronutrients. In: Allen, L, de Benoist B, Dary O, Hurrell R, eds; 2006.
5. Coates J, Colaiezzi B, Fiedler JL, Wirth J, Lividini K, Rogers B. A program needs-driven approach to selecting dietary assessment methods for decision-making in food fortification programs. Food Nutr Bull. 2012;33:S146-56.
6. World Health Organization (WHO). Estimating Appropriate Levels of Vitamins and Minerals for Food Fortification Programmes: The WHO Intake Monitoring, Assessment and Planning Program (IMAPP): meeting report. Available:http://www.who.int/nutrition/publications/micronutrients/ (accessed February 4 2015).
7. Wrieden W, Peace H, Armstrong J, Barton K. A short review of dietary assessment methods used in National and Scottish Research Studies. Briefing Paper Prepared for: Working Group on Monitoring Scottish Dietary Targets Workshop; 2003. Available:http://www.food.gov.uk/sites/default/files/multimedia/pdfs/scotdietassessmethods.pdf (accessed February 4 2015)
8. Abdeen Z, Ramlawi A, Qaswari R, et al. Predicted efficacy of the Palestinian wheat flour fortification programme: complementary analysis of biochemical and dietary data. Public Health Nutr. 2014;1-11.
9. Gadgil M, Joshi K, Pandit A, et al. Imbalance of folic acid and vitamin B12 is associated with birth outcome: An Indian pregnant women study. Eur J Clin Nutr. 2014;68:726-9.
10. Engle-Stone R, Nankap M, Ndjebayi AO, Brown KH. Simulations based on representative 24-h recall data predict region-specific differences in adequacy of vitamin A intake among Cameroonian women and young children following large-scale fortification of vegetable oil and other potential food vehicles. J Nutr. 2014;144:1826-34.
11. Engle-Stone R, Ndjebayi AO, Nankap M, Brown KH. Consumption of potentially fortifiable foods by women and young children varies by ecological zone and socio-economic status in Cameroon. J Nutr. 2012;142:555-65.

12. Tooze JA, Kipnis V, Buckman DW, et al. A mixed-effects model approach for estimating the distribution of usual intake of nutrients: The NCI method. Stat Med. 2010;29:2857-68.
13. Engle-Stone R, Haskell MJ, Nankap M, Ndjebayi AO, Brown KH. Breast milk retinol and plasma retinol-binding protein concentrations provide similar estimates of vitamin A deficiency prevalence and identify similar risk groups among women in Cameroon but breast milk retinol underestimates the prevalence of deficiency among young children. J Nutr. 2014;144:209-17.
14. Allen LH, Haskell M. Estimating the potential for vitamin A toxicity in women and young children. J Nutr. 2002;132: 2907S-19S.
15. Haskell MJ, Lembcke JL, Salazar M, Green MH, Peerson JM, Brown KH. Population-based plasma kinetics of an oral dose of [2H4] retinyl acetate among preschool-aged, Peruvian children. Am J Clin Nutr. 2003;77:681-6.
16. Olson JA. Serum levels of vitamin A and carotenoids as reflectors of nutritional status. J Natl Cancer Inst. 1984;73:1439-44.
17. Arroyave G, Mejia LA, Aguilar JR. The effect of vitamin A fortification of sugar on the serum vitamin A levels of preschool Guatemalan children: A longitudinal evaluation. Am J Clin Nutr. 1981;34:41-9.
18. Institute of Medicine (IOM), ed. Dietary Reference Intakes for Vitamin A, Vitamin K, Arsenic, Boron, Chromium, Copper, Iodine, Iron, Manganese, Molybdenum, Nickel, Silicon, Vanadium, and Zinc. Washington, DC: National Academy Press; 2000.
19. European Food and Safety Authority (EFSA). Tolerable Upper Intake Levels for Vitamins and Minerals. Available:http://www.efsa.europa.eu/en/ndatopics/docs/ndatolerableuil.pdf (accessed 4 February 2015).
20. Otten JJ, Hellwig JP, Meyers LD, eds. Institute of Medicine (IOM); Dietary Reference Intakes Essential Guide Nutrient Requirements. Washington DC, USA: The National Academies Press; 2006.
21. Rothman KJ, Moore LL, Singer MR, Nguyen US, Mannino S, Milunsky A. Teratogenicity of high vitamin A intake. N Engl J Med. 1995;333:1369-73.
22. World Health Organization. Global prevalence of vitamin A deficiency in populations at risk 1995-2005. WHO Global Database on Vitamin A Deficiency; 2009;
23. Guatemala. Ministerio de Salud Publica y Asistencia Social. Encuesta Nacional de Micronutrientes; 1996.
24. Smith FR, Goodman DS. Vitamin A transport in human vitamin A toxicity. N Engl J Med. 1976;294:805-8.
25. Krasinski SD, Russell RM, Otradovec CL, et al. Relationship of vitamin A and vitamin E intake to fasting plasma retinol, retinol-binding protein, retinyl esters, carotene, alpha-tocopherol, and cholesterol among elderly people and young adults: Increased plasma retinyl esters among vitamin A-supplement users. Am J Clin Nutr. 1989; 49:112-20.
26. Ballew C, Galuska D, Gillespie C. High serum retinyl esters are not associated with reduced bone mineral density in the Third National Health And Nutrition Examination Survey, 1988-1994. J Bone Miner Res. 2001;16:2306-12.
27. Blomhoff R, Blomhoff HK. Overview of retinoid metabolism and function. J Neurobiol. 2006;66:606-30.
28. Edwards IR, Lindquist M. Understanding and communicating key concepts in risk management: What do we mean by benefit and risk? Drug Saf. 2009;32:449-52.
29. Kalogeras N, Odekerken-Schroder G, Pennings JM, et al. State of the art in benefit-risk analysis: Economics and marketing-finance. Food Chem Toxicol. 2012;50:56-66.
30. Baghurst P, Beaumont-Smith N. The relationship between the consumption of fruits and vegetables and health status: Report to Department of Health and Aged Care and the Strategic Intergovernmental Nutrition Alliance. CSIRO Health Sciences & Nutrition; 1999. Available:http://www.health.vic.gov.au/archive/archive2014/nphp/publications/signal/fvstat.pdf (accessed 4 February 2015)
31. Hoekstra J, Verkaik-Kloosterman J, Rompelberg C, et al. Integrated risk-benefit analyses: Method development with folic acid as example. Food Chem Toxicol. 2008;46:893-909.
32. Hoekstra J, Hart A, Boobis A, et al. BRAFO tiered approach for benefit-risk assessment of foods. Food Chem Toxicol. 2012;50(4):S684-98.

33. Gyles CL, Lenoir-Wijnkoop I, Carlberg JG, et al. Health economics and nutrition: A review of published evidence. Nutr Rev. 2012;70:693-708.
34. European Commission. Discussion Paper on the setting of maximum and minimum amounts for vitamins and minerals in foodstuffs. Available:http://ec.europa.eu/food/food/labellingnutrition/supplements/discus_paper_amount_vitamins.pdf (accessed 2014 August).
35. European Food Safety Authority (EFSA). Guidance on human health risk-benefit assessment of foods. EFSA Journal. 2010;8:1673-8.
36. Tijhuis MJ, De Jong N, Pohjola MV, et al. State of the art in benefit-risk analysis: Food and nutrition. Food Chem Toxicol. 2012;50:5-25.
37. Allen LH. New approaches for designing and evaluating food fortification programs. J Nutr. 2006;136:1055-8.
38. Bruins MJ, Mugambi G, Verkaik-Kloosterman J, et al. Addressing the risk of inadequate and excessive micronutrient intakes: traditional versus new approaches to setting adequate and safe micronutrient levels in foods. Food Nutr Res. 2015;59: 26020.

5

Prevalence and Characterization of *Salmonella* Isolated from Beef in Namibia

Renatus P. Shilangale[1,2*], Godwin P. Kaaya[3] and Percy M. Chimwamurombe[2]

[1]*Central Veterinary Laboratory, Private Bag 13187, Windhoek, Namibia.*
[2]*Department of Biological Sciences, Faculty of Science, University of Namibia, Private Bag 13301, Windhoek, Namibia.*
[3]*School of Biological Sciences, University of Nairobi, P.O.Box 30197-00100, Nairobi, Kenya.*

Authors' contributions

This work was carried out in collaboration between all authors. Author RPS designed the study, performed the statistical analysis and wrote the first draft of the manuscript. Authors GPK and PMC managed the review of the manuscript. All authors read and approved the final manuscript.

ABSTRACT

Aim: The purpose of this research was to determine the prevalence of *Salmonella* in raw beef produced from selected commercial abbatoirs in Namibia.
Methodology: A total of 9508 of beef samples from three different types of samples; meat cuts, carcass swabs and meat fluid were collected from the three local abattoirs over a period of two years starting from January 2008 to December 2009. Pre-enrichment for isolation of *Salmonella* was done in Buffered peptone water followed by enrichment in the Rappaport-Vassiliadis and selenite cystine broth. The isolation of *Salmonella* was done on Xylose Lysine Desoxycholate and Brilliant Green agar followed by biochemical confirmation and serotyping according to Kauffman-White scheme.
Results: The overall prevalence of *Salmonella* was 0.85% for beef samples derived from meat cuts, meat fluid and carcass swabs. The prevalence of *Salmonella* in carcass swabs (2.67%) was significantly different (P = 0.05) from that of meat cuts (0.50%) and meat fluid (0.43%). No significant difference (P = 0.05) on the prevalence of *Salmonella* existed between the meat cuts and meat fluid. A total of 34 different types of *Salmonella* serovars were identified with *S.* Chester being the most frequently isolated serovars (n = 12), followed by *S.* Reading and *S.* Bredeney (n = 6) and *S.* Typhimurium (n = 5).
Conclusions: The prevalence of *Salmonella* in raw beef found in this survey was lower than those observed in Sub Sahara Africa with *S.* Chester being the most prevalent serovar.

**Corresponding author: Email: ShilangaleR@mawf.gov.na*

Keywords: Namibia; Salmonella; beef meat cuts; beef carcass swab; beef meat fluid; serovar.

1. INTRODUCTION

Salmonella bacteria cause foodborne illness in humans commonly known as salmonellosis. The outbreaks of salmonellosis are commonly associated with consumption of food of animal origin which is contaminated with *Salmonella* [1,2]. The link between *Salmonella* infections and food consumption is supported by the findings of other researchers elsewhere [3,4]. According to a study in Australia, a wide range of *Salmonella* serovars, all of which have been isolated from humans, were found in both cattle and sheep [5]. In the US and Canada, *S.* Heidelberg is among the most frequently isolated serovar both in clinical cases of salmonellosis and from retail meats and food animals [6]. Another study on food animals, meat products and slaughterhouses personnel in Ethiopia suggested the likely link of human *Salmonella* infections and food of animal origin [3]. According to the Center for Diseases Control and Prevention (CDC) beef contributed up to 11% of the foodborne outbreaks for the year 2009 – 2010 in the United States [2].

In order to reduce the threat of *Salmonella* to food safety, the sources of contamination need to be identified in the food chain. This is because scientific investigation on the sources and distribution of *Salmonella* contamination in food are a basic step to mitigate the spread of *Salmonella* in food that contributes to food safety [7].

Hazard Analysis Critical Control Point (HACCP) system, a food safety management tool is currently being used in Namibia to manage the risk of pathogens in the abattoirs that produce beef for export. These abattoirs test for *Salmonella* on beef carcasses as part of the quality control criteria in order to meet the stringent export requirements for the European Union, Norway and South Africa. Norway, one of the major markets of Namibia's beef has a zero tolerance policy on *Salmonella* based on the Norwegian National Food Law [8]. Although Namibia has managed to export beef to Norway, there is limited information on the prevalence of *Salmonella* in Namibia. In most of the developing countries including Namibia the lack of epidemiological studies on foodborne diseases are probably due to poor coordination and lack of resources [7].

The purpose of this study was to determine the prevalence and to characterize *Salmonella* in raw beef (meat cuts, carcass swabs and meat fluid) produced from the beef export abbatoirs in Namibia.

2. MATERIALS AND METHODS

2.1 Sample Collection

A total of 9508 of beef samples were collected from the three beef export abattoirs over a period of two years starting from January 2008 to December 2009. Three different types of samples were collected; meat cuts, carcass swabs and meat fluid. Of these, 3424 of meat cuts, 1688 of carcass swabs and 4396 of meat fluid samples were collected. Carcass swabs were collected before chilling and meat cuts and meat fluid were collected from the packaged finished product after deboning. Samples were collected by the State veterinary officials stationed at the three abattoirs for routine microbiological analysis using sterile dilution bags and media bottles.

2.2 Sample Collection Techniques

Meat cuts were collected from frozen packaged beef prepared for export to South Africa and the European markets. Approximately 5 g of meat was sampled from 10 randomly selected packages of frozen meat using the excision method. These samples were then pooled together into a sterile stomacher bag to make at total of approximately 50 g samples per bag. For carcass swabs sampling, two gauze swabs (wet and dry) were used to sample the surface area of 100 cm^2 per site of a carcass of which four sites were sampled. The wet swab was used first then the dry swab to sample the same sampling site. The 8 swabs per carcass were pooled together into a 500 ml media bottle containing 200 ml buffered peptone water (BPW). The four sites sampled were rump, flank, brisket and neck. Meat fluid samples were taken from vacuum packaged deboned beef stored for 24 h at 0 to 2°C. Approximately 10 ml of meat fluid from 10 randomly selected vacuum packed meats was sampled and pooled (approximately 100 ml in total). The sampling was done by draining the meat fluid from the vacuum packaged meat using a sterile needle with a syringe and then pooled into sterile 250 ml media bottles.

Samples were transported to the laboratory for analysis on the same day. During transportation samples were kept at refrigeration temperatures (2 to 8°C) using a cooler box with ice bricks. When received at the laboratory, samples were stored in the refrigerator (1 to 5°C) before the analysis. The isolation of *Salmonella* was done within 24 h from the time when samples were received.

2.3 Isolation and Identification

As a routine microbial analysis, samples corrected were analyzed for the presence of *Salmonella* at the Central Veterinary Laboratory of the Ministry of Agriculture, Water and Forestry in Windhoek, Namibia. The pre-enrichment stage for all type of samples was done using BPW. For meat cuts, 25 g sample was pre-enriched into 225 ml of buffered peptone water (Merck, Darmstadt) and incubated at 37°C for 24 h. For the carcass swabs, 300 ml of BPW (Merck, Darmstadt, Germany) at ambient temperature was added into a sample containing 200 ml of BPW in a 500 ml media bottle. The samples were then incubated at 37°C for 18 to 24 hours. For meat fluid, approximately 50 ml of the meat fluid sample was transferred into a 500 ml media bottle. Then 450 ml of BPW (Merck, Darmstadt, Germany) was added to the sample and incubation was maintained at 37°C for 18 to 24 hours.

Subsequently, 0.1 ml of the pre-enrichment culture was added to 10 ml of Rappaport-Vassiliadis (RV) broth (Merck, Darmstadt) and 10 ml to 100 ml of selenite cystine (SC) broth (Merck, Darmstadt) and incubated for 24 h at 41.5°C and 37°C, respectively. The culture was then streaked onto two selective agar: Xylose Lysine Desoxycholate (XLD) (Merck, Wadeville) and Brilliant Green Agar (BGA) (Scharlau Chemie SA, Barcelona) and incubated at 37°C for 24 h. The presumptive *Salmonella* colonies were then confirmed serologically and biochemically. On XLD agar, presumptive *Salmonella* colonies appear black or pink with or without black center while on BGA *Salmonella* change the medium from pink to red. For serological confirmation of *Salmonella*, Omnivalent anti-sera (Siemens, Marburg) were used. For biochemical confirmation, the following tests were performed: triple sugar iron reactions, urea production, the Voges-Proskauer reaction, the indole reaction, the lysine decarboxylase reaction and the detection of β-galactosidase.

The confirmed strains of *Salmonella* were then further identified biochemically using miniaturized commercial systems (Vitek-BioMérieux, Marcy l'Etoile). Serological identification with commercially available antisera (State Serum Institute, Copenhagen) for detection of somatic and flagellar antigens was performed and the isolates were identified and named in accordance with the Kauffman-White scheme [9]. Further biochemical and serological identification were done at the Istituto 'G. Caporale', Teramo, Italy.

2.4 Data Analysis

The prevalence of *Salmonella* strains was evaluated in terms of percentage of occurances on the total number of *Salmonella* isolates within a sample type. The differences between observations were analyzed using a Fisher Exact test with the confidence interval of 95%. The differences were considered significant at $P = 0.05$.

3. RESULTS AND DISCUSSION

The prevalence of *Salmonella* serovars in meat cuts, meat fluid and carcass swabs samples are reported in Tables 1, 2 and 3 respectively. The overall prevalence of *Salmonella* was 0.85% for beef samples derived from meat cuts, meat fluid and carcass swabs.

The prevalence of *Salmonella* in meat cuts (n = 3424) was 0.50%, whereas in meat fluid samples (n = 4396) the prevalence was 0.43%. The prevalence of *Salmonella* in carcass swabs (n = 1688) was 2.67%. The present study found that the prevalence of *Salmonella* in carcass swabs was significantly different (P = 0.05) from meat cut and meat fluid. On the other hand, there was no significant difference (P = 0.05) on the prevalence of *Salmonella* between meat cuts and meat fluid. This was based on the Fisher's exact test.

The low prevalence of *Salmonella* found in the present study was not comparable to the once reported elsewhere in the Region [3,10,11]. A study done in Ethiopia on the prevalence of *Salmonella* from beef carcasses in abattoirs found the prevalence of *Salmonella* to be 13.3% [10]. Another study in Ethiopia found the prevalence of *Salmonella* contamination in minced beef to be 14.4% [3]. In Botswana, a study done in raw beef sausages found the prevalence of *Salmonella* to be as high as

25.3% [11]. However, the two latter studies were done on naturally contaminated raw beef from retail outlets where the level of bacterial contamination is likely to be higher than the levels at the abattoir due to post handling or processing conditions.

The lower prevalence of *Salmonella* in beef found in the present study could be due to the effective implementation and maintenance of high hygiene standards together with the HACCP system in the abattoirs as part of the EU export requirements. Namibia together with Botswana and Swaziland are the only African countries that have access to the EU beef market where implementation of the HACCP system is mandatory. The EU requires the third countries exporting to the EU to implement the same requirements as the EU Member States. Nevertheless, the differences in the isolation procedures used in the present study could have resulted to the low isolation rate of *Salmonella* because of the use of SC and RV combination as selective media. According to Hammack et al. [12] findings, the SC broth with a recovery rate of 96% and RV medium with a recovery rate of 90% have less proportion of *Salmonella* recovery as compared to tetrathionate broth (TT) of 97%. This study recommended a combination of RV and TT for recovery of *Salmonella* from food with low microbial load. The recommendation was based on the fact that the selectivity of RV medium and TT broth increase the recovery of *Salmonella* by inhibiting the growth of the competitors in favor of *Salmonella* [12]. Unlike the present study, the study on the natural contaminated carcasses in Ethiopia used a combination of RV and TT as selective media [10]. This could probably have resulted to the higher prevalence of *Salmonella* in the carcass as compared to the present study.

Table 1. Distribution of *Salmonella* serovars isolated in meat cuts and the proportion of positive isolates

Salmonella serotype	Number of positive samples	(%) (*N* = 17)
S. Schwarzengrund	1	5.88
S. Bredeney	1	5.88
S. Saintpaul	1	5.88
S. Parkroyal	1	5.88
S. Chester	5	29.41
S. Winston	1	5.88
S. Bahrenfeld	1	5.88
Salmonella group K*	1	5.88
S. Dublin	1	5.88
S. Lamberhurst	1	5.88
S. Uganda	1	5.88
S. Anatum	1	5.88
S. Sao	1	5.88
Total	17	100.00

* *non-typeable isolates of Salmonella*

Table 2. Distribution of *Salmonella* serovars isolated in meat fluid and the proportion of positive isolates

Salmonella serotype	Number of positive samples	(%) (*N* = 19)
S. Braenderup	2	10.53
S. Ball	1	5.26
S. Chester	1	5.26
S. Fischerkietz	1	5.26
S. Cannstatt	2	10.53
S. Petahtikve	1	5.26
S. Bredeney	3	15.79
S. Anatum	3	15.79
Salmonella enterica subsp. *salamae**	1	5.26
S. Winston	1	5.26
Salmonella group K*	1	5.26
S. Vaertan	1	5.26
S. Cerro	1	5.26
Total	19	100.00

* *non-typeable isolates of Salmonella*

Table 3. Distribution of *Salmonella* serovars isolated in carcass swabs and the proportion of positive isolates

Salmonella serotype	Number of positive samples	(%) (*N* = 45)
S. Typhimurium	5	11.11
S. Saint-paul	2	4.44
S. Chester	6	13.33
S. Minnesota	1	2.22
S. Braenderup	3	6.67
S. Parkroyal	1	2.22
S. Djermaia	1	2.22
S. Petahtikve	1	2.22
S. Fischerkietz	1	2.22
S. Kaapstad	1	2.22
Salmonella group I*	1	2.22
S. Cannstatt	1	2.22
S. Reading	6	13.33
S. Bredeney	2	4.44
S. Chichiri	1	2.22
S. Newport	1	2.22
S. Kintambo	1	2.22
S. Banana	1	2.22
Salmonella enterica subsp. salamae *	5	11.11
S. Anatum	1	2.22
Salmonella Group D2*	1	2.22
Salmonella Group C1*	1	2.22
S. Schwarzengrund	1	2.22
Total	45	100.00

**non-typeable isolates of Salmonella*

The lower prevalence of *Salmonella* in meat cuts and meat fluid as compared to carcass swabs in the present study was probably a result of different handling, sampling techniques used and the processing stage where samples were taken. This is because before packaging the meat had to be trimmed during deboning in order to remove the fat and dry part of the meat from the surface. Trimming the outer layer could have had reduced the population of bacterial contamination on the surface of the final product. Bacterial contamination and growth occurs on the surface of the meat while the inner part of the carcass is usually considered sterile. Furthermore, the lower prevalence of *Salmonella* in vacuum packaged beef could be the result of the preservation conditions which the beef was subjected to. Vacuum packaging has been reported to reduce the population of *Salmonella* stored at 4°C by 1 log cycle in one day [13]. A different study has shown that vacuum packaging could significantly reduce *S.* Typhimurium counts in beef stored at 4°C [14].

In general, a total of 34 different types of *Salmonella* serovars were identified with *S.* Chester being the most frequently isolated with 12 isolates, followed by *S.* Reading and *S.* Bredeney with 6 serovars each and *S.* Typhimurium with 5 isolates. Six strains were found to be positive for *S. enterica* subsp. Salamae, two for *Salmonella* group K, one *Salmonella* Group I, one each for *Salmonella* Group C1 and D2. These groups could not be identified further because they did not express the phase 2 'H' antigens. Nevertheless, there is little information in the Sub Sahara Africa on the prevalence of *Salmonella* in beef from the slaughter houses which can be used to make a comparison with this study. Most of the studies in Sub Sahara Africa are pertaining to clinical cases [5,15,16] and retail markets [3,11]. Unlike Namibia, a study done in Botswana found *Salmonella enterica* subsp. salamae II, *S.* Thompson and *S.* Anatum to the most frequently isolated *Salmonella* strains from different raw meat sausages made from beef with the prevalence of 13.8, 6.2 and 6.2% respectively [11].

Of the four most frequently isolated serovars found in this study, *S.* Typhimurium is the only serovar that is often being reported to be among the most frequently isolated elsewhere as compared to other serovars. A study in Sweden done between 1993 and 1997 on the isolation of *Salmonella* from animals and animal feed reported *S.* Typhimurium to be the second most frequently isolated serotype in cattle [17]. However, 78 of the 115 isolates originated from infected herds where the remaining isolates were collected at autopsies, sanitary slaughter and surveillance at slaughterhouses. In a different study in Ireland on the prevalence of *Salmonella* in cattle carcass *S.* Typhimurium was the third most frequently isolated serotype after *S.* Dublin and *S.* Agona [18]. In the US, *S.* Typhimurium was the second most common isolate after *S.* Montevideo [19]. In a different study in meat and meat products, *S.* Typhimurium was among the top five most frequently isolated in Algeria [20].

Moreover, the differences on the type of strains found in the present study and those of other studies can probably be due to the differences in the geographical locations where these studies were carried out. This idea may be supported by the findings of other researchers who have suggested that the distribution of different types of *Salmonella* vary based on the geographical location [18,21]. According to Hendriksen et al. [21], the majority of salmonellosis cases in humans are caused by a limited number of *Salmonella* serovars which may vary over time from one country to another. According to McEvoy et al. [18], the higher prevalence in the incidence of *Salmonella* may be among other factors due to geographical variation. These suggestions imply that different types of *Salmonella* serovars may be expected to be more prevalent in this region and Namibia in particular as compared to other regions. Namibia's weather pattern is different from many countries and is among of the driest countries in the world which means it may have different growth conditions when compared to other countries.

Nevertheless, *S.* Typhimurium was the second most frequently isolated serotype in carcass swabs and was among the top four when the three beef products were analyzed together. This observation suggests that *S.* Typhimurium is an important bacterium of public concern in Namibia just as it is in other parts of the world. In the US, 43% of all *Salmonella* isolates from human sources are only from three types of strains; *S.* Enteritidis, *S.* Newport and *S.* Typhimurium, and of which the latter contributes to 11% of all *Salmonella* outbreaks [2]. However, there is no information that can be used to estimate the potential public health threat of *S.* Typhimurium in Namibia.

The most notable *Salmonella* strain in this study was *S.* Chester. Although this strain is infrequently reported in foods worldwide, it is the most common stain isolated in this study. In a different study by Shilangale et al. [22], *S.* Chester was the most prevalent serovar of *Salmonella* in Namibia in animal feed made from bovine and ovine byproducts. In this study the prevalence of *S.* Chester was found to be as much as 19.7% [22].

Unlike Namibia, in Canada *S.* Chester is considered to be a rare serotype of *Salmonella* with regard to human infections [23]. Despite being considered as a rare serovar in human infections, *S.* Chester can still be one of the *Salmonella* serovars of public concern worldwide due to some few outbreaks which have been reported [24,25]. This is because of its recent links to several salmonellosis outbreaks in North America and Australia. In 2010, *S.* Chester caused a multistate outbreak of human infections in 18 States in the US [24]. In the same year, another outbreak of *S.* Chester in Canada which involved 26 people was due to the consumption of luncheon meat, sausage and hard cheese [25]. Another notable case of *S.* Chester happened in Australia where 36 people of Aboriginal community were infected due to consumption of turtle meat in 1998 [26]. However, there are no clinical records to determine if *S.* Chester is a major cause of foodborne salmonellosis in Namibia apart of being the most frequently isolated strain in beef. Unlike in the developed countries, in Namibia sources of attribution of most of the *Salmonella* infections are unknown because most of these cases are treated based on the clinical diagnosis.

4. CONCLUSION

The prevalence of *Salmonella* in raw beef found in this survey was lower than those observed in Africa with *S.* Chester being the most prevalent serovar. Although the prevalence of *Salmonella* contamination is low in beef in Namibia the prevalence of *Salmonella* in beef suggest that there is still a risk of *Salmonella* infection through beef consumption as well as possible export of *Salmonella* strains to other countries through trade. Because of scant *Salmonella* studies in the Region, the information on the prevalence in the present study may be useful in future epidemiological studies of *Salmonella* in Namibia and in the Region. The information of the present study may further be useful in the beef production in their efforts to improve the hygiene and safety standards in the food chain as part of the requirements to meet the high demanding export requirements.

ACKNOWLEDGEMENTS

The authors wish to acknowledge the work of the Food Hygiene personnel at the Central Veterinary laboratory for microbiological analyses of sample. Further acknowledgement is extended to the Istituto "G. Caporale" of Teramo, Italy for serotyping of *Salmonella*.

COMPETING INTERESTS

Authors have declared that no competing interests exist.

REFERENCES

1. Center for Diseases Control and Prevention (CDC). Multistate outbreak of *Salmonella* Typhimurium infections associated with eating ground beef-United States, 2004. MMWR Morb. Mortal. Wkly. Rep. 2006;55:180-182.
2. Morbidity and Mortality Weekly Report (MMWR). Surveillance for Foodborne Disease Outbreaks - United States, 2009 – 2010. Morbidity and Mortality Weekly Report. 2013;62(3):41–47. Available:http://www.cdc.gov/mmwr/pdf/wk/mm6203.pdf
3. Ejeta G, Molla B, Alemayehu D, Muckle A. *Salmonella* serotypes isolated from minced meat beef, mutton and pork in Addis Ababa, Ethiopia. Revue de Medicine Veterinaire. 2004;155(11):547-551.
4. Vanselow BA, Hornitzky MA, Walker KH, Eamens GJ, Bailey GD, Gill PA, et al. *Salmonella* and on-farm risk factors in healthy slaughter-age cattle and sheep in eastern Australia. Aust Vet Journal. 2007; 85 (12):498-502.
5. Kariuki S, Revathi G, Gakuya F, Yamo V, Muyodi J, Hart A. Lack of clonal relationship between non-typhi *Salmonella* strains types from humans and those from animals living in close contact. FEMS Immunol Med Microbiol. 2002;33:165-171.
6. Zhao S, White DG, Friedman SL, Glenn A, Blickenstaff K, Ayers SL, et al. Antimicrobial resistance in *Salmonella enterica* Serovar Heidelberg isolates from retail meats, including poultry, from 2002 to 2006. Appl Env Micro. 2008;74(21): 6656- 6662.
7. Molla B, Alemayehu D, Salah W. Sources and distribution of *Salmonella* serotypes isolated from food animals, slaughterhouse personnel and retail meat products in Ethiopia: 1997-2002. Ethiop J Health Dev. 2003;17(1):63-70.
8. Isakbaeva E, Lindstedt BA, Schimmer B, Vardund T, Stavnes TL, Hauge K, et al. *Salmonella* Typhimurium DT104 outbreak linked to imported minced beef, Norway, October – November 2005. Euro Surveillance. 2005;10(45):2829.
9. Grimont PAD, Weill FX. Antigenic formulae of the *Salmonella* serovars. 9th Edition. WHO collaborating centre for reference and research on *Salmonella*. Institut Pasteur, 28 rue du Dr. Roux, 75724 Paris Cedex 15, France; 2007.
10. Dabassa A, Bacha K. The Prevalence and Antibiogram of *Salmonella* and *Shigella* Isolated from abattoir, Jimma town, South West, Ethiopia. Int J Pharm and Biol Res. 2012;3(4):143-148.
11. Samaxa RG, Matsheka MI, Mpoloka SW, Gashe BA. Prevalence and antimicrobial susceptibility of *Salmonella* Isolated from a variety of raw meat sausages in Gaborone (Botswana) retail stores. J Food Prot. 2012;75(4):637-642.
12. Hammack T, Amaguaña RM, June GA, Sherrod PS, Andrews WH. Relative effectiveness of Selenite Cystine Broth, Tetrathionate Broth, and Rappaport-Vassiliadis medium for the recovery of *Salmonella* spp. from foods with a low microbial load. J Food Prot. 1999;62(1): 16–21.
13. Kim J, Song JB. Effect of vacuum packaging on the microbiological profile of chilled chicken during storage. Agric. Chem. Biotechnol. 2004;47(1):35-37.
14. Tu L, Mustapha A. Reduction of *Brochothrix thermosphacta* and *Salmonella* Serotype Typhimurium on vacuum-packaged fresh beef treated with Nisin and Nisin combined with EDTA. J Food Science. 2002;67(1):302-306.
15. Gordon MA, Graham SM, Walsh AL, Wilson L, Phiri A, Molyneux E, et al. Epidemics of invasive *Salmonella enterica* serovar Enteritidis and *S. enterica* serovar Typhimurium infection associated with multidrug resistance among adults and children in Malawi. Clin Infect Dis. 2008;46: 963-969.
16. Morpeth SC, Ramadhani HO, Crump JA. Invasive non-Typhi *Salmonella* disease in Africa. Clin Infct Dis. 2009;49:606-611.
17. Boqvist S, Hansson I, Bjerselius UN, Hamilton C, Wahlström H, Noll B, et al.

Salmonella Isolated from Animals and Feed Production in Sweden between 1993 and 1997. Acta Veterinaria Scandinavica. 2003;44(3-4):181-197.
18. McEvoy JM, Doherty AM, Sheridan JJ, Blair IS, McDowell DA. The prevalence of *Salmonella* spp. in bovine faecal, rumen and carcass samples at a commercial abattoir. J Appl Microbiol. 2003;94(4): 693-700.
19. Schlosser W, Hogue A, Ebel E, Rose B, Umholtz R, Ferris K, et al. Analysis of *Salmonella* serotypes from selected carcasses and raw ground products sampled prior to implementation of the Pathogen Reduction; Hazard Analysis and Critical Control Point Final Rule in the US. Int J Food Microbiol. 2000;58:107-111.
20. Mezali L, Hamdi TM. Prevalence and Antimicrobial Resistance of *Salmonella* Isolated from Meat and Meat Products in Algiers (Algeria). Foodborne Path and Dis. 2012;9(6):522-529.
21. Hendriksen RS, Vieira AR, Karlsmose S, Lo Fo Wong DM, Jensen AB, Wegener HC et al. Global monitoring of *Salmonella* serovar distribution from the world health organization global foodborne infections network country data bank: Results of quality assured laboratories from 2001 to 2007. Foodborne Pathogens and Disease. 2011;8:887-900.
22. Shilangale RP, Di Giannatale E, Chimwamurombe PM, Kaaya GP. Prevalence and antimicrobial resistance pattern of *Salmonella* in animal feed produced in Namibia. Vet Ital. 2012;48(2): 125-132.
23. Taylor J, Galanis E, Wilcott L, Hoang L, Stone J, Ekkert J et al. An outbreak of *Salmonella* Chester infection in Canada: Rare serotype, uncommon exposure, and unusual population demographic facilitate rapid identification of food vehicle. J Food Prot. 2012;75(4):738–742.
24. Center for Diseases Control and Prevention (CDC). What is Salmonellosis? 2012. Accessed 14 May 2013.
Available:http://www.cdc.gov/salmonella/general/
25. Foodborne Illness Outbreak Database (FIOD). Canada Freybe Gourmet Foods, Ltd. and G. Brandt Meat Products Ltd., Head Cheese; 2010. Accessed 14 May 2013.
Available:http://www.outbreakdatabase.com/details/canada-freybe-gourmet-foods-ltd.-and-g.-brandt-meat-products-ltd.-head-cheese-2010/?
26. O'Glady K, Krause V. An outbreak of salmonellosis linked to a marine turtle. Southeast Asian J Trop Med Public Health. 1999;30(2):324–327.

6

Food and Nutrition Hazards Related to Public Health, Food Technology and Ecology: Perceptions of Experts and Non-experts in the United Kingdom

Hurjus Bahra[1] and Anne Mullen[1*]

[1]*Diabetes and Nutritional Sciences Division, King's College London, Franklin-Wilkins Building, 150 Stamford Street, London, UK.*

Authors' contributions

This work was carried out in collaboration between both authors. Authors HB and AM designed the study. Author HB completed the data collection and data analysis. Author AM compiled the manuscript. Both authors read and approved the final manuscript.

ABSTRACT

Aim: This study investigated perceptions of risk to self, risk to society, risk denial, control, responsibility, trust and knowledge of food and nutrition-related hazards among expert and non-expert groups with an aim of better understanding barriers to effective communication between experts and the public.
Design: Participants completed a questionnaire housed on an online platform.
Methodology: Experts were recruited from nutritionists in the food industry in the UK. Non-experts were recruited from a voluntary public panel with access to the online platform. Questions documented demographic variables and scores for the risk that food and nutrition-related hazards pose to self and society. Participants scored perceptions of expert and personal control of hazards, personal responsibility for averting the hazards, trust in experts for managing the hazards and personal knowledge of the hazards. The hazards were categorised for analysis into public nutrition, food technology and food ecology to reflect degrees of citizen participation in risk management.
Results: Experts scored perception of risk to self and risk to society from food technology hazards significantly lower than non-experts. Both groups had greatest risk denial, scores of personal control, personal responsibility and personal knowledge, and lowest scores for expert control, for public nutrition hazards. Trust in experts was higher among experts than it was among non-experts. Gender, personal responsibility and knowledge, but not trust in experts, were significant predictors of risk perception.
Conclusions: There were similarities in perception of risk of public nutrition and food ecology

**Corresponding author: Email: anne.mullen@kcl.ac.uk, a.mullen@DairyCouncil.org.uk*

hazards between experts and non-experts, but differences in the perception of risk and trust in experts relating to food technology. Both groups perceived higher personal control of and personal responsibility for, and lower expert control for, public nutrition hazards.

Keywords: Public health nutrition; food technology; food ecology; risk perception; expert; non-expert.

1. INTRODUCTION

Effective communication in public health relies on a lateral appreciation of risk perception between the 'expert' and 'non-expert'. Early studies in the field of risk perception focused on hazardous environmental activities and novel technologies, in nuclear power and commercial transport for example [1]. Comment and study on risk perception relating to food and nutrition hazards have emerged in recent times in reaction to high-profile, acute food safety events including the BSE crisis in the UK, dioxins in animal feed in Belgium, bacterial contamination of infant formula in France and polychlorinated biphenyls in Scottish salmon [2].

Despite efforts in communication and mitigation of food and nutrition hazards, many such concerns in the United Kingdom (UK) remain chronic and persistent. The Health Survey for England in 1991 reported 53% of men and 44% of women were overweight or obese; data published two decades later, in 2011, shows 65% of men and 58% of women were overweight or obese. Similarly, prevalence of overweight and obesity among boys and girls aged 2 to 15 years has increased from 24% and 26%, respectively, in 1995 to 31% and 28%, respectively, in 2011 [3]. Recent National Diet and Nutrition Survey data shows total fat intake is 34-36% food energy among adults and children, which was similar to previous studies, although saturated fat intakes ranged from 13 to 15% food energy, which was lower to previous surveys, but still above the dietary reference value [4]. Trends in alcohol consumption in the UK show increases among women, middle and older age groups and very young adolescents [5]. Data from 2009 indicates that 37% of men and 29% of women in the UK were reported to consume greater the recommended limit of alcohol per day, with 20% of men and 13% of women exceeding twice the recommended limit [6].

As well as these persistent challenges in public health, The Giessen Declaration identifies additional areas of concern for food and nutrition experts to address in the 21st century [7]. Food waste, the carbon footprint of the food supply and depletion of natural reserves are of present concern in public nutrition [8,9], although there has been some positive changes reported in recent years. Food wastage in the UK is reported to have fallen from 8.3 million tonnes to 7.2 million tonnes per year between 2006/7 and 2010 [10]. Indications that stocks of exploited fish in the EU have improved have recently been reported [11]. The complexity of food supply in the EU was drawn into sharp focus following the European investigation into authenticity of beef products; it is proposed that genetic technologies may be used more widely in the future to authenticate the origin of food ingredients [12]. Other developments in the use of genetic technologies for food and nutrition include harnessing it to address food shortage. The planet's population is expected to increase by 2 to 4 billion people over the next 3 to 4 decades, and food shortages are compounded by slowed development of arable land, energy costs, climate change and economic depression [13]. By 2008, 10% of cropland globally was planted with genetically modified crops, and 90% of the 13 million producers engaged were small-holder, resource poor farmers [13]. Genetic technology is seen by many as a viable approach to improving food security, particularly in developing countries [14], although dialogue in the public relating to genetic technologies remains contentious [15-17].

Food and nutrition experts must effectively communicate about persistent public health concerns, but also for moving forward with strategies for addressing global issues on climate change and food supply. Reconciling the perception of risk that hazards pose between experts and the public they communicate to is, therefore, of importance. It is reported that technical, quantitative and objective risk measurement is taken as the key informant of risk perception among experts [18]; non-experts incorporate broader issues including dread, catastrophic potential, controllability, equity and risk to future generations [19], as well as ethics [15] in their perception of risk. The expert-lay discrepancy in the perception of risk has been established in a variety of fields, including toxicology, biotechnology [20,21] and nutrition

[22,23], and may impact risk communication between expert and non-expert. This survey hopes to inform the communication of nutrition and health messages between experts and the public. As a reconciled perception of risk is key to effective communication, the main aim of this survey is to investigate the perception of risk to self and society of food and nutrition hazards among expert and non-expert groups. Perceptions of control, responsibility trust and knowledge of the food and nutrition related hazards were also investigated, as these can offer insights into what influences how risk is perceived.

2. METHODOLOGY

The survey was conducted according to the guidelines laid down in the Declaration of Helsinki and all procedures involving human subjects were approved by the Biomedical Sciences, Dentistry, Medicine, Natural and Mathematical Sciences Research Ethics Sub-Committee of King's College London (BDM/11/12-18).

A questionnaire investigating perceptions of risk to self and society, control, trust in experts, responsibility for and knowledge of nine food and nutrition related hazards was developed in line with those used in previous studies [21,22]. Hazards were chosen subjectively by the researchers to reflect Arnstein's three levels on the citizen 'ladder of participation' [24]: degrees of citizen participation (obesity, alcohol consumption, high fat diet), tokenism (carbon footprint of food supply, depletion of fish stocks, food shortage) and non-participation (salt in manufactured foods, genetically modified food, *E. coli* contamination of food) in their mitigation. The hazards were subsequently categorised for analysis as relating to 'public nutrition' (obesity, alcohol, high fat diet), 'food technology' (genetically modified food, salt in manufactured food, *E. coli* contamination) and 'food ecology' (food shortage, carbon footprint from food supply and fish stock depletion). The hazards were randomly presented, and not categorised, on the questionnaire. Participants were asked to score on a scale from 1 to 10 (low to high) their perceptions of risk, control, trust in experts, responsibility for and knowledge of each food and nutrition related hazard. Scores for perceived risk to self and risk to society were used determine risk denial. Questions on demography were administered. The questionnaire was housed on an online survey platform (SurveyShack, http://live.surveyshack.com/).

Experts were recruited from the Nutritionists in Industry (NII) group representing nutritionists working in the food industry in the United Kingdom. There are 100 members of this group. The response rate was 20%. A non-expert population was recruited from panellists pre-recruited to the online survey platform. There were no inclusion criteria for participant through SurveyShack. A small fee was paid to SurveyShack for the survey to be opened and promoted on the platform until 200 responses were retrieved. This response rate was achieved in under one week. Both the NII and SurveyShack distributed information sheets including the link to the online survey to their members by email.

Data was analysed using SPSS 20.0. Categorical data defining the subject characteristics of the expert and non-expert group was analysed by Chi-squared test. Scores of perception of risk, control, personal responsibility, trust in experts and knowledge of hazards and categorised hazards were compared between the expert and non-expert groups using Mann-Whitney U tests. Variance in scores for categorised hazards within expert and non-expert groups was compared using Friedman's two way analysis of variance by rank. Linear regression was used to investigate potential predictors of risk perception. Data is presented as median and 95% confidence interval (CI) of the median. A P value of ≤0.05 was considered statistically significant.

3. RESULTS AND DISCUSSION

The characteristics of the expert and non-expert groups are presented (Table 1). Twenty experts and 200 non-experts responded to the survey.

3.1 Expert and Non-expert Perceived Risk to Self and Risk to Society

Expert and non-expert scores of perceived risk to self and risk to society for each hazard are presented (Table 2). The expert group scores of perceived risk to self and risk to society of genetically modified foods (P=0.008 and P=0.000, respectively) and salt in manufactured foods (P=0.006 and P=0.023, respectively) were significantly lower than non-expert scores. The expert group score of perceived risk to self from fish stock depletion was significantly higher than the non-expert score (*P*=0.012). For all other hazards, there were no significant differences between expert and non-expert scores of perceived risk to self and risk to society.

Expert and non-expert scores of perceived risk to self and risk to society of categorised hazards (ie: public nutrition, food technology and food ecology) are presented (Fig. 1). As indicated in Table 2, the expert group scored perceived risk to self and risk to society of food technology hazards significantly lower than non-experts (*P*=0.024 and *P*=0.014, Fig. 1a and Fig. 1b respectively). Experts did not attribute significantly different scores for perceived risk to self from public nutrition, food technology and food ecology hazards. Non-experts attributed decreasing scores of perceived risk to self from public nutrition, food technology and food ecology hazards (*P*=0.000; Fig. 1a). Experts and non-experts attributed significantly higher scores of perceived risk to society from public nutrition hazards than food technology or food ecology (*P*=0.000 for both; Fig. 1b).

3.2 Expert and Non-expert Risk Denial

Expert and non-expert risk denial, calculated as the difference between scores of perceived risk to self and perceived risk to society, is presented (Fig. 2). The expert group had significantly greater risk denial regarding obesity and food shortage than the non-expert group (*P*=0.016 and *P*=0.015, respectively) and significantly lower risk denial regarding genetically modified foods than the non-expert group (*P*=0.027; Fig. 2a). Both expert and non-expert risk had significantly greater risk denial for public nutrition hazards than for food technology and food ecology hazards (*P*=0.000 for both; Fig. 2b).

Table 1. Subject characteristics (column %) of expert and non-expert groups

Variable	Expert (n 20)	Non-expert (n 200)	P
Gender			0.001
Male	0	36.0	
Female	100.0	64.0	
Ethnicity			NSD
White	95.0	90.5	
Mixed	0	1.5	
Asian or Asian British	5.0	4.5	
Black or Black British	0	1.0	
Chinese or other ethnicity	0	2.5	
Age			NSD
18-24 years	0	4.0	
25-34 years	45.0	23.0	
35-44 years	25.0	26.0	
45-54 years	25.0	23.5	
55+ years	0	23.5	
Unanswered	5.0	0	
Education			0.000
No formal/primary/secondary	0	58.5	
Third level	95.0	41.5	
Unanswered	5.0	0	
Employment			0.000
Manual/skilled labour	0	10.0	
Non-science/non-health/admin/business	26.3	44.0	
Science/healthcare	73.7	9.0	
Student/unemployed	0	37.0	

(a)

(b)

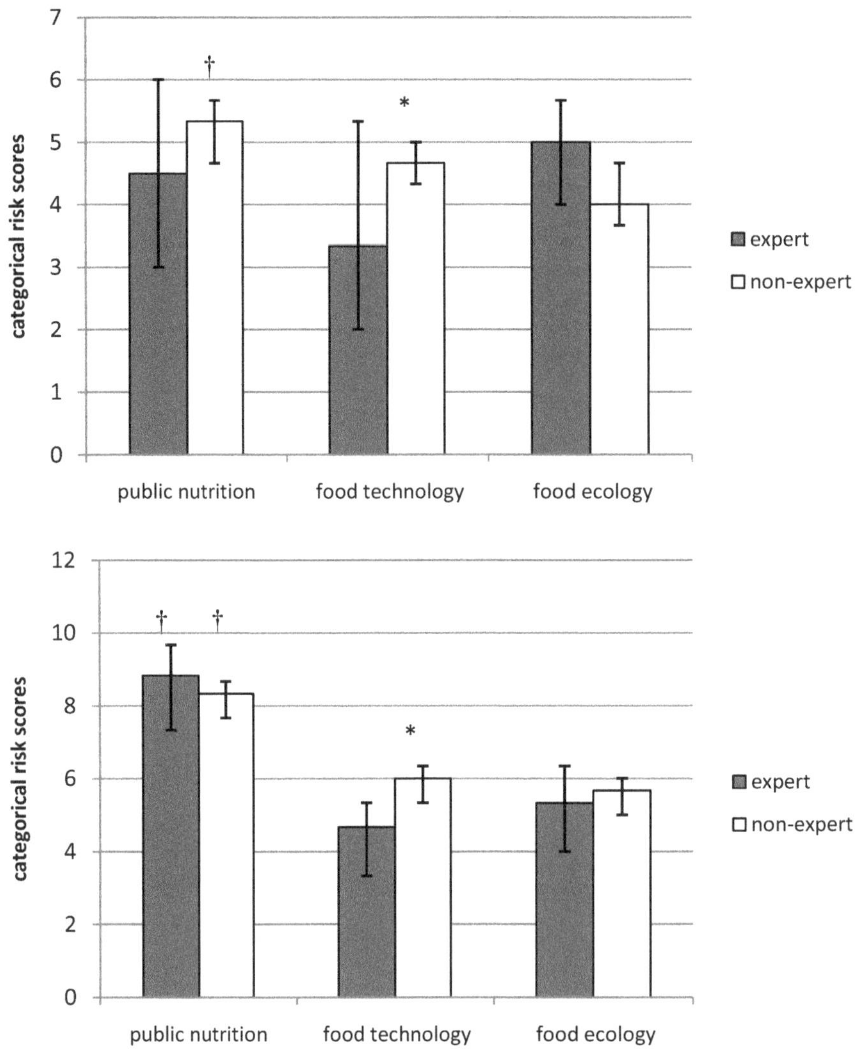

Fig. 1. Categorical risk scores of food and nutrition related hazards to (a) self and (b) society among expert and non-expert groups; Data represents median and 95% CI median; * *P*<0.05 within category between expert and non-expert groups; † *P*<0.05 across categories within expert or non-expert group

3.3 Personal and Expert Control

Expert and non-expert perceptions of personal and expert control of each food and nutrition related hazard (Table 3) and categorised hazards (Fig. 3) were investigated. The expert group scores of personal control of obesity and high fat diet were significantly higher than non-expert scores (P=0.014 and P=0.002, respectively). For all other hazards there were no significant differences between scores of personal control and expert control between expert and non-expert groups. Experts attributed significantly higher scores for personal control of public nutrition hazards than non-experts (P=0.024; Fig. 3a). Both expert and non-expert risk had significantly higher scores for personal control of public nutrition hazards than for food technology and food ecology hazards (P=0.000 for both; Fig. 3a). There were no differences between scores attributed by exerts and non-experts to expert control of each category of hazard, but both expert and non-expert groups scored expert control of public nutrition hazards significantly lower than food technology and food ecology (P=0.002 and P=0.000, respectively; Fig. 3b).

Table 2. Risk scores of food and nutrition related hazards to self and to society among expert and non-expert groups; data represents median and 95% CI median

	Risk to self				P	Risk to society				P
	Expert		Non-expert			Expert		Non-expert		
	Median	**95% CI**	**Median**	**95% CI**		**Median**	**95% CI**	**Median**	**95% CI**	
Public Health										
Obesity	3.5	2.0-7.0	6.0	5.0-6.0	NSD	9.0	8.0-10.0	9.0	8.0-9.0	NSD
Alcohol	4.5	2.0-5.0	4.0	3.0-5.0	NSD	9.0	7.0-10.0	8.0	8.0-8.0	NSD
High fat diet	6.0	2.0-8.0	6.0	6.0-7.0	NSD	8.5	8.0-9.0	8.0	8.0-9.0	NSD
Food technology										
Genetically modified food	2.0	1.0-3.0	4.0	3.0-5.0	0.008	2.0	1.0-3.0	5.0	5.0-6.0	0.000
Salt in manufactured food	4.5	2.0-5.0	6.0	6.0-7.0	0.006	5.0	3.0-7.0	7.0	7.0-8.0	0.023
E. coli contamination	5.0	3.0-5.0	5.0	4.0-5.0	NSD	6.5	5.0-7.0	6.0	5.0-6.0	NSD
Food ecology										
Foot shortage	2.5	1.0-4.0	3.0	2.0-4.0	NSD	4.0	3.0-7.0	5.0	3.0-5.0	NSD
Carbon footprint of food supply	4.5	3.0-6.0	4.0	3.0-5.0	NSD	3.0	2.0-6.0	6.0	6.0-6.0	NSD
Fish stock depletion	7.5	5.0-8.0	5.0	4.0-6.0	0.012	7.0	6.0-9.0	6.0	6.0-7.0	NSD

(a)

(b)

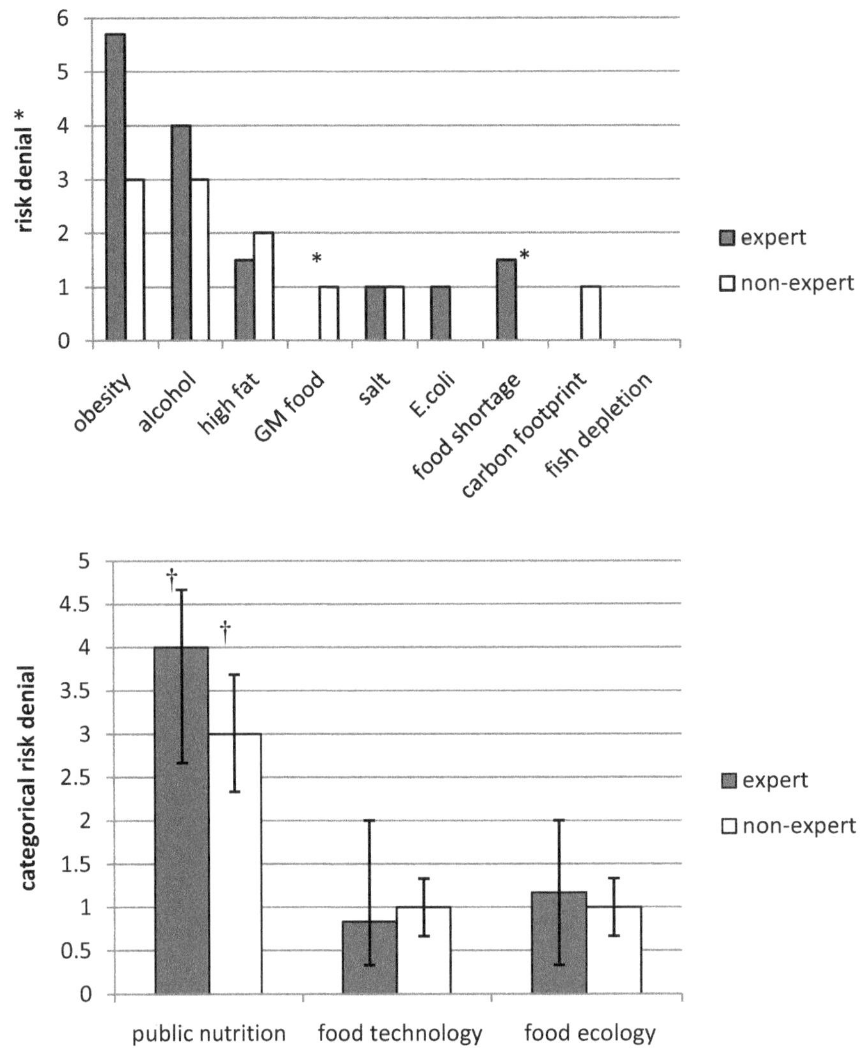

Fig. 2. Risk denial among expert and non-expert groups for (a) food and nutrition related hazard and (b) categorical hazards. Data represents median values in Fig. 2a and median and 95% CI of median in Fig. 2b. *p<0.05 between expert and non-expert groups; †p<0.05 across categories within expert or non-expert group

3.4 Personal Responsibility, Trust in Experts and Personal Knowledge

Expert and non-expert perceptions of personal responsibility for each food and nutrition related hazard (Table 4) and category of hazard (data not shown) were investigated. Experts had a significantly lower personal responsibility score than non-experts for carbon footprint of food (*P*=0.014). For all other hazards there were no significant differences between expert and non-expert scores of personal responsibility. There were no significant differences in scores of personal responsibility for categorised hazards between expert and non-expert groups. Both expert and non-expert groups attributed scores in descending order for personal responsibility of public nutrition, food technology and food ecology hazards (*P*=0.000, for both; data not shown).

Expert and non-expert perceptions of trust in experts for each food and nutrition related hazard (Table 4) and category of hazard (data not shown) were investigated. Experts had a significantly higher trust in expert score than non-experts for high fat diet and salt in manufactured food (*P*=0.039 and *P*=0.001). For all other hazards there were no significant differences between expert and non-expert scores of trust in experts. Experts attributed significantly higher scores for trust in experts for food technology hazards than non-experts (*P*=0.034; data not shown). For all other comparsions there were no significant differences in scores for trust in experts attributed by experts and non-experts.

Expert and non-expert personal knowledge for each food and nutrition related hazard (Table 4) and category of hazard (data not shown) were investigated. Experts reported significantly higher personal knowledge than non-experts for high fat diet (*P*=0.001), genetic modification (*P*=0.034) and salt in manufactured foods (*P*=0.000; Table 4). For all other hazards there were no significant differences between expert and non-expert reported personal knowledge. Experts attributed significantly higher scores for personal knowledge for public nutrition and food technology hazards than non-experts (*P*=0.013 and *P*=0.026, respectively; data not shown), and there were no significant difference between the experts and non-experts for personal knowledge of food ecology hazards. Both expert and non-expert groups attributed scores in descending order for personal knowledge of public nutrition, food technology and food ecology hazards (*P*=0.000 for both; data not shown).

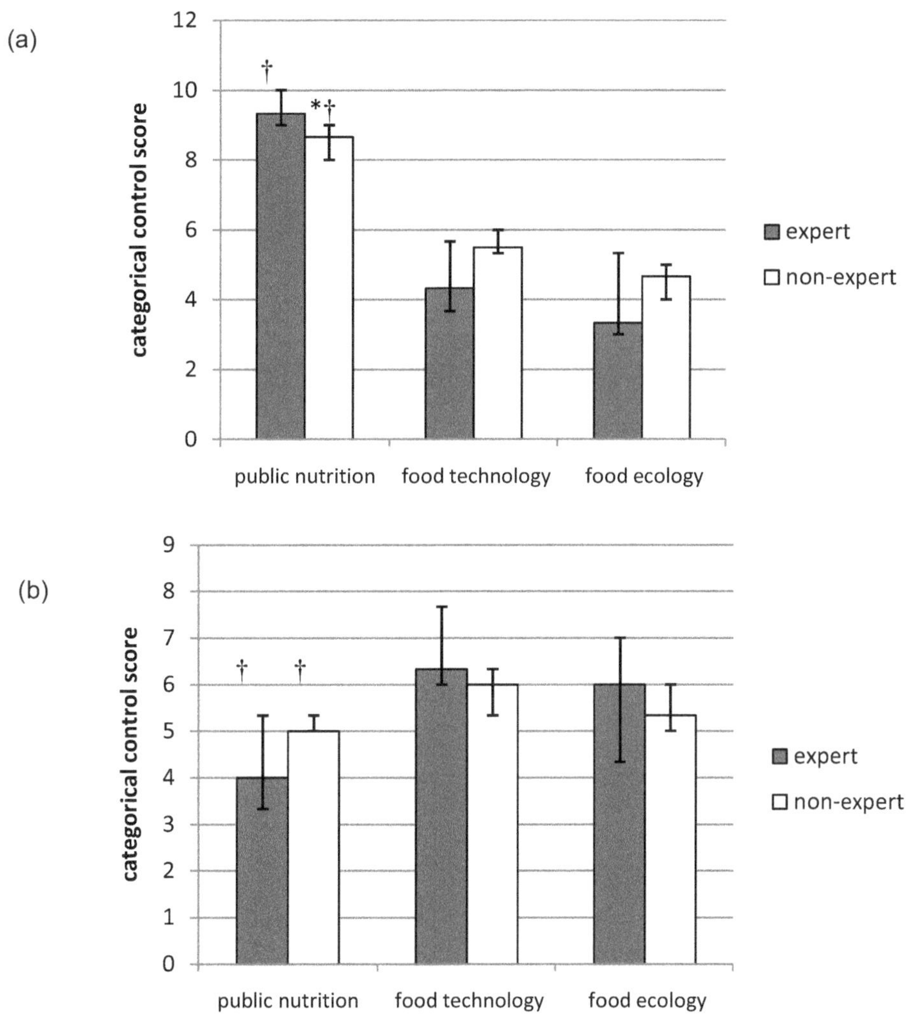

Fig. 3. Categorical scores of control food and nutrition related hazards by(a) the person and (b) experts among expert and non-expert groups; Data represents median and 95% CI of median; *p<0.05 within category between expert and non-expert groups; †p<0.05 across categories within expert or non-expert group

Table 3. Personal control and expert control scores of food and nutrition related hazards to self and to society among expert and non-expert groups; data represents median and 95% CI median

	Personal control				P	Expert control				P
	Expert		Non-expert			Expert		Non-expert		
	Median	95% CI	Median	95% CI		Median	95% CI	Median	95% CI	
Public Health										
Obesity	9.0	9.0-10.0	9.0	8.0-9.0	0.014	5.0	4.0-6.0	5.0	5.0-5.0	NSD
Alcohol	9.0	7.0-10.0	10.0	9.0-10.0	NSD	5.0	2.0-6.0	5.0	5.0-5.0	NSD
High fat diet	10.0	7.0-10.0	9.0	8.0-9.0	0.002	4.5	3.0-5.0	5.0	5.0-6.0	NSD
Food technology										
Genetically modified food	3.0	3.0-5.0	5.0	5.0-6.0	NSD	6.0	5.0-8.0	6.0	5.0-7.0	NSD
Salt in manufactured food	6.0	4.0-7.0	6.0	6.0-7.0	NSD	7.0	6.0-8.0	6.0	5.0-7.0	NSD
E. coli contamination	5.0	3.0-5.0	6.0	5.0-6.0	NSD	7.0	6.0-8.0	6.0	5.0-6.0	NSD
Food ecology										
Foot shortage	5.0	2.0-6.0	5.0	4.0-5.0	NSD	6.0	5.0-7.0	5.0	5.0-6.0	NSD
Carbon footprint of food supply	3.0	2.0-6.0	5.0	5.0-6.0	NSD	6.0	5.0-8.0	5.0	5.0-6.0	NSD
Fish stock depletion	2.0	2.0-5.0	4.0	3.0-5.0	NSD	6.0	5.0-8.0	6.0	5.0-6.0	NSD

Table 4. Personal responsibility, trust in experts and personal knowledge of food and nutrition related hazards among expert and non-expert groups

	Personal responsibility				P	Trust in experts				P	Personal knowledge				P
	Expert		Non-expert			Expert		Non-expert			Expert		Non-expert		
	Median	95% CI	Median	95% CI		Median	95% CI	Median	95% CI		Median	95% CI	Median	95% CI	
Public health															
Obesity	8.0	7.0-9.0	9.0	8.0-9.0	NSD	6.0	3.0-8.0	5.0	4.0-5.0	NSD	9.0	8.0-9.0	8.0	8.0-8.0	NSD
Alcohol	9.0	9.0-9.0	9.0	9.0-9.0	NSD	5.0	3.0-8.0	5.0	4.0-5.0	NSD	8.0	8.0-9.0	8.0	8.0-8.0	NSD
High fat diet	8.0	7.0-9.0	8.0	8.0-9.0	NSD	6.0	4.0-7.0	5.0	4.0-5.0	0.039	9.0	9.0-10.0	8.0	7.0-8.0	0.001
Food technology															
Genetically modified food	4.0	1.0-7.0	5.0	4.0-5.0	NSD	6.0	4.0-8.0	5.0	5.0-5.0	NSD	7.0	5.0-7.0	5.0	5.0-6.0	0.034
Salt in manufactured food	6.5	5.0-7.0	6.0	5.0-7.0	NSD	7.0	6.0-8.0	5.0	4.0-5.0	0.001	9.0	9.0-9.0	6.5	6.0-7.0	0.000
E. coli contamination	4.0	2.0-5.0	5.0	4.0-5.0	NSD	6.0	4.0-7.0	5.0	5.0-5.0	NSD	5.0	4.0-6.0	5.0	5.0-6.0	NSD
Food ecology															
Foot shortage	2.0	1.0-4.0	3.0	2.0-4.0	NSD	6.0	4.0-7.0	5.0	5.0-5.0	NSD	5.0	4.0-6.0	5.0	4.0-5.0	NSD
Carbon footprint of food supply	2.0	2.0-4.0	5.0	4.0-5.0	0.014	6.0	4.0-7.0	5.0	4.0-5.0	NSD	5.0	4.0-6.0	5.0	5.0-5.0	NSD
Fish stock depletion	3.0	1.0-5.0	4.0	3.0-5.0	NSD	6.0	4.0-6.0	5.0	4.0-5.0	NSD	6.0	4.0-6.0	5.0	5.0-6.0	NSD

3.5 Predictors of Perceived Risk to Self and Risk to Society

From the variables investigated (gender, ethnicity, age, education level, being an expert/non-expert, perceptions of risk to society, personal control, expert control, personal responsibility, trust in experts and personal knowledge), none were significant predictors of perceived risk to self from public nutrition hazards (adjusted $r^{2=}0.045$; data not shown); gender, perceptions of risk to society and expert control were predictors of perceived risk to self from food technology hazards ($P=0.010$, $P=0.000$ and $P=0.013$, respectively; adjusted $r^{2=}0.588$; data not shown); perceived risk to society was the only predictor of perceived risk to self from food ecology hazards ($P=0.000$; adjusted $r^{2=}0.507$; data not shown). From the variables investigated (gender, ethnicity, age, education level, being an expert/non-expert, perceptions of risk to self, personal control, expert control, personal responsibility, trust in experts and personal knowledge), perceptions of personal knowledge, expert control and personal control were predictors of perceived risk to society from public nutrition hazards ($P=0.000$, $P=0.025$ and $P=0.000$, respectively; adjusted $r^{2=}0.296$; data not shown); gender, being an expert/non-expert, perceptions of personal responsibility, personal knowledge and risk to self were predictors of perceived risk to society from food technology hazards ($P=0.000$, $P=0.001$, $P=0.023$, $P=0.002$ and $P=0.000$, respectively; adjusted $r^{2=}0.662$; data not shown); gender, perceptions of risk to self and personal knowledge were predictors of perceived risk to society from food ecology hazards ($P=0.003$, $P=0.000$ and $P=0.004$, respectively; adjusted $r^{2=}0.607$; data not shown).

4. CONCLUSION

Non-experts had significantly higher scores of risk to self and society from food technology hazards than experts, with genetically modified food and salt in manufactured food being significant. Non-experts had significantly lower scores of risk to self from food ecology hazards than experts. Both groups gave higher scores for risk to society to public nutrition hazards than to food technology or food ecology hazards. Non-experts had higher risk denial for genetically modified foods than experts, but experts had significantly higher risk denial for obesity and food shortage than non-experts. Risk denial for public health hazards was higher among both expert and non-expert groups than for food technology or food ecology hazards. These results indicate that non-experts have a higher perception of risk to self and society from technological hazards, and that both groups have high risk denial for public nutrition hazards, with obesity a significant example for experts. Savadori et al. [20] investigated risk judgements relating to food biotechnology applications among 58 expert and 58 non-experts. For 7 applications of genetic modification relating to food, plants and animals (non-medical), the public had significantly higher risk judgements than experts. The group studied the influence of four factors relating to risk from food biotechnology applications (harmful and dreadful application, useful application, science knowledge and new application). The four factor model accounted for 44% of the variance of risk judgement among experts and 30% of the variance among non-experts. Non-experts scored more highly on scales measuring 'harmful and dreadful application' and lower on scales measuring 'useful application' for biotechnology applied to food. Just 2 of the factors (harmful and dread application, and useful application) were significantly associated with expert risk judgement, whereas all 4 factors were for the non-expert. The authors concluded that non-experts have a wider perception of risk associated with food biotechnology than experts. Our survey supports the work of Savadori et al. [20] showing a disparity in the perception of risk to self and society among experts and non-experts relating to genetic modification. There have been a number of discursive publications concerning the 'democratisation of risk' in areas where policy is driven largely by scientific expertise, such as the genetic modification of food [15-17]. The expert-lay discrepancy in risk perception of food technology may highlight the need for such an approach to interpretation and communication of risk.

Control has been shown to be a determinant of risk perception [25] and the degree of control an individual has over their exposure to a hazard is a key predictor of risk denial [26]. Non-experts had significantly lower scores for personal control of public nutrition hazards than experts, particularly for obesity and high fat diet, but both groups had higher scores for personal control of public nutrition related hazards than for food technology and food ecology hazards. Information from untrusted sources amplifies risk perception [27]. Wynne cautions that trust should not be assumed to be a functional substitute for information in risky technologies [15], and elsewhere it has been shown that trust is only weakly associated with risk perception [28]. In the present survey, non-experts had lower scores for trust in experts regarding high fat diet and salt in manufactured food than experts, and attributed significantly lower scores for trust in experts relating to food technology hazards than experts did. Autonomy and personal responsibility attenuate risk perception [27]. In this survey, there were no differences in how the groups perceived personal responsibility. Both experts and non-experts scored personal responsibility for public nutrition hazards higher than for food technology and food ecology hazards. Experts had significantly greater scores for personal knowledge of public nutrition and food technology hazards than non-experts, with high fat diet, salt in manufactured food and genetic modification being significant examples.

The perception of risk is attenuated by characteristics of familiarity, individual control, naturalness, statistical probability, clear benefits, fair distribution of exposure, voluntariness of exposure, information by trusted sources and coverage in the media, as well as freedom of choice, perception of autonomy and personal responsibility [27]. Risk perception may be amplified by characteristics of exoticness/novelty, control by others, manmade, catastrophic potential, little or no benefit, unfair distribution of exposure, imposition, information by untrusted sources and lack of coverage in the media [27]. Risk perception may also be shaped by gender, ethnicity and socio-economic distinctions [29].Our model for investigation of risk perception among the entire cohort, incorporating factors such as gender, age, education, perceptions of control, responsibility, trust and knowledge, showed poor to moderate ability to predict variance in the scores of risk to self and risk to society. None of the predictors was associated with risk to self from public nutrition hazards. Gender, risk to society and expert control were associated with risk to self from food technology. Only risk to society was associated with risk to self from food ecology hazards. Personal knowledge, expert control and personal control were associated with perception of risk to society from public nutrition hazards. Gender, being an expert/non-expert, personal responsibility, knowledge and perception of risk to self were associated with perception of risk to society from food technology hazards. Gender, risk to self and personal knowledge were associated with risk to society from food ecology hazards variance. Our survey showed that control, personal responsibility and knowledge, as well as gender, were significant predictors of risk perception. Trust in experts was not a predictor.

There are several limitations to the present study. Significant differences in the characteristics of the expert and non-expert groups were evidenced, and were anticipated: the experts were comprised exclusively of females, a higher proportion of the expert group had a third level education, and a higher proportion of the expert group was employed in the science-based/health care profession section. However, the response rate from the experts approached was poor, at 20%, and we recruited solely from experts in the food industry. A power calculation was not conducted; previous studies have detected significant difference between expert and non-expert group perceptions of risk, including58 experts in biotechnology and 58 members of the public [20], 46 experts in nanotechnology and 375 members of the public [21], 40 experts in nutrition and 40 members of the public [22] and 21 experts in food safety and 29 members of the public [23]. We subjectively chose the hazards to investigate and the categorisation of the hazards as public health, food technology and food ecology. Furthermore, the terms 'hazard' and 'risk' were not defined on the questionnaire and were open to interpretation by the participant. The terms 'alcohol', 'genetic modification', 'salt in manufactured food' and 'carbon footprint of food' are evocative of, but not strictly, hazards; whereas 'obesity', 'high fat diet', '*E. coli* contamination', 'food shortage' and 'fish stock depletion' are more evidently hazardous. Additional factors in the psychometric paradigm for determining risk perception among non-experts [1,21] may have been incorporated into the study, including 'dread', 'unknown effects', 'voluntariness of risk' and 'ethics'. To build on the present work, we would suggest focus groups as a means of collecting richer, qualitative data on the perception of risk among expert and non-expert groups. Online health information searches have been used as a proxy for public health risk perception [30], and this may be a useful tool for understanding the perception of risk related to food and nutrition in the public.

The key message in this study is that disparity exists between expert and non-expert perceptions of risk to self and society, particularly with respect to food technology. Both experts and non-experts had more risk denial for public nutrition hazards than for food technology or food ecology. Both groups had higher scores for personal control, personal responsibility and personal knowledge of public nutrition hazards, compared with food technology and food ecology hazards. Both groups gave higher scores to expert control for food technology and food ecology hazards than for public nutrition hazards. Trust in experts was not associated with risk perception, and did not vary depending on the nature of the hazard. This study, we believe, is the first to investigate risk perception of expert and non-expert groups on food and nutrition hazards relating to public nutrition, food technology and food ecology. The results signify the need for further research into reconciling, or levelling, the perception of risk between experts and the public they address, in an effort to optimise risk communication.

ETHICAL APPROVAL

The study was conducted according to the guidelines laid down in the Declaration of Helsinki and all procedures involving human subjects were approved by the Biomedical Sciences, Dentistry, Medicine, Natural and Mathematical Sciences Research Ethics Sub-Committee of King's College London (BDM/11/12-18).

ACKNOWLEDGEMENTS

We wish to thank all those who participated in this study.

COMPETING INTERESTS

The authors have no conflicts of interest to declare.

REFERENCES

1. Slovic P. Perception of risk. Science. 1987;236:280-285.
2. McGloin A, Delaney L, Hudson E, Wall P. Symposium on 'The challenge of translating nutrition research into public health nutrition'. Session 5: Nutrition communication. The challenge of effective food risk communication. Proc Nutr Soc. 2009;68:135-41.
3. Oyebode O, Mindell J. Use of data from the health survey for England in obesity policy making and monitoring. Obes Rev. 2013;14:463-76.
4. Pot GK, Prynne CJ, Roberts C, Olson A, Nicholson SK, Whitton C, et al. National Diet and Nutrition Survey: fat and fatty acid intake from the first year of the rolling programme and comparison with previous surveys. Br J Nutr. 2012;107:405-15.
5. Smith LA, Foxcroft DR. The effect of alcohol advertising, marketing and portrayal on drinking behaviour in young people: systematic review of prospective cohort studies. BMC Public Health. 2009;9:51.
6. Office for National Statistics. Social Trends, No. 41, Health. 2011. Accessed 26 April 2014. Available:http://www.ons.gov.uk/ons/rel/social-trends-rd/social-trends/social-trends-41/index.html.
7. The Giessen Declaration. Public Health Nutr. 2005;8:783-86.
8. Friel S, Dangour AD, Garnett T, Lock K, Chalabi Z, Roberts I, et al. Public health benefits of strategies to reduce greenhouse-gas emissions: food and agriculture. Lancet. 2009;374:2016-25.
9. Sulda H, Coveney J, Bentley M. An investigation of the ways in which public health nutrition policy and practices can address climate change. Public Health Nutr. 2010;13:304-13.
10. Quested TE, Parry AD, Easteal S, Swannell R. Food and drink wasted from households in the UK. Nutr Bull. 2011;36:460-46.
11. Cardinale M. Fishery reform: many stocks secure. Nature. 2013;476:282.
12. O'Mahony PJ. Finding horse meat in beef products-a global problem. QJM. 2013;106:595-97.
13. Fedoroff NV. The past, present and future of crop genetic modification. N Biotechnol. 2010;27: 461-65.
14. Ruane J, Sonnino A. Agricultural biotechnologies in developing countries and their possible contribution to food security. J Biotechnol. 2010;156:356-63.
15. Wynne B. Creating public alienation: Expert cultures of risk and ethics on GMOs. Sci Cult. 2001;10:445-81.
16. Ferretti MP. Why public participation in risk regulation? The Case of Authorizing GMO Products in the European Union. Sci Cult. 2007;16:377-95.
17. Kurian P, Wright J. Science, governance, and public participation: an analysis of decision making on genetic modification in

Aotearoa/New Zealand. Public Underst Sci. 2012;21:447-64.
18. Hansen J, Holm L, Frewer L, Robinson P, Sandoe P. Beyond the knowledge deficit: recent research into lay and expert attitudes to food risks. Appetite. 2003;41:111-21.
19. Slovic P. Trust, emotion, sex, politics, and science: surveying the risk-assessment battlefield. Risk Anal. 1999;19:689-701.
20. Savadori L, Savio S, Nicotra E, Rumiati R, Finucane M, Slovic P. Expert and public perception of risk from biotechnology. Risk Anal. 2004;24:1289-99.
21. Siegrist M, Keller C, Kastenholz M, Frey S, Wiek A. Laypeople's and experts' perception of nanotechnology hazards. Risk Anal. 2007;27:59-69.
22. Mullen A, Lambe J, Gibney MJ. Risk perception in expert and non-expert groups. Proc Nutr Soc. 2002;61:81A.
23. Webster K, Jardine C, Cash SB, McMullen LM. Risk ranking: investigating expert and public differences in evaluating food safety hazards. J Food Prot. 2010;73:1875-85.
24. Arnstein SR. A ladder of citizen participation. J AmInst Plann. 1969;35:216-44.
25. Brandimarte L, Acquisti A, Loewenstein G. Misplaced confidences: privacy and the control paradox. Soc Psychol Pers Sci. 2012;4:340-47.
26. Sjoberg L. Factors in risk perception. Risk Anal. 2001;20:1-11.
27. Schmidt M. Investigating risk perception: a short introduction. Chapter 3 of Loss of agro-biodiversity in Vavllov centres with a special focus on the risk of GMOs, PhD Thesis, Austria; 2004. Accessed 26 April 2014.
Available: http://faculty.mercer.edu/butler_aj/documents/Intro_risk_perception_Schmidt.pdf.
28. Sjoberg L. Limits of knowledge and the limited importance of trust. Risk Anal. 2001;21:189-98.
29. Frewer L. Risk Perception and risk communication about food safety issues. Nutr Bull. 2000; 25:31-3.
30. Liang B, Scammon DL. Incidence of online health information search: a useful proxy for public health risk perception. J Med Internet Res. 2013;15:e114.

7

Health Risks Associated with Freshwater Fish Consumption

Angelika A. Linowska[1*] and Ewa Sobecka[1]

[1]Division of Hydrobiology, Ichthyology and Biotechnology of Breeding, West Pomeranian University of Technology, K. Królewicza 4, Szczecin, Poland.

Authors' contributions

This work was carried out in collaboration between both authors. Author ES headed the project and drafted the manuscript. Author AAL conducted the analysis and interpretation. Author AAL was involved in the early work and assisted with the interpretation of the result and was in charge of data management. Both authors were involved in the critical revision of the manuscript for important intellectual content. Both authors read and approved the final manuscript.

ABSTRACT

Aims: To determine the prevalence and density of the larval digenean trematode *Paracoenogonimus ovatus* in fish muscle tissue. To determine larval survival under selected thermal conditions.

Place and Duration of Study: Division of Hydrobiology, Ichthyology and Biotechnology of Breeding, West Pomeranian University of Technology, Szczecin, October 2009 and 2014.

Methodology: Twelve fish species from Lake Dąbie (Poland) that are exploited by both commercial and recreational fisheries were selected. Samples of muscle tissue were weighed and then viewed under a trichinoscope where encysted *P. ovatus* larvae (metacercariae) could be counted. Then, they were subjected to artificial digestive juice (0.1% activated pepsin and 5% citric acid) to facilitate isolating the cysts from the muscle tissues. The parasites that were isolated were removed from their shells, and the species was identified. Further, ten freshwater bream fillets with encysted larvae were placed under refrigeration (4°C) and frozen (-18°C); larval viability was checked after 24, 48, and 78 h.

Results: The prevalence of *P. ovatus* larvae increases (76.5% in 2009 and 80.5% in 2014). The highest density was noted in roach (*Rutilus rutilus*), freshwater bream (*Abramis brama*), and white bream (*Blicca bjoerkna*). It was confirmed that metacercariae are resistant to decreases in temperature to 4°C, but that further reducing the temperature to -18°C results in larval death in 24 h.

Conclusion: Many species of freshwater fishes are infected by *P. ovatus* metacercariae. Eating

**Corresponding author: Email: angelika.linowska@zut.edu.pl*

improperly prepared fish and fisheries products may provide serious health problems. Further studies are needed to neutralize the living metacercariae in raw materials and food products.

Keywords: Fish; parasites; Paracoenogonimus ovatus larvae; viability.

1. INTRODUCTION

Over fifty helminth parasite species, including digenean trematodes, that infect fish and shellfish are known to be pathogenic to humans. Usually illnesses are rare and present minor to moderate symptoms; however, some parasites can potentially pose serious health risks [1]. Domesticated and wild animals can also become infected accidentally. Many traditional recipes across the world call for either raw or semi-cooked fish or shellfish meat. Since larval parasites can be located in muscle tissues, consumption of it can pose real risks to human health.

Many parasite species infect muscle tissues. The most frequently noted is the nematode *Anisakis simplex*, which is found in the muscle tissues of gadids and herring. Consuming live larval *A. simplex* leads to a strong reaction from the digestive system, and it can produce symptoms similar to those of ulcerative diseases or stomach cancer [2]. Larval tapeworms of the genus *Diphyllobothrium*, which infect the muscles of freshwater fish, are also pathogenic. Many species of digenetic trematodes have been listed as transmissible to man through the consumption of fish, crustacea or molluscs, but a few represent notable zoonotic threats [2]. The familly Heterophyidae comprises very small trematodes which inhabit the intestine of birds and mammals. The metacercariae can be found in a wide variety of fresh and marine fish muscles [2].

There are few studies of the prevalence and density of trematode metacercariae infection in different fish species. Sometimes fish exhibit external signs of infection including muscle cahexia, "sunken sides", and protruding eyes. However, in most instances, there are no visible indications of the presence of larvae. This is why it is difficult immediately after catching fish to determine whether they present a potential health risk to human consumers.

The aim of this study was to determine which of the most abundantly occurring and frequently caught fish species occurring in Lake Dąbie (Poland) are most infected with *P. ovatus* metacercariae. An additional aim was to verify the viability of isolated larval trematodes during the storage of raw materials.

2. MATERIALS AND METHODS

2.1 Fish Species

The initial study was performed in the fall of 2009 when 34 fish were examined, including 24 fish from the family Cyprinidae (10 freshwater bream - *Abramis brama*, 10 roach - *Rutilus rutilus*, 3 crucian carp - *Carassius carassius*, 1 rudd - *Scardinius erythrophthalmus*) and 9 species of Percidae (5 European pearch - *Perca fluviatilis*, 4 ruffe -*Gymnocephalus cernua*), and one northern pike - *Esox lucius* (Esocidae).

The next study was conducted in fall 2014, when 113 fish were examined: 5 species from the family Ciprinidae (25 white bream - *Blicca bjoerkna*, 42 freshwater bream, 2 tench - *Tinca tinca*, 28 roach, 1 asp - *Aspius aspius*), 12 fish from the family Percidae (8 European pearch, 2 pike-perch - *Sander lucioperca*, 1 ruffe), and one from the family Gobidae (3 round goby - *Neogobius melanostomus*).

The fish were caught from fishing cutters operating in Lake Dąbie, which is one of the largest lakes in Poland, and the species selected reflect the species structure of commercial and recreational catches. The fish species used in this study occur commonly in Europe and in north, west, and central Asia.

2.2 Digestion and Tissue Compression

The initial study focused on evaluating prevalence (% of fish infected) and the density (the number of parasites per 1 g of mass of muscles) of encysted digenean trematode metacercariae in the muscle tissues of various fish species. The locations in muscle tissues that were chosen most frequently by the metacercariae were also determined. To do this, each fish was divided as shown in Fig. 1. The fish were filleted, and the gently minced skinned muscle was compressed under a trichinoscope to permit counting the number of metacercariae in the tissues. Based on the results, it was noted that metacercariae occur most frequently and most abundantly in the anterior dorsal section of fillets (sector I).

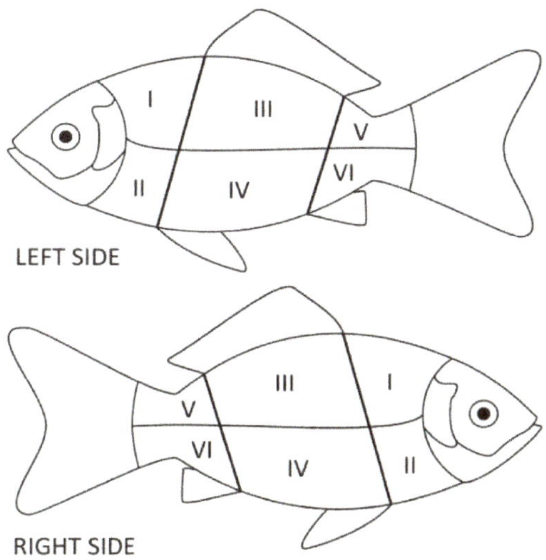

Fig. 1. Pattern of fillet sectors

In both studies (2009 and 2014), muscle tissue samples were collected from the anterior dorsal section of skinless fillets. The mean weight of these samples was 2.162 g (SD 0.558 g), and these were also analyzed with a trichinoscope. The encysted metacercariae were counted, and then to facilitate isolation, the muscle tissues were subjected to a 0.1% solution of activated pepsin and 5% citric acid at a temperature of 20°C [3]. Pepsin is activated by lowering the pH of the solution to less than 5, and it is most active within a pH range of 1.8 to 3.5 [4]. In the human stomach this role is played by 0.2-0.5% hydrochloric acid. It was determined experimentally that the pH of the 5% citric acid solution with its value of about 2 corresponds to the pH value achieved when a 0.5% hydrochloric acid solution is used. Thus, it is possible to use citric acid as a pepsin activator in artificial digestive juice. The metacercariae were removed from their cysts using a preparatory needle. Once freed of the cysts, the larvae were measured and identified under an Olympus BX 50 microscope coupled with a camera running AxioVs40 V 4.8.2.0 software.

2.3 Lethal Conditions

Attempts were also made to determine lethal conditions for *P. ovatus* metacercariae. To do this, 10 freshwater bream fillets with a mean mass of 23 g were refrigerated (at 4°C) or frozen (at -18°C). Because the fillets used in the experiment had a low mass and thickness (about 1 cm), the internal temperature of the fillets quickly reached that of storage conditions. The experiment lasted for 72 h. On each successive day, muscle tissue samples were collected from each sample, the metacercariae were isolated, and the locomotive activity of the encysted larvae was verified under a microscope [5].

3. RESULTS AND DISCUSSION

The metacercariae of the trematode *P. ovatus* form spherical, double-layered cysts with an approximate diameter of 500 µm that usually have a yellow-brown pigmentation. The outer wall of the cyst comes away easily from the internal layer, which is highly compact. The oral sucker is clearly visible, the ventral sucker is smaller and is considerably distant from the oesophagus. Sometimes the metacercariae gonad set is visible in the cyst, and it resembles the shape of the letters M or W (Fig. 2).

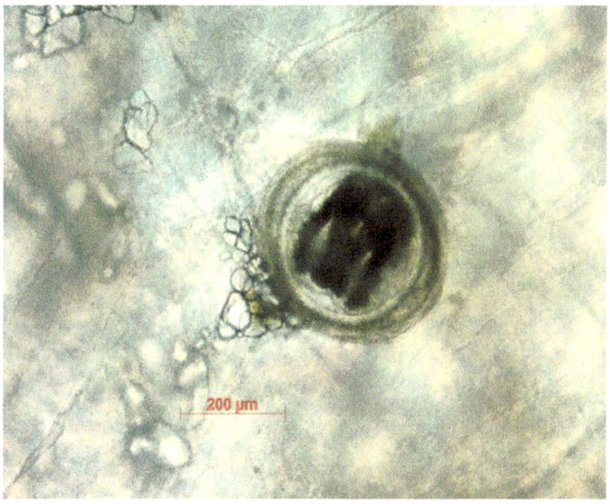

Fig. 2. Metacercariae of *Paracoenogonimus ovatus* in the fish muscles
M/W shape of gonads visible

These morphological characters permit differentiating this species from others, which also, although less frequently, occur in muscle tissues. However, the cysts in muscle tissues are only visible under a microscope at a high magnification of 40 X (Fig. 3). Metacercariae are grouped primarily in the anterior segment of the dorsal segment at depths ranging from just beneath the skin to about 50% of fillet thickness.

The prevalence of *P. ovatus* metacercariae in the fish analyzed in 2009 was $\bar{x}$ 76.47 (SD 10.44), while in 2014 it was $\bar{x}$ 80.53 (SD 13.46). Both the initial study and the one performed in 2014 indicated that the highest density of larvae in 1 g of muscle tissue mass was noted in roach and freshwater bream (Table 1). In addition, a large number of metacercariae was also observed in 2014 white bream. Parasites were not found in the muscles of crucian carp caught in 2009, or in the round goby, ruffe, or European pearch examined in 2014. Metacercariae were noted to occur singly, as in northern pike (4), or in fairly large numbers in pike-perch (15), but the insufficient number of this fish did not permit conducting a proper evaluation.

The density per 1 g of muscle tissue mass in total sample of fishes examined in 2009 was $\overline{X}$ 8.97 (SD 10.44), density range from 0 to 42 *P. ovatus* metacercariae, and in 2014 it was $\overline{X}$ 8.73 (SD 13.46), density range from 0 to 84 *P. ovatus* metacercariae.

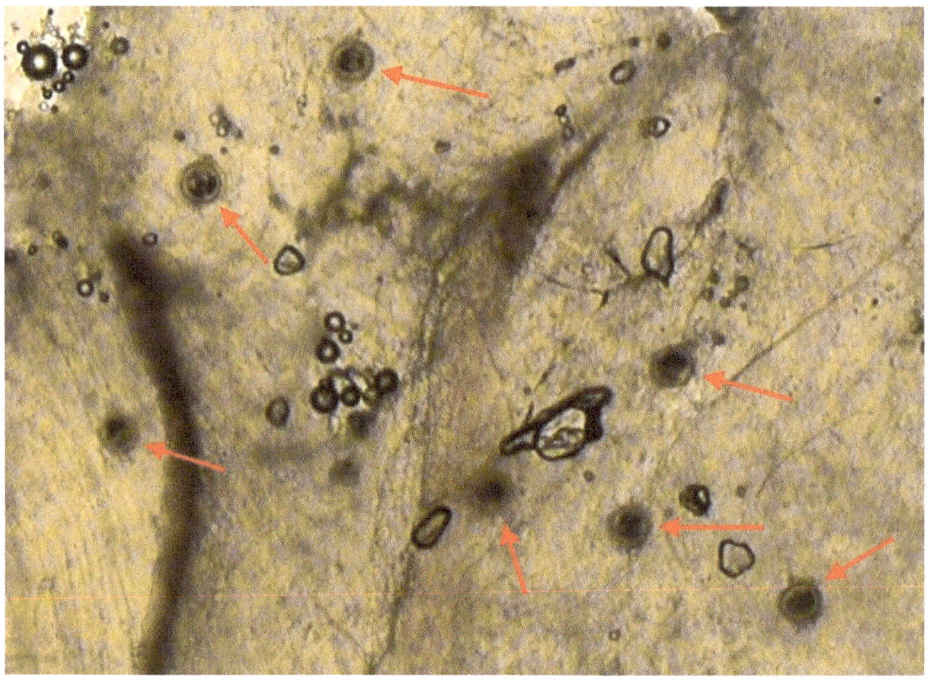

Fig. 3. Metacercariae of *Paracoenogonimus ovatus*
Mmagnification 40 X Rred arrows shows the location in the muscles

Table 1. *Paracoenogonimus ovatus* infection parameters of fish studied

Species	2009			2014		
	Density (in 1 g)		Density range	Density (in 1 g)		Density range
	$\overline{X}$	SD		$\overline{X}$	SD	
Asp				(+)		2
Crucian carp	0	0	0			
European pearch	0.53	0.74	0-2	0	0	0
Freshwater bream	11.31	5.66	4-22	5.17	6.98	0-35
Northern pike	(+)		4			
Pike-perch				7.30	10.32	0-15
Roach	16.38	13.90	1-42	17.80	21.83	0-84
Round goby				0	0	0
Rudd	(+)		18			
Ruffe	0.94	1.25	0-3	(+)		0
Tench				2.31	0.26	2-3
White bream				9.58	7.53	0-30

(+) - one fish was examined

The larvae (50 metacercariae – 10 larvae from 5 samples of 5 fishes) were viable after 24, 48, and 72 h of refrigeration (4°C); however, the metacercariae from fillets that had been frozen (-18°C) exhibited no locomotion after 24 h under these conditions.

In most instances, humans become infected with digenean trematodes by consuming the intermediate hosts of these parasites either raw or semi-cooked, partially marinated, inappropriately smoked, or from fish and/or sea food that is poorly canned. Consumers that are at particularly high risk are those who live in regions where fish is the main staple food and also where the risk of infection is linked to traditional foods or culinary trends. This is why the probability of infection in Europe has increased in recent years.

When some trematodes infect hosts incidentally they can only survive for a very short period of time and produce only a few eggs; however, they can cause serious pathology in the host body. Human infection with trematodes from the family Opisthorchiidae in Europe is attributed to *Opisthorchis felineus*, but *Pseudamphistomum truncatum* and trematodes from the genus *Metorchis*, which have been confirmed in foxes in the vicinity of Berlin and in otters in England, Wales, Ireland, and Poland [6-9], can also infect humans [10,11]. The first human cases of metorchiasis were described in 1993. To date *Metorchis*, which the larvae can also be found in the muscles of fish, has been identified in carnivores in the North America, and in the human, ducks, cats and dogs in China. *Metorchis bilis* has been implicated in human infection in Siberia [12]. Experimental infections with *P. ovatus* metacercariae isolated from cyprinid muscle tissue have been described in mice [13] and dogs [14,15]. Consequently, there is also a great potential for *P. ovatus* to be a concern for human health.

The recommendation for procedures can be found in the Guidelines for fish and fishery product [16]. The Codex Alimentarius provides the guidance for standards of good and safe food practice including different methods of conservation used in the fish processing technology [17].

To date, not much has been done to control the occurrence of metacercariae in fish products. Most processing parameters (temperature, pH, salinity) applied in fish processing have only undergone limited study with regard to their capability of eliminating the activity of trematodes [1]. Additionally, since *P. ovatus* is not on the FAO list of potential human pathogens, no measures are implemented in Europe to prevent the consumption of live larvae of this trematode species.

In addition, the *P. ovatus* complex life cycle makes it difficult to prevent this infection using biological methods (elimination of intermediate host of the aquatic environment). Instead, adequate fish processing could help protect human health.

4. CONCLUSION

Our study on the occurrence of *P. ovatus* in the muscle tissues of different fish species indicated that metacercariae most frequently settle in the anterior dorsal segment of the fish and are not visible to the naked eye in fish fillets. Physically removing encysted larvae is impossible because of their small size and possible high density of occurrence. What is essential is to study metacercariae viability in fish muscle tissues and their resistance to different food processing methods (salting, pickling, smoking and freezing). The results of this study indicate that *P. ovatus* metacercariae do not tolerate negative temperatures; refrigerated storage (4°C) alone does not guarantee the death of the parasite; meanwhile the temperature of -18°C gives expected effect. It is also essential to inform more fully communities associated with aquaculture, the food industry, food safety monitoring, public health, education, and the general public, especially consumers and fishing enthusiasts.

COMPETING INTERESTS

Authors have declared that no competing interests exist.

REFERENCES

1. Huss HH, Ababouch L, Gram L. Assessment and management of seafood safety and quality. FAO Fisheries Technical Paper. 2003;444.
2. Adams AM, Murrell KD, Cross JH. Parasites of fish and risks to public health. Revue scientifique et technique. International Office of Epizootics. 1997; 16(2):652-660.

3. Kim MK, Pyo KH, Hwang YS, Chun HS, Park KH, Ko SH, Chai JY, Shin EH. Effect of citric acid on the acidification of artificial pepsin solution for metacercariae isolation from fish. Veterinary Parasitology. 2013; 198:111-115.
4. Szewczuk A, Szczeklik A. Hydrolazy peptydylopeptydów. In: Szczeklik A, editor. Enzymologia kliniczna. 3rd ed. Warszawa: Państwowy Zakład Wydawnictw Lekarskich; 1974.
5. Scientific Opinion on risk assessment of parasites in fishery products. EFSA Journal. 2010;8(4):1543.
6. Sherrard-Smith E, Cable J, Chadwick EA. Distribution of Eurasian otter biliary parasites, *Pseudamphistomum truncatum* and *Metorchis albidus* (Family Opisthorchiidae), in England and Wales. Parasitology. 2009;136(9):1015-22.
7. Simpson VR, Gibbons LM, Khalil LF, Williams JL. Cholecystitis in otters (*Lutra lutra*) and mink (*Mustela vison*) caused by the fluke *Pseudamphistomum truncatum*. Veterinary Record. 2005;157(2):49-52.
8. Hawkins CJ, Caffery JM, Stuart P, Lawton C. Biliary parasite *Pseudamphistomum truncatum* (Opistorchiidae) in American mink (*Mustela vison*) and Eurasian otter (*Lutra lutra*) in Ireland. Parasitology Research. 2010;107(4):993-997.
9. Hildebrand J, Popiołek M, Zaleśny G, Piróg A. A record of *Pseudamphistomum truncatum (*Rudolphi, 1819) (Digenea, Opistorchiidae) in the Eurasian otter (*Lutra lutra* L.) from Poland. Wiadomości Parazytologiczne. 2011;57(3):151-154.
10. MacLean JD, Arthur JR, Ward BJ, Gyorkos TW, Curtis MA, Kokoskin E. Common-source outbreak of acute infection due to the North American liver fluke *Metorchis conjunctus*. Lancet. 1996;347(8995):154-158.
11. Khamidullin RI, Lyubina S, Khamidullin IR, Medinsky BL. Trematodoses in tataria. Meditsinkatya Parazitologiya i Parazitarnye Bolezni. 1991;2:60–61.
12. Berger S. Infectious diseases of the world. GIDEON Informatics, Inc. Los Angeles, California, USA; 2015.
13. Katsurada F. Biology and pathogenic significance of trematodes. Papers on Helminthology, Commemorating 30 Year Jubilee of Prof. K. I. Skriabin. 1937; 281-286.
14. Shaapan RM. Parasites of fishes and its effect on public health. M.V. Sc. Thesis, Fac. Vet. Med., Cairo Univ; 1997.
15. Mahdy OA, Sheaheed B. Studies on metacercarial infection among Tilapia species in Egypt. Helminthologia. 2001; 38(1):35-42.
16. Fish and Fishery Products Hazards and Controls Guidance. U.S. Department of health and human services, public health service, food and drug administration. Center for Food Safety and Applied Nutrition, Office of Food Safety; 2011. Fourth Edition. Available:http://www.fda.gov/downloads/Food/GuidanceRegulation/UCM251970.pdf
17. Proposed draft code of practice for fish and fishery products (Section of Smoked Fish). Codex alimentarius commission, joint FAO/WHO standards programme. Codex Commitee on Fish and Fishery Products. Thirty-first Session: CX/FFP 11/31/5; 2011. Available:ftp://ftp.fao.org/codex/Meetings/CCFFP/ccffp31/fp31_05_Add1e.pdf

8

Nutritional Risk Assessment of Eleven Minerals and Trace Elements: Prevalence of Inadequate and Excessive Intakes from the Second French Total Diet Study

Esther Kalonji[1*], Véronique Sirot[1], Laurent Noel[2], Thierry Guerin[2], Irène Margaritis[1] and Jean-Charles Leblanc[1]

[1]Risk Assessment Department, French Agency for Food, Environmental and Occupational Health & Safety (ANSES), 14 Rue Pierre et Marie Curie, F-94701 Maisons-Alfort, France.
[2]CIME Unit, Food Safety Laboratory, French Agency for Food, Environmental and Occupational Health & Safety (ANSES), 23, Avenue du Général de Gaulle, F-94706 Maisons-Alfort, France.

Authors' contributions

This work was carried out in collaboration between all authors. Author EK analyzed the data and wrote the manuscript. Author VS was in charge of the scientific coordination of the TDS, calculated the intakes and wrote the manuscript. Authors LN and TG were in charge of the analytical part of the TDS: they analyzed the TDS samples and provided composition data. Author IM read and amended the manuscript. Author JCL was the main coordinator of the TDS, he read and amended the manuscript. All authors read and approved the final manuscript.

ABSTRACT

Aims: Adequate coverage of nutrient requirements is a real health concern and surveillance of the nutritional status of a population is a key element for public policies. This study aimed at providing a reliable nutritional risk assessment of the French population based on prevalences of inadequate and excessive intakes of eleven minerals and trace elements.
Methodology: Intakes from foods (dietary supplements excluded) were estimated by combining composition data from the second national Total Diet Study (TDS2/2007-2009) and consumption data from the Individual and National Study on Food Consumption. Results were compared with those from other TDSs.
Results: Sodium intakes exceeded World Health Organization (WHO) guidance values, respectively for 74% of adults, and for 76% of children. For calcium and magnesium, the prevalence of inadequate intakes in adults and children ranged from approximately 50 to 70% to

**Corresponding author: Email: esther.kalonji@anses.fr*

over 80% in teenagers. Prevalences of inadequacy were 13% in adults and 18% in children for selenium, and 40% in children and 74% in 16-17 year-old girls for iron.
Conclusion: These substantial risks of inadequate intakes should be considered in the light of nutritional status biomarkers. Furthermore, effort to reduce excessive intakes of sodium in the French population should be maintained.

Keywords: Minerals; trace elements; total diet study; prevalence of inadequacy.

1. INTRODUCTION

Minerals and trace elements play numerous biological roles in humans, and today there is clear evidence that both deficiencies and excessive intakes can undermine physiological functions and lead to certain conditions including numerous metabolic disturbances (impairments of the hormonal and immune system, impaired wound healing and nervous system development, musculoskeletal disorders, growth retardation, appetite and digestive disorders, etc.), chronic diseases (osteomalacia and osteoporosis, diabetes, cardiovascular disease, cancers, etc.) and even an increase in mortality rates [1-3].

Consequently, fulfilling mineral and trace element requirements is a real concern from a public health viewpoint. Roman-Vinas et al. [4] recently showed, applying the same nutritional values to analyse national surveys, that the mean prevalence of inadequacy in adults is below 11% for zinc and iron in European Union countries. In addition, calcium, selenium, and iodine are among nutrients showing a higher prevalence of inadequate intakes in Europe (above 20%). Mean sodium intakes of European adult populations range from about 3-5 g/day [5], which is over the amounts required for normal function. In France, studies on nutritional status concern few minerals and trace elements [6,7] and their results seem not systematically corroborate inadequacy estimations [8].

Under these conditions, appropriate knowledge of the nutritional status of different population groups and the identification of critical nutrients are key elements for building regulatory and public health policies to insure adequate nutrient intakes. In addition, the general surveillance over time of the nutritional status of these populations helps evaluate the effectiveness of public measures and adjust nutritional recommend-dations (for example the setting of food-based dietary guidelines) [9].

Data from nutritional and clinical surveys are used to describe the nutritional status of a population or of a specific group of individuals. For a given nutrient, this assessment is based on the existence of a clinical deficiency or insufficiency as evidenced by reliable biomarkers [10-13]. In the absence of consistent data to characterize a population's nutritional status, the development of nutritional indicators is useful to identify situations of inadequate intakes, and is now considered as a reliable method to estimate the probability of physiological inadequacy, thanks to the quality of composition and consumption data [14]. In this respect, Total Diet Studies (TDS) have the advantage of providing a realistic estimation of nutrient intakes via foods by yielding more refined data representing an average dietary pattern for the population (baskets representing food consumption) and by considering foods in a "table-ready" state in order to take into account the impact of home cooking [15,16].

The first French TDS (TDS1) was undertaken between 2000 and 2004 by the French National Institute for Agricultural Research (INRA), in collaboration with the French Food Safety Agency (AFSSA) and gave a comprehensive appraisal of dietary exposure to or intakes of mycotoxins, as well as inorganic contaminants and minerals [17]. In 2006, AFSSA undertook a second TDS (TDS2) which included a larger number of target nutrients and substances, some of which had already been analyzed in the first study.

The aim of the present work was to provide a reliable nutritional risk assessment of the French population, including prevalences of inadequacy and excessive intakes, based on data from this second TDS. Eleven minerals and trace elements were considered: calcium (Ca), magnesium (Mg), sodium (Na), iron (Fe), potassium (K), selenium (Se), manganese (Mn), copper (Cu), zinc (Zn), lithium (Li), and molybdenum (Mo). Some of these nutrients required increased surveillance, for example Na for which TDS1 and other surveys (Individual and

National Food Consumption survey, INCA1 and INCA2) had revealed that intakes exceed the reference value, or for example Ca, Mg and Fe for which intakes lower than the nutritional reference values have been identified.

This work also presented a comparison between TDS1 and TDS2 data.

2. MATERIALS AND METHODS

2.1 Food Sampling

The methodology of food sampling has already been described [18]. To summarize, core foods (n=212) were selected as representative of the population's diet. The selection of these foods was based on two main criteria: (i) the foods most consumed by adults and/or children with a consumer rate of at least 5% according to the INCA2 survey, and (ii) foods which are the main known or assumed contributors to exposure, if they were not selected by the first criterion. The core foods covered about 90% of the whole diet.

Metropolitan France was divided into eight major regions and food samples (n=1319) were collected in the regions where they were most consumed, according to the INCA2 data. When available, each food item was collected during two different seasons to take account of possible differences in composition.

In order to be representative of French food consumption habits, each food sample was a composite sample of up to 15 subsamples of equal weight of the same food, taking into account the market share of the different brands, the origin, species, processing and packaging, if relevant, flavouring, etc [18]. Subsamples were prepared as consumed by the population (i.e. peeled, cooked, etc.) according to the cooking habits recorded in the consumption survey (including salt added during cooking), then pooled in composite samples that were analyzed.

2.2 Analysis of the Food Samples/ Composition Data

The analytical methods used and the occurrence data have been described in detail [19-22]. Analysis were performed by the French National Reference Laboratory for heavy metals in foodstuffs of animal origin (ANSES Laboratory for Food Safety) in compliance with good laboratory practice, internal quality procedures and the ISO/IEC 17025 standard [23].

To ensure and confirm the analytical accuracy of the multi-elemental method used, the laboratory regularly participated in proficiency test schemes as external quality controls (EQC). The EQC results have already been discussed in detail elsewhere [24].

2.3 Food Consumption Data

The consumption data used came from the second individual and national study on food consumption (INCA2). This survey was carried out from December 2005 to May 2007 to take into account consumption differences over year's periods. A total of 2624 adults aged 18-79 years and 1455 children aged 3-17 years were recruited as representative of the population in France through stratification [25-27]. To assess food consumption, subjects were asked to complete a 7-day food diary to describe their intakes: food name, home-made or industrial origin, brands, etc. Portion sizes for each food consumed were estimated through photographs compiled in a manual [28] or with household measurements (such as a spoon). Moreover, certain individual characteristics were also recorded, including body weight, age, gender, region and city, socio-professional status, etc.

Individual intakes from potential under- and over-reporters were excluded from the analyses. Under- and over-reporters were identified using the Goldberg cut-off value for energy intake [29]. Basal metabolic rate for each subject was calculated by the Schofield equations [30], using individual age, sex, height and weight [30]. A total of 1918 adults and 1444 children were considered in this work.

2.4 Nutritional Intake Assessment

The assessment of nutritional intakes (excluding food supplements) consisted in combining the national consumption data with the composition data from the analyses for each subject of the INCA2 survey according to the following formula:

$$I_{i,j} = \sum_{k=1}^{n} C_{i,k} \times L_{k,j}$$

where $I_{i,j}$ is the daily intake of nutrient j for subject i, $C_{i,k}$ is the mean daily consumption level of food k by subject i (k=1 to n), $L_{k,j}$ is the level of nutrient j in food k. Each food consumed by a subject was assigned a mean composition level calculated for the two samples from each sampling wave from its region. If the food was not sampled in the region, then it was assigned the average of the mean levels from all the regions sampled.

Mean intakes were calculated for different population subgroups (4-6 year old, 7-10, 11-14, 15-17, 18 and over, women of childbearing age (18-45 year old), subjects over 65) as well as 5th and 95th percentiles (P5, P95) for the whole adult (aged 18 to 79 years) and child (aged 4 to 17 years) populations.

Comparisons of the results with the first French TDS were performed using T-tests, using SAS version 9.2 (SAS Institute Inc., Cary, NC, USA).

2.5 Nutritional Situation Assessment: Comparisons with Reference Values

2.5.1 Prevalence of inadequate intake estimation

In this study, individual intakes were compared to French nutritional reference values, namely, estimated average requirement (EAR), population reference intake. French population reference intakes (ANC - *apports nutritionnels conseillés*) have been established for a given nutrient as the level of intake necessary to cover the requirements of virtually the entire healthy population (97.5%) [10]. By convention, the ANC is defined as the estimated average requirements (EAR)[1] plus almost two standard deviations of 15% each. Thus, nutrient intakes below the ANC are not necessarily a reflection of an inadequacy. Consequently, the EAR, which applies to individuals, has been used as the relevant criterion for the estimation of the prevalence of inadequate intakes in a population. The EAR cut-point method was used to approximate the proportion of the population with adequate intakes. The EAR values used were calculated using the following formula: EAR = 0.77 x ANC for all nutrients except magnesium, for which the EAR = 0.83 x ANC [14], and iron, for which a probabilistic method was used with a risk curve derived from the estimation of the requirement distribution [31]. This method allows to assess the probability for each subject to cover its needs using the risk curve and the estimated intake. Then the probability for the whole population to be at risk of inadequacy is calculated. When only a range for the population reference intake or EAR was found in the literature, the mean value of this range was used for comparison with intake. For zinc, two EAR values were defined depending on the composition of the diet, as diet is known to modify intestinal absorption of this trace element [32]. Our calculations considered the EAR corresponding to a 30% intestinal absorption rate, which is the value proposed in the case of a diet relatively rich in animal products.

Percentages of the population whose intakes were below the requirements (or who had a prevalence of inadequate intakes) were calculated with their confidence interval (CI_{95}).

2.5.2 Prevalence of excessive intake estimation

Individual intakes were compared to tolerable upper intake levels (UL) when they exist. The values defined at European level (EFSA) or if not available, in France, were given preference [10, 33]. If neither France nor EFSA had defined an UL for a nutrient, similar values proposed by agencies abroad were considered.

Concerning sodium, the intake limit recommended by the WHO (5 g of salt per day, or about 2000 mg of sodium for adults) [34] has been used, as data are insufficient to establish an UL for this mineral. For children, WHO recommends to extrapolate this value according to the energy requirements. Hence, for a 3 year old child whose energy needs are 1200 kcal/day, the intake limit recommended will be 1.2 g/day (3 g/day NaCl).

Percentages of the population whose intakes exceeded the reference value (UL or guidance) were calculated with their CI_{95}.

3. RESULTS

For each mineral or trace element studied, the results were displayed considering an estimation of nutrient intakes, allowing the identification of main food contributors (Tables 1 and 2), and estimation of prevalence of insufficient and excessive intakes (Table 3). Specific data concerning different age groups were also presented (Tables 4 and 5).

[1] *The estimated average requirements result from values acquired from an experimental group consisting of a limited number of individuals and corresponding to average individual requirements.*

Table 1. Mean estimated intake of minerals by the French adult population (18-79 y) from the different food groups

Food group	Ca (mg/day)	Mn (µg/day)	Mg (mg/day)	Cu (µg/day)	Zn (µg/day)	Li (µg/day)	Na (mg/day)	Mo (µg/day)	Se (µg/day)	K (mg/day)	Fe (µg/day)
Bread and dried bread products	29	623	34	125	609	0.74	802	15.1	2.9	229	1255
Breakfast cereals	3.8	13	1.1	3	17	0.02	4	0.4	0.1	8.2	48
Pasta	6.2	114	10	111	139	1.55	81	3.3	1.0	18	250
Rice and wheat products	2.5	60	4.1	35	81	1.04	38	2.8	0.6	12	75
Croissant-like pastries	3.6	36	2.9	11	62	0.06	52	1.0	0.2	20	156
Sweet and savoury biscuits and bars	2.9	35	4.1	15	49	0.08	32	1.3	0.2	29	122
Pastries and cakes	11	54	6.2	24	106	0.25	72	2.3	0.8	53	315
Milk	88	2	10	8	324	0.27	36	3.8	2.1	138	35
Ultra-fresh dairy products	92	15	11	9	318	0.40	38	4.0	2.0	145	47
Cheese	150	5	8.2	15	541	0.10	172	2.0	0.7	35	28
Eggs and egg products	7.5	6	2.4	11	197	0.16	32	1.1	0.8	29	265
Butter	2.0	0	0.3	2	8	0.14	21	0.5	0.4	3.6	6
Oils	0.1	0	0.02	0	2	0.00	0.07	0.1	0.2	0.1	2
Margarine	0.5	0	0.07	0	3	0.01	7.6	0.1	0.1	1.6	3
Meat	5.7	5	14	36	2018	0.18	34	0.8	1.2	179	786
Poultry and game	4.4	4	11	23	395	0.09	31	1.2	1.1	112	259
Offal	0.2	4	0.4	162	91	0.01	2	1.4	0.5	5.1	78
Delicatessen meats	3.7	11	7.1	37	612	0.18	297	1.9	0.9	106	417
Fish	2.4	6	4.6	9	55	0.13	35	0.3	2.0	42	65
Crustaceans and molluscs	4.1	18	2.7	20	194	0.22	17	1.0	0.8	8.7	81
Vegetables (excluding potatoes)	36	170	20	70	251	2.36	93	8.4	2.8	279	562
Potatoes and potato products	8.0	61	14	50	137	0.89	48	3.0	1.4	220	198
Pulses	1.8	26	2.3	14	56	0.31	7.7	6.8	0.1	14	80
Fruit	14	115	17	80	110	1.09	2.0	5.2	3.4	256	180
Dried fruits, nuts and seeds	1.4	39	3.3	21	62	0.03	7.3	1.9	0.1	20	41
Ice creams, sorbets and frozen desserts	6.2	15	2.9	13	34	0.09	5.0	0.7	0.2	22	338
Chocolate	6.2	45	5.7	32	60	0.04	3.7	0.8	0.1	29	352
Sugars and sugar derivatives	1.5	26	1.0	4	11	0.09	1.0	0.7	0.5	13	29
Water	102	4	18	72	56	16.72	16	2.3	17.3	4.6	36
Non-alcoholic beverages	8.1	44	6.9	16	20	0.72	4.5	1.3	2.9	84	60
Alcoholic beverages	15	109	14	14	121	1.09	2.8	4.1	3.6	103	296
Coffee	54	125	26	698	73	8.31	4.5	1.2	5.5	254	54

Other hot beverages	31	138	5.5	58	64	6.76	7.1	1.1	3.2	33	95
Pizzas, quiches and savoury pastries	21	36	4.9	15	180	0.24	104	1.8	0.4	41	146
Sandwiches and snacks	6.0	37	3.8	12	204	0.16	79	1.2	0.4	32	135
Soups and broths	11	45	6.5	37	75	2.49	156	1.9	1.8	103	119
Mixed dishes	15	63	9.3	43	481	0.79	145	4.6	0.9	89	363
Dairy-based desserts	18	23	5.5	19	89	0.12	13	1.2	0.5	43	259
Compotes and cooked fruit	6.7	14	0.9	7	6	0.04	0.2	0.6	0.4	16	17
Seasonings and sauces	2.9	11	2.6	6	23	0.18	152	0.4	0.3	24	60
Miscellaneous foods	0.05	0	0.07	0	1	0.00	0.1	0.0	0.0	0.1	2
Total	786	2158	304	1936	7935	48.2	2653	93.9	64.4	2854	7715

Table 2. Mean estimated intake of minerals by the French child population (3-17 y) from the different food groups

Food group	Ca (mg/day)	Mn (µg/day)	Mg (mg/day)	Cu (µg/day)	Zn (µg/day)	Li (µg/day)	Na (mg/day)	Mo (µg/day)	Se (µg/day)	K (mg/day)	Fe (µg/day)
Bread and dried bread products	17	290	16	59	286	0.36	388	7.2	1.4	109	593
Breakfast cereals	17	33	3.4	9	55	0.06	15	1.2	0.3	27	254
Pasta	6.9	127	11	125	154	1.72	91	3.7	1.1	20	276
Rice and wheat products	2.5	59	4.1	33	78	0.96	35	2.6	0.6	12	76
Croissant-like pastries	6.0	59	4.7	18	103	0.10	86	1.7	0.4	33	263
Sweet and savoury biscuits and bars	5.3	71	8.0	31	93	0.16	51	2.5	0.4	51	255
Pastries and cakes	13	61	7.4	29	139	0.26	84	2.6	0.8	59	387
Milk	174	5	20	14	645	0.56	71	7.6	4.1	274	70
Ultra-fresh dairy products	86	12	10	9	293	0.38	35	3.8	1.9	132	44
Cheese	85	3	4.6	8	309	0.07	97	1.2	0.4	20	17
Eggs and egg products	5.0	4	1.6	7	131	0.11	21	0.7	0.5	20	173
Butter	1.4	0	0.2	2	5	0.10	15	0.3	0.3	2.5	4
Oils	0.1	0	0.01	0	2	0.00	0.05	0.0	0.2	0.04	1
Margarine	0.2	0	0.032	0	2	0.01	4.1	0.1	0.1	0.8	2
Meat	4.0	4	11	28	1618	0.14	26	0.6	0.9	138	639
Poultry and game	3.1	3	7.4	14	260	0.05	20	0.8	0.7	77	153
Offal	0.1	1	0.2	53	31	0.00	0.6	0.5	0.1	2.0	34
Delicatessen meats	3.2	8	5.2	25	451	0.15	214	1.2	0.6	79	289
Fish	1.9	9	4.0	7	48	0.14	36	0.5	1.3	37	53
Crustaceans and molluscs	1.7	4	0.9	7	48	0.07	6.0	0.4	0.3	2.9	29
Vegetables (excluding potatoes)	21	102	12	40	159	1.38	59	5.7	1.6	155	334
Potatoes and potato products	8.3	58	13	45	132	0.85	53	2.8	1.3	203	187
Pulses	1.6	23	2.0	13	50	0.27	6.9	6.2	0.1	12	74

Fruit	6.5	69	9.2	39	58	0.50	1.1	2.3	1.6	133	92
Dried fruits, nuts and seeds	0.6	16	1.3	9	25	0.01	3.0	0.8	0.0	7.9	16
Ice creams, sorbets and frozen desserts	8.0	19	3.8	16	44	0.11	6.3	0.9	0.3	28	432
Chocolate	13	104	10	54	105	0.08	7.6	1.4	0.3	56	453
Sugars and sugar derivatives	1.0	11	0.8	2	7	0.04	1.5	0.5	0.2	6.3	24
Water	56	3	7.5	53	38	6.78	9.8	1.5	11.0	3.3	27
Non-alcoholic beverages	12	69	9.7	21	27	1.11	6.2	1.5	4.1	123	83
Alcoholic beverages	0.2	1	0.2	0	1	0.01	0.046	0.0	0.1	1.5	3
Coffee	1.5	3	0.6	17	2	0.20	0.1	0.0	0.1	6.0	1
Other hot beverages	16	24	4.3	19	78	0.48	7.2	0.8	0.6	37	128
Pizzas, quiches and savoury pastries	20	33	4.5	14	167	0.22	97	1.7	0.4	38	135
Sandwiches and snacks	5.1	31	3.2	10	178	0.13	66	1.0	0.3	27	116
Soups and broths	5.6	23	3.4	19	40	1.19	85	1.0	0.9	52	64
Mixed dishes	19	67	10	41	420	0.72	163	4.4	1.0	94	321
Dairy-based desserts	23	30	7.9	29	123	0.16	17	1.6	0.7	62	389
Compotes and cooked fruit	7.5	19	1.2	8	9	0.06	0.2	0.9	0.5	22	23
Seasonings and sauces	2.2	9	2.0	5	20	0.15	113	0.4	0.2	26	53
Miscellaenous foods	0.002	0	0.003	0	0	0.00	0.003	0.0	0.0	0.003	0
Total	659	1465	227	929	6431	19.84	1998	74.7	41.5	2186	6566

Table 3. Intakes of minerals by the French adult and child populations, and percentage of the populations not reaching the requirements or exceeding the limitation values

Nutrient	Nutrients Intakes						Reference values	% Inadequate intakes	
	Adults			Children				Adults	Children
	P5*	Mean	P95†	P5	Mean	P95		%<EAR or %>UL [CI 95%]	%<EAR or %>UL [CI 95%]
Calcium (mg/d)	345	786	1419	298	659	1118	RDA=700 to 1200 [10] UL = 2500 [33]	50 [48 ; 52] 0.05 [-0.05 ; 0.1]	68 [66 ; 70] 0
Manganese (mg/d)	1.07	2.16	3.55	0.71	1.46	2.56	no requirement value UL = 10 [10]	- 0	- 0
Magnesium (mg/d)	180	304	457	136	227	342	RDA = 6 (mg/kg bw/d) [10] UL = 700 [10]	65 [63 ; 67] 0.05 [-0.06 ; 0.2]	53 [51 ; 56] 0
Copper (mg/d)	0.73	1.94	4.11	0.5	0.93	1.73	RDA = 0.8 to 2 [10] UL = 5 [33]	33 [31 ; 35] 2.6 [1.9 ; 3.3]	73 [71 ; 76] 0.8 [0.4 ; 1.3]
Zinc (mg/d)	4.09	7.93	13.3	3.5	6.43	10.7	RDA = 3 to 19 [10] UL = 25 [33]	26 [24 ; 28] 0.08 [-0.05 ; 0.2]	48 [46 ; 51] 0.5 [0.1 ; 0.8]
Lithium (µg/d)	14.9	48.2	93.6	9	19.8	38.6	no requirement value no UL	- -	- -

Sodium (g/d)	1.3	2.65	4.5	1.08	2	3.41	no requirement value WHO guidance value = 2[34]	- 74 [72 ; 76]	- 76 [74 ; 76]
Molybdenum (µg/d)	49.1	93.9	155	40.3	74.7	130	no requirement value UL = 600 [33]	- 0	- 0.08 [-0.07 ; 0 2]
Selenium (µg/d)	34.6	64.4	100.7	23.8	41.5	66.2	RDA = 20 to 70 [10] UL = 90 to 300 [33]	13 [12 ; 15] 0	18 [17 ; 20] 0.08 [-0.07 ; 0 2]
Potassium (mg/d)	1635	2854	4333	1303	2186	3278	no requirement value no UL	- -	- -
Iron (mg/d)	4.03	7.71	12.6	3.4	6.57	11.1	EAR = 2.73 to 17.51 [31] UL = 28 [10]	36 [34;37] 0	40 [38 ; 42] 0

RDA, recommended daily allowance;
*EAR, estimated average requirement (for each individual, depending on their gender and age, intakes were compared to EAR calculated as 0.77*RDA except for magnesium: 0.83*RDA);*
UL, upper safety level.
**, fifth percentile;†,ninety-fifth percentile.*

Table 4. Mean total intakes of minerals in the different subgroups of the French population

	Ca, mg/day	Mn, mg/day	Mg, mg/day	Cu, mg/day	Zn, mg/day	Li, µg/day	Na, g/day	Mo, µg/day	Se, µg/day	K, g/day	Fe, mg/day
4-6 years old	633	1.13	192	0.70	5.24	16.25	1.52	63.1	35.3	1.91	5.18
7-10 years old	680	1.44	226	0.91	6.35	18.53	1.96	76.3	40.5	2.19	6.60
11-14 years old	676	1.62	244	1.03	7.01	21.65	2.21	78.8	44.5	2.32	7.19
15-17 years old	635	1.68	245	1.08	7.17	23.48	2.32	80.5	46.1	2.33	7.30
Women of childbearing age (18-45 years old)	719	1.78	257	1.62	6.57	41.99	2.17	77.6	58.0	2.41	6.53
Older subjects (>64 years old)	742	2.24	297	1.89	7.44	49.31	2.74	96.4	64.4	2.86	7.37

Table 5. Mean iron intakes in different subgroups of the French population and prevalence of inadequate intake

	n	Intake, mg/day	Prevalence of inadequate intake, %	CI 95%
Adults				
Men 18-64 y	649	9.12	13	[11;15]
Women 18-54 y	792	6.61	67	[65;69]
Men 65-75 y	102	8.65	17	[12;22]
Women 55-75 y	323	6.50	28	[25;31]
> 75 y	52	6.98	29	[20;37]
Children				
4 to 6 y	242	5.18	33	[31;35]
7 to 9 y	236	6.67	26	[23;29]
10 to 12 y	295	6.74	36	[32;40]
Boys 13-15 y	164	8.17	39	[32;45]
Girls 13-15 y	212	6.53	64	[58;69]
Boys 16-17 y	144	8.37	47	[40;55]
Girls 16-17 y	151	6.44	74	[70;79]

CI, confidence interval; y, years old.

3.1 Main Food Contributors to Intakes

Bread and bread products appeared to be one of the major contributors to the intake of iron (16% in adults, 9% in children), manganese (29% in adults, 20% in children), magnesium (11% in adults, 7% in children), molybdenum (16% in adults, 10% in children), sodium (30% in adults, 19% in children), and zinc (8% for adults only). Milk also appeared to be a major contributor to different nutritional intakes. For children, it contributed to 26% of calcium intake. For adults, milk brought 11% of the total intake of calcium after cheese (19%) and other dairy products (12%), and water (13%). For children, milk was also a major vector of potassium (13%), zinc (10%), selenium (10%), molybdenum (10%), and magnesium (9%). Among beverages, coffee was estimated to contribute significantly to intake of copper (36%), lithium (17%), potassium (9%), selenium (9%), and magnesium (9%) in adults, and water appeared to contribute significantly to intake of lithium (35% in adults, 34% in children) and selenium (27% in adults, 26% in children). For adults and children, meats, including beef, mutton and lamb, and pork, contributed to 10% of total iron intake and 25% of total zinc intake. For adults and children, delicatessen meats appeared to contribute to sodium (11%) and zinc (8%) intakes. Of note, vegetables (excluding potatoes) contributed to potassium (10%) and magnesium (7%) intakes in adults.

3.2 Prevalence of Excessive Intake Estimation

Generally increasing with age (Table 4), sodium intakes exceeded WHO recommended intake limit, respectively for 74% [72-76] of adults, and for 76% [74-79] of children. Intakes reached 2.65 g/d on average in adults (P95=4.5 g/d) and 2 g/d in children (P95=3.41 g/d). Moreover, 2.6% [1.9-3.3] of adults and 0.8% [0.4-1.3] of children had copper intakes exceeding the upper safe limit. In addition, 0.5% [0.1-0.8] of children had zinc intakes higher than the upper safe limit.

For the other nutrients, the prevalence of excessive intakes was equal to zero or statistically insignificant (Ca, Mn, Mg, Mo, Se, Fe).

3.3 Prevalence of Inadequate Intake Estimation

Mean calcium intakes were assessed at 786 mg/d in adults and 659 mg/d in children (P5 at 345 and 298 respectively). For magnesium, mean intakes were 304 mg/d in adults and 227 mg/d in children (P5 at 180 and 136 respectively).

Then for both nutrients, prevalences of inadequate intakes in adults and children ranged from about 50 to 70%, whereas prevalences of inadequate intakes of iron and selenium remained lower (13 to 40%). Indeed, intakes differed greatly between population subgroups (Table 4) and, due to different EARs, the highest prevalences affected specific subgroups of the population. For instance, 60% [47-72] of elderly people (>75 years) did not reach the EAR defined for the selenium intake (data not shown). About 86% [83-88] of teenagers aged 10-17 had a calcium intake lower than their EAR, and more than 80% of teenagers aged 12-17 had a magnesium intake lower than their EAR. For those three nutrients, the prevalence increased with age in children, and was higher for girls than boys.

Concerning iron, prevalence of inadequate intake ranged from 64 to 74% in females aged 13 to 54 (Table 5), with mean intakes estimated between 6.44 and 6.61 mg/d. Among other adult and child subgroups, the prevalence remained between 13 and 47%. Highest mean intakes were reached by 18-64 y old men (9.12 mg/d).

For copper and zinc, prevalences of inadequate intakes varied from 26% (adult zinc intakes) to 73% (child copper intakes).

For sodium, lithium, manganese, potassium and molybdenum no estimation of inadequate intakes was calculated because no requirement values have been set for these nutrients.

4. DISCUSSION

Results of this 2nd French TDS can be compared with the results of the 1st French TDS conducted in 2001 [17] as the methods used were the same for both consumption surveys, for the sampling method and the analysis performed. Compared with the results of the 1st French TDS, the intakes of lithium, copper and magnesium appeared to be significantly higher (until 2 times higher, P<0.0001 to 0.003). Concerning copper and lithium this may be due mainly to higher concentrations measured in foods. For magnesium, this trend was already observed with the INCA1 and INCA2 surveys [26,27]. In adults, calcium intake also significantly increased since TDS1 (P<0.0001), in accordance with the consumption level of dairy products [26], whereas in children calcium intake significantly decreased (P<0.0001) with dairy product consumption [27]. Manganese intakes were slightly but significantly higher in TDS2, for adults as well as children (P<0.0001), in spite of equivalent concentrations in foods. For some nutrients, differences in intakes in the two studies might be explained by differences in food group coverage within the diet. For instance, in TDS1, almost all kinds of pulses, which are a major vector food for molybdenum, were included in the sampling plan (and in the intake assessment) while in TDS2 only about 62% of the pulses consumed in France were covered [18]. Then intake of molybdenum could have been underestimated in the present study. In both studies, zinc and sodium intakes of children are similar. Iron intakes were not estimated in TDS1; in TDS2, the mean intakes were lower (around 40%) than those estimated by combining the INCA2 consumption data with national nutritional composition data (not published), certainly due to incomplete dietary coverage and most likely also due to lower levels in foods or to analytical uncertainty. For sodium, the mean intake levels were in the same range as those estimated during TDS1, given the analytical uncertainty related to sample preparation, and were slightly below those estimated by combining the INCA2 consumption data with national nutritional composition data (not published), due to incomplete coverage of the total diet. The mean intakes of zinc are around the same levels as those estimated during TDS1, but slightly lower than those estimated by INCA2 (20-25%), which is consistent with incomplete coverage of the diet.

Mean adult intakes of calcium, sodium, molybdenum, potassium and selenium were in the same range of those reported in other European and non-European TDSs (Table 6). For manganese and zinc, TDS results were heterogeneous, with French TDS2 intakes being the median. Magnesium intakes were about 30% higher in French TDS2 than in the American TDS [35] and the Italian TDS [36]. Copper intakes in French TDS2 were up to 100% higher than those reported for other TDSs. Iron intakes in French TDS2 were 30-40% lower than those reported for other TDSs. These differences may be explained by differences in food consumption levels and iron concentrations in the different food groups. For instance, the iron content of corn flakes is about 10-20 times higher in the USA than in Europe (France and Spain) [37]. Moreover, consumption of corn flakes is also higher (12-13 g/day for adults in the USA vs 4.9 in France) [26]. Moreover, the iron content of alcoholic beverages reported in the Spanish study [38], which contributed to 13.9% of the total intake, was 6.5 mg/kg, whereas in the French TDS2 it was equal to 1.1 mg/kg, with a 4% contribution to the total intake. Finally, while the intakes of meat, seafood products and eggs were similar between adults in the Italian and French populations, their contributions to the total iron intake were higher in Italy than in France (48% versus 23%). These differences might be due to higher iron content in animal products in Italy.

Table 6. Comparison of mean intakes of minerals and trace elements (mg/day) in adults according to total diet studies from different countries

	France Leblanc et al. (2005) [17]	Ireland (2011) [39]	UK Rose et al. (2010) [40]*	USA Egan et al. (2002)[35]	Australia FSANZ (2011) [41]	New Zealand NZFSA (2009) [42]	Italy Lombardi-Boccia et al. (2003) [36]	Italy Turconi et al. (2009) [43]	Japan Aung et al. (2006)[44]	Lebanon Nasreddine et al. (2010)[45]	Spain Urieta et al. (1996)[38]
Calcium	786		-	512-796	695-1170	-	738	-	-		-
Magnesium	304		-	184-262	-	-	262	-	-		-
Manganese	2.16		67* 4.69	[LB 1.88; UB 2.93]	3.94-5.38	-	-	1.38 [0.45-2.19]	2.72	2.04	-
Copper	1.94		17.23* 1.21	[LB 0.73 ; UB 1.36]	1.22-1.89	-	1.2	1.14 [0.73-1.96]	0.78	1.10	-
Zinc	7.93		140.7* 9.85	7.6-12.7	7.8-14.8	-	10.6	12.0 [7.5-19.8]	6.75	10.97	11.6
Lithium	0.48		-	-	-	-	-	-	-		-
Sodium	2653		-	1682-2739	-	2049-3405	3812	-	-		-
Molybdenum	0.09		1.61-1.64* 0.11	-	0.076-0.12	-	-	0.079 µg [0.032-0.106]	0.13		-
Potassium	2854		-	1853-2683	2960-4440		2913	-	-		-
Selenium	0.06	0.05-0.09	0.83-0.95* 0.06-0.07	[LB 0.074 ; UB 0.126]	0.097-0.17	0.056-0.082	0.104	-	-		-
Iron	7.71		-	9.0-.13.9	8.7-14.8	-	12.7	11.0 [6.41-19.45]	-	13.0	11.3

**, values specifically expressed in µg/kg bw/day, assuming a mean body weight of 70 kg*

In the absence of data related to the physiological and nutritional status of a population, an estimation of the prevalence of adequate intakes (probability of adequacy) only is possible thanks to statistical methods. Among the several available methods, the EAR cut-point method was chosen as the most relevant to estimate the inadequacy of most nutrient intakes [46-48]. Furthermore, this approach requires a decision to be made on acceptable or non-acceptable prevalence of inadequate intakes. A maximum desirable prevalence of inadequate intakes (or acceptable prevalence of inadequacy) was set by WHO and the Food and Agriculture Organization (FAO) at 2-3% of a population; in other words, the population is in a satisfactory nutritional situation when most individuals in the population, i.e. 97-98%, fulfil the EAR values [49]. In line with these recommendations, Food Standards Australia New Zealand (FSANZ) considers that when 3% or less of a population has nutrient intakes below the EAR, this population group as a whole can be considered to have adequate intakes of this nutrient [50]. In addition, prevalence of inadequate intake values were put into perspective with regard to the type of data used to define the EARs (endpoint criteria, uncertainties, etc.). Touvier et al. chose a high threshold for prevalence of inadequacy to qualify a population group as "at risk" [14,51]. Indeed, the population groups for which the CI_{95} of the prevalence of inadequacy for a nutrient included a value equal to or above 70%, and did not include the 50% value, were considered as "particularly at risk" for this nutrient. Hence, the same value of prevalence of inadequacy may be considered as a reflection of a risk of inadequacy or not according to the interpretation of the criteria used. Thus, the heterogeneity of the methods used to estimate the prevalence of inadequacy and of the interpretation criteria shall be kept in mind when comparing conclusions in terms of public health objectives drawn on prevalence of inadequacy data from different studies and different countries [13]. Consequently, there is a need to harmonise methodologies worldwide, or at least at European level as recently stressed by some authors [4,48], even if a scientific consensus on the judgment criteria to be used to interpret these prevalences of adequate intake is difficult to reach, considering the nutritional specificities of the populations and population subgroups concerned and the public health goals pursued.

The results of TDS2 regarding dietary intakes of trace elements showed that some prevalences of inadequate intakes may appear to be relatively high, notably for calcium, iron, magnesium and copper. This can be partly due to food sampling in this study only covering about 90% of the French population's diet. Moreover, due to the short observation period of 7 days, the total variance has been over-estimated [52]. We can hypothesize that the dietary intakes at the low percentiles may be slightly under-estimated and the prevalences of inadequacy slightly overestimated.

Moreover, it should be noted that conclusions cannot be drawn from prevalences of inadequate intakes, which are only probabilities of inadequacy, as to the risk or to the lack of risk for the general population. They varied greatly between subgroups and according to gender (Table 4). For example, for calcium, only children over 9, adolescents and women over 55 were particularly at risk of inadequacy (data not shown). Several data showed that teenagers consumed less dairy products than younger children [27,53]. Moreover, women in the perimenopause period may not adapt their diet to their increased needs in calcium. However, as useful indicators, these prevalences of inadequate intakes highlight the need to acquire biological data based on appropriate biomarkers in order to fully assess a population or population subgroup's nutritional status, in particular for magnesium and copper.

The estimated prevalence of inadequate intakes for iron ranged from 13% to 74%, depending on the age group and gender. Prevalences were notably higher among young girls (64% [58,69] for 13-15 year-olds and 74% [70,79] for 16-17 year-olds) and pre-menopausal women (67% [65,69]). These high prevalences should be compared with the much lower incidences of iron deficiency-induced anaemia and iron reserve depletion in the general population (around 2 to 11% for adults) [9]. These data suggest that it is necessary to identify causes of anaemia (intake-related and/or physiological). Indeed, the prevalences of inadequate intakes were calculated mainly on the basis of population reference intakes published in 2001. These values were established on the basis of the relevant literature available at the time, which sometimes relied on observations in the population. For some nutrients, such as magnesium, which do not seem to pose any particular public health risk, recent and relevant data suggest that the requirements should be reassessed. For instance, Hunt and Johnson suggest lower needs, of about 180 mg/day (versus an EAR of 6 mg/kg bw/day, i.e. 360 mg/day assuming a body weight of 60 kg), estimated on the basis of balance studies [54]. However, the relationship between magnesium intakes, status biomarkers and pathologies (cardiovascular disease, type 2 diabetes and osteoporosis) is not fully understood and a diagnosis of moderate deficiency is hard to establish [55]. Consequently, the high prevalences of inadequacy in magnesium in TDS2, concordant with the literature [14,51,55], must be considered in the light of factors likely to influence needs (genetics, whole diet, specific nutrients, physical activity, etc.).

However, the nutritional intakes from food supplement consumption were not included in the TDS2 analysis. While food supplement use is liable to increase intakes for several minerals, food supplement consumers are not necessarily those with the lowest intakes from food sources [56-58]. The cumulated intakes from the whole diet, including fortified foods and supplements, may lead to exceeding the upper safety limits, as recently shown by several authors [57,58], especially for minerals and trace elements for which high intakes are already close to the upper safe limit (copper, zinc, selenium…).

Some excessive intakes have already been detected in the general population without taking into account supplements (Table 4). In the few cases where observed tolerable upper intake levels were exceeded, the excess levels were extremely low and sometimes insignificant. Concerning sodium, the high excess levels of the WHO guideline value underline the need to continue efforts to reduce intake. Such objectives can be achieved by general modifications in consumption habits (food choices, added salt, etc.) and the gradual reduction of salt in manufactured products, as recommended by public health programs, namely the PNNS in France [59].

5. CONCLUSION

The data compiled from this second TDS showed that substantial prevalences of inadequacy have been identified in some sub-groups, such as children and the elderly. Our conclusions are based on data using a reliable and robust estimation approach, insofar as they confirm the conclusions of previous studies. However, there is a considerable need to put these estimated intakes into perspective with regard to biological data. The TDS results might be useful to encourage examination of certain nutrients in more detail with a view to policy development. In TDS2, only the general population over 3 years of age were included. Undeniably, studies focusing on physiological sub-groups (infants, toddlers, pregnant and lactating women, etc.) as well as on sub-groups with various specificities (geographic location, low income, dietary habits, etc.) would be needed in the medium term in terms of risk evaluation in a public health objective.

ACKNOWLEDGEMENTS

This study was supported by public funds from the French Ministries in charge of Food, the Environment, Health and Consumer Affairs.

The authors would like to thank the experts from the ANSES Expert Committee on Human nutrition for their comments during the risk assessment process, namely J.M. Bard, J. Belegaud, O. Bruyère, J.M. Chardigny, S. Charrières, V. Coxam, P. Crenn, J. Delarue, F. Feillet, J.P. Girardet, A. Lapillonne, P. Legrand, F. Mariotti, A. Martin, C. Moulis, A. Quignard-Boulangé, D. Rieu, P. Sauvant, S. Schneider, A. Servin, D. Tomé, D. Turck (Chair) & M.P. Vasson. We are particularly grateful to F. Mariotti & A. Martin for their valuable comments and suggestions on aspects of method and data analysis.

COMPETITING INTERESTS

Authors have declared that no competing interests exist.

REFERENCES

1. WHO, Trace elements in human nutrition and health (A report of a re-evaluation of the role of trace elements in human health and nutrition), 1996; Geneva, Switzerland.
2. Institute of Medecine, Dietary reference intakes: The essential guide to nutrient requirements; 2006. Available:http://books.nap.edu/openbook.php?record_id=11537
3. Martin A. The "apports nutritionnels conseilles (ANC)" for the French population. Reprod Nutr Dev. 2001;41(2): 119-28.
4. Vinas BR, Barba LR, Ngo J, Gurinovic M, Novakovic R, Cavelaars A, et al., Projected

prevalence of inadequate nutrient intakes in Europe. Ann Nutr Metab. 2011;59(2-4): 84-95.
5. EFSA, Opinion of the Scientific Panel on Dietetic Products, Nutrition and Allergies on a request from the Commission related to the Tolerable Upper Intake Level of Sodium. The EFSA Journal. 2005;209:1-26; 2005. EFSA: Parma, Italy.
6. Castetbon K, Vernay M, Malon A, Salanave B, Deschamps V, Roudier C, et al., Dietary intake, physical activity and nutritional status in adults: The French nutrition and health survey (ENNS, 2006-2007). Br J Nutr. 2009;102(5):733-43.
7. Castetbon K, Mejean C, Deschamps V, Bellin-Lestienne C, Oleko A, Darmon N, et al., Dietary behaviour and nutritional status in underprivileged people using food aid (ABENA study, 2004-2005). J Hum Nutr Diet. 2011;24(6):560-71.
8. ANSES. Avis de l'Agence nationale de sécurité sanitaire de l'alimentation, de l'environnement et du travail relatif à l'évaluation des apports en vitamines et minéraux issus de l'alimentation non enrichie, de l'alimentation enrichie et des compléments alimentaires dans la population française : Estimation des apports usuels, des prévalences d'inadéquation et des risques de dépassement des limites de sécurité. In press; 2015.
9. INVS. Étude nationale nutrition santé ENNS, 2006 - Situation nutritionnelle en France en 2006 selon les indicateurs d'objectif et les repères du. Programme National Nutrition Santé (PNNS), INVS: Saint-Maurice. France. 2007;77.
10. Martin A, Azaïs-Braesco V, Bresson JL, Couet C, Cynober L, Guéguen L, et al., Apports nutritionnels conseillés pour la population française, ed. Tec & Doc; 2001, Paris: Lavoisier.
11. Arnaud MJ, Update on the assessment of magnesium status. Br J Nutr. 2008;99(3): 24-36.
12. Zimmermann MB, Methods to assess iron and iodine status. Br J Nutr. 2008;99(3):2-9.
13. Mensink GBM, Fletcher R, Gurinovic M, Huybrechts I, Lafay L, Serra-Majem L, et al., Mapping low intake of macronutrients across Europe. Br J Nutr. 2013;110:755-773.
14. Touvier M, Lioret S, Vanrullen I, Bocle JC, Boutron-Ruault MC, Berta JL, et al., Vitamin and mineral inadequacy in the French population: estimation and application for the optimization of food fortification. Int J Vitam Nutr Res. 2006;76(6):343-51.
15. EFSA. Joint Guidance of EFSA, FAO and WHO - Towards a harmonised Total Diet Study approach: A guidance document, 2011;1-66.
16. Moy G, Vannort RW. Total Diet Studies. Springer; 2013.
17. Leblanc JC, Guerin T, Noel L, Calamassi-Tran G, Volatier JL, Verger P. Dietary exposure estimates of 18 elements from the 1st French Total Diet Study. Food Addit Contam. 2005;22(7):624-41.
18. Sirot V, Volatier JL, Calamassi-Tran G, Dubuisson C, Menard C, Dufour A, et al., Core food of the French food supply: second Total Diet Study. Food Addit Contam Part A Chem Anal Control Expo Risk Assess. 2009;26(5):623-39.
19. Chekri R, Noel L, Millour S, Vastel C, Kadar A, Sirot V, et al., Calcium, magnesium, sodium and potassium levels in foodstuffs from the second French Total Diet Study. J Food Comp Anal. 2012;25: 97-107.
20. Millour S, Noel L, Chekri R, Vastel C, Kadar A, Sirot V, et al., Strontium, silver, tin, iron, tellurium, gallium, germanium, barium and vanadium levels in foodstuffs from the second french total diet study. J Food Comp Anal. 2012;25:108-129.
21. Noel L, Chekri R, Millour S, Vastel C, kadar A, Sirot V, et al., Li, Cr, Mn, Co, Ni, Cu, Zn, Se and Mo levels in foodstuffs from the Second French TDS. Food Chem. 2012;132:1502-1513.
22. Millour S, Noël L, Kadar A, Chekri R, Vastel C, Sirot V, et al., Pb, Hg, Cd, As, Sb and Al levels in foodstuffs from the 2nd French total diet study. Food Chem. 2011;126:1787–1799.
23. ISO/IEC. General requirements for the competence of calibration and testing laboratories. ISO/IEC 17025. Geneva, Switzerland: International Organization for Standardisation; 2005.
24. Millour S, Noël L, Chekri R, Vastel C, Kadar A, Guérin T. Internal quality controls applied in inductively coupled plasma mass spectrometry multi-elemental analysis for the 2nd french total diet study. Accredit Qual Assur. 2010;15:503-513.

25. Dufour A, Lafay L, Volatier JL. La mesure des consommations alimentaires par l'éude INCA2 (The measurement of food consumption by the INCA2 study), in méthodes de sondage (Sampling Methods), D.H. P Guilbert, A Ruiz-Gazen and Y Tille, Editor. Dunod: Paris, France. 2008;132-137.
26. Dubuisson C, Lioret S, Touvier M. Dufour A, Calamassi-Tran G, Volatier JL, et al., Trends in food and nutritional intakes of French adults from 1999 to 2007: results from the INCA surveys. Br J Nutr. 2010; 103(7):1035-48.
27. Lioret S, Dubuisson C, Dufour A, Touvier M, Calamassi-Tran G, Maire B, et al., Trends in food intake in French children from 1999 to 2007: results from the INCA (étude Individuelle Nationale des Consommations Alimentaires) dietary surveys. Br J Nutr. 2010;103(4):585-601.
28. Hercberg S, Deheeger M, Preziosi P. Portions Alimentaires : Manuel Photos pour l'Estimation des Quantites (Food Portions: Photo Manual for Quantity Estimation); 1994. Paris: PolyTechnica.
29. Goldberg GR, Black AE, Jebb SA, Cole TJ, Murgatroyd PR, Coward WA, et al., Critical evaluation of energy intake data using fundamental principles of energy physiology. 1. Derivation of cut-off values to identify under-recording. Eur J Clin Nutr. 1991;45(569-581).
30. Schofield WN. Predicting basal metabolic rate, new standards and review of previous work. Hum Nutr Clin Nutr. 1985;39:5-41.
31. Institute of Medicine, Iron, in Dietary Reference Intake for vitamin A, vitamin K, arsenic, boron, chromium, copper, iodine, iron, manganese, molybdenum, nickel, silicon, vanadium, and zinc, N. A. Press, Editor. National Academy Press: Washington DC, USA. 2001;290-393.
32. Arnaud J. Zinc, in apports nutritionnels conseillés pour la population française, 3e édition, Tec & Doc, Editor. Lavoisier: Paris, France. 2001;155-158.
33. SCF. Tolerable upper intake levels for vitamins and minerals; 2006.
34. WHO, Reducing salt intake in populations. Report of a WHO forum and technical meeting, Paris, France; 2007. Paris, France.
35. Egan SK, Tao SS, Pennington JA, Bolger PM. US Food and Drug Administration's Total Diet Study: Intake of nutritional and toxic elements, 1991-96. Food Addit Contam. 2002;19(2):103-25.
36. Lombardi-Boccia G, Aguzzi A, Cappelloni M, Di Lullo G, Lucarini M. Total-diet study: dietary intakes of macro elements and trace elements in Italy. Br J Nutr. 2003;90 (6):1117-21.
37. USFDA. National Nutrient Database for Standard Reference. Available:http://ndb.nal.usda.gov/ndb/foods/list. 2012.
38. Urieta I, Jalon M, Eguilero I. Food surveillance in the Basque Country (Spain). II. Estimation of the dietary intake of organochlorine pesticides, heavy metals, arsenic, aflatoxin M1, iron and zinc through the total diet study, 1990/91. Food Addit Contam, 1996;13(1):29-52.
39. FSA. Report on a Total Diet Study carried out by the food safety authority of Ireland in the period. 2001-2005;2011.
40. Rose M, Baxter M, Brereton N, Baskaran C. Dietary exposure to metals and other elements in the 2006 UK Total Diet Study and some trends over the last 30 years. Food Addit Contam Part A Chem Anal Control Expo Risk Assess. 2010;27(10): 1380-404.
41. FSANZ. The 23rd Australian Total Diet study; 2011. Available:http://www.foodstandards.gov.au/_srcfiles/FSANZ%2023rd%20ATDS_v8_.pdf,
42. NZFSA. 2009; New Zealand Total Diet Study; 2009. Wellington, New Zealand. Available:http://www.nzfsa.govt.nz/science/research-projects/total-diet-survey/2009.htm
43. Turconi G, Minoia C, Ronchi A, Roggi C. Dietary exposure estimates of twenty-one trace elements from a Total Diet Study carried out in Pavia, Northern Italy. Br J Nutr. 2009;101(8):1200-8.
44. Aung NN, Yoshinaga J, Takahashi JI. Dietary intake of toxic and essential trace elements by the children and parents living in Tokyo Metropolitan Area, Japan. Food Addit Contam. 2006;23(9):883-94.
45. Nasreddine L, Nashalian O, Naja F, Itani L, Parent-Massin D, Nabhani-Zeidan M, et al., Dietary exposure to essential and toxic trace elements from a total diet study in an adult lebanese urban population. Food Chem Toxicol. 2010;48(5):1262-9.
46. Medicine Io. Dietary Reference Intakes : Applications in dietary assessment.

Washington, DC: National Academy Press; 2000.
47. de Lauzon B, Volatier JL, Martin A. A Monte Carlo simulation to validate the EAR cut-point method for assessing the prevalence of nutrient inadequacy at the population level. Public Health Nutr. 2004; 7(7):893-900.
48. Vinas BR, Majem LS, Barba LR, Ngo J, Alvarez AG, Wijnhoven TMA, et al., Overview of methods used to evaluate the adequacy of nutrient intakes for individuals and populations. Brit J Nut. 2009; 101(2): S6-S11.
49. WHO/FAO, Guidelines on food fortification with micronutrients; 2007: WHO.
50. FSANZ. 22nd Australian total diet study. A total diet study of five trace elements: iodine, selenium, chromium, molybdenum, and nickel; 2008. FSZANZ. Available:http://www.foodstandards.gov.au/scienceandeducation/publications/22ndaustraliantotaldietstudy/,
51. Moshfegh A, Goldman J, Ahuja J, TR hodes D, La Comb R. What we eat in America, NHANES 2005-2006: Usual nutrient intakes from food and water compared to 1997 dietary reference intakes for vitamin D, calcium, phosphorus, and magnesium. U.S. department of Agriculture, Agricultural Research Service; 2009.
52. Goedhart PW, van der Voet H, Knüppel S, Dekkers ALM, Dodd KW, Boeing H, et al., A comparison by simulation of different methods to estimate the usual intake distribution for episodically consumed foods. Supporting Publications. 2012;299:65. 2012. Available:www.efsa.europa.eu/publications
53. Baird DL, Syrette J, Hendrie GA, Riley MD, Bowen J, Noakes M. Dairy food intake of Australian children and adolescents 2-16 years of age: 2007 Australian National Children's Nutrition and Physical Activity Survey. Public Health Nutr. 2012;1-14.
54. Hunt CD, Johnson LK. Magnesium requirements: new estimations for men and women by cross-sectional statistical analyses of metabolic magnesium balance data. Am J Clin Nutr. 2006;84(4):843-52.
55. Rosanoff A, Weaver CM, Rude RK. Suboptimal magnesium status in the United States: are the health consequences underestimated? Nutr Rev. 2012;70(3):153-64.
56. Robson PJ, Siou GL, Ullman R, Bryant HE. Sociodemographic, health and lifestyle characteristics reported by discrete groups of adult dietary supplement users in Alberta, Canada: findings from the tomorrow project. Public Health Nutr. 2008; 11(12):1238-47.
57. Bailey RL, Fulgoni VL, 3rd Keast DR, Dwyer JT. Dietary supplement use is associated with higher intakes of minerals from food sources. Am J Clin Nutr. 2011; 94(5):1376-81.
58. Shakur YA, Tarasuk V, Corey P, O' Connor DL. A comparison of micronutrient inadequacy and risk of high micronutrient intakes among vitamin and mineral supplement users and nonusers in Canada. J Nutr. 2012;142(3):534-40.
59. Ministère du travail de l'emploi et de la santé. Programme National Nutrition Santé 2011-2015. Available:http://www.sante.gouv.fr/IMG/pdf/PNNS_2011-2015.pdf. 2011.

9

Chemical Sensitivity in the Elderly: Lessons Learned from Micronutrient Consumption in the Dutch Elderly Population

Lya G. Soeteman-Hernández[1*], Eugène H. Jansen[1], Hans Verhagen[1,2], Elly J. M. Buurma-Rethans[1] and Jan van Benthem[1]

[1]National Institute for Public Health and the Environment (RIVM), P.O.Box 1, 3720 BA, Bilthoven, The Netherlands.
[2]University of Ulster, Northern Ireland Centre for Food and Health (NICHE), Cromore Road, Coleraine, BT52 1SA, Northern Ireland.

Author's contribution

This work was carried out in collaboration between all authors. Author JVB headed the project. Authors HV, EHJ and EJMBR conducted the analysis and interpretation and author LGSH drafted the manuscript. All authors were involved in the early work and assisted with the interpretation of the results. Authors HV, EHJ and EJMBR were in charge of data management. All authors were involved in the critical revision of the manuscript for important intellectual content. All authors read and approved the final manuscript.

ABSTRACT

A food consumption survey in the Dutch elderly population (51-69 years of age) showed an increased trend in micronutrient supplement intake (36.4%; 120/347 participants). Because data on chemical sensitivity in the elderly is lacking, evaluation as to whether the current uncertainty factor (UF) of 10 is sufficient to protect the elderly was investigated using the micronutrient consumption data in the elderly Dutch population as a case study. Theories of ageing, and differences in toxicokinetic and toxicodynamics are briefly discussed in the context of chemical sensitivity in the elderly. Evidence suggests that for the healthy elderly, no additional default UFs are recommended because the present UF of 10 is probably sufficient. However, more research is needed to ensure that there is no additional risk, particularly in the not-so healthy elderly population. Although there is a trend of increased consumption of micronutrient supplements (i.e. vitamins and minerals) by the Dutch population, the existing European legislation for micronutrients in fortified foods

**Corresponding author: Email: Lya.Hernandez@rivm.nl*

(Regulation 1925/2006) and food supplements (Directive 2002/46) is now being translated to simultaneously set maximum levels of micronutrients in foods and in supplements. For the healthy elderly, no foreseeable risk is expected due to the consumption of micronutrients. For the unhealthy elderly, the effects of micronutrient consumption are not yet known and therefore, dietary supplement intakes need to be continuously monitored with detailed questioning on health status, supplement and prescription drug use. In addition, the generation of an international and up-to-date database on the composition of available dietary supplements is needed to fill the current data gaps.

Keywords: Elderly; ageing; chemical toxicity; risk assessment; micronutrients.

1. INTRODUCTION

Traditionally, human risk assessment consists of selecting the no observed adverse effect level (NOAEL) or the benchmark dose (BMD) from a pivotal study with a critical effect and adding uncertainty factors (UFs) to account for inter-species kinetics (UF = 4), inter-species dynamics (UF = 2.5), variability in human kinetics (UF = 3.16) and variability in human dynamics (UF = 3.16); all combined to the traditional uncertainty factor of 100 used [1]. Given that in this modern and western society new chemical compounds and pharmaceuticals are constantly being developed, their potential hazardous effect(s) to the human population needs to be continuously investigated. Of particular importance is whether the intra-species UF derivation used to extrapolate from the average human to susceptible populations is sufficient to protect sensitive populations such as the elderly. In this review focus was given on the elderly population and their micronutrient consumption in The Netherlands.

Recent studies aimed at measuring human variability using a database of 60 therapeutic drugs to represent a variety of metabolic and elimination pathways to derive pathway-related UFs for each metabolic route and subgroup (i.e. neonates, elderly and healthy adults) suggests that in the elderly population, drug metabolism is significantly impaired, in comparison to healthy adults [2]. Another therapeutic drug pharmacokinetic database consisting of 46 drugs encompassing a variety of clearance pathways (phase I metabolism, phase II conjugation and renal excretion) and more than 4500 subjects for the evaluation of *in vivo* drug clearance, showed that individuals over the age of 60 had a greater drug half-life (up to 60% increase) and slower clearance (up to 50% reduction) [3]. This decrease in elimination of toxicants is possibly due to factors such as the reduction in liver mass, reduction in the activity of hepatic metabolizing enzymes, decreased liver blood flow and modifications in plasma drug binding; all associated with natural ageing process [4,5].

The elderly are heterogeneous with respect to the ageing processes and, in most instances, the deficit in the elderly relates more to the lifestyle and diseases than to the ageing processes itself. The inability to distinguish between intrinsic ageing, natural disease and toxic responses makes the question whether the elderly should be considered a sensitive group for risk assessment very challenging.

Even though there are clear indications that the elderly respond differently to chemical exposure compared to the average population due to physiological differences, the point to address for the Dutch government is whether a new policy should be set to protect the elderly. The aim of this paper is therefore to find a scientific basis for whether the elderly (> 55 years of age) should be considered a sensitive group and whether the traditional UF of 100 is sufficient to protect the elderly population against adverse effects.

In the Netherlands, the healthy life expectancy for men is 70 years of age and the healthy life expectancy for women is 73 years [6], in comparison to the life expectancy of 78 and 82 years of age, respectively. The period of 65 years of age up to 75 seems to be a critical period, given that surveys indicate that during this period men and women in the Netherlands have the highest reports that they don't feel well (e.g. don't feel healthy) [7]. In general, approximately 75, 70 and 65% of 45-64, 65-74 and 75^+ year-old men, respectively felt healthy; and approximately 70, 62 and 50% of 45-64, 65-74 and 75^+ year-old women, respectively felt healthy [7]. In 2005, in the Netherlands, the healthy life expectancy in men and women aged 65 has been estimated to be 10.4 and 10.9 years, respectively [8]. Therefore, in general when dealing with the healthy elderly population, we are dealing with approximately 70-75% of 46-64 year-old individuals, with approximately 62-70% of 65-74 year-old individuals, and with approximately 50-65% of 75^{+} year-old individuals [7].

A recent food consumption survey in 2007/2008 of the Dutch elderly population (51-69 years of age) showed that micronutrient supplements were consumed on a regular basis by 36.4% (120/347) of the participants (25.8% of the men and 42.9% of the women). The supplements were composed of a single component (18%) or a multi-component supplement (42%). A market inventory in the Netherlands on 120 nutrient supplements for elderly revealed that the mean composition (P50) for vitE was 1.5-fold higher than the recommended daily allowance (RDA), the 95th percentile intake (P95) value was up to 13.5-fold higher than the RDA and a maximum value was 26.8 times the RDA; although none exceeded the tolerable upper intake level of 300 mg/day but were still on the high end [9]. For vitC, the P95 value was found to be 4.38 times the RDA and the maximum value was 14 times the RDA [10]. These data illustrate that consumption of supplements by the elderly population may contribute to high nutrient densities. On the other hand, surveys dealing with micronutrient intakes in Europe indicate that the elderly are usually under-represented. Nevertheless, a survey from the UK of institutionalized and non-institutionalized individuals over 64 years of age, showed that in the majority of instances, the elderly intakes were below the lower reference nutrient intake (LRNI) [11]. Although the elderly-specific nutrient data is limited, the available data indicate that the elderly generally suffer from micronutrient deficiencies and that there is a trend of increased consumption of micronutrients, which, in some instances, may contribute to high nutrient densities, as in the case of vitE and vitC in the Netherlands. For this reason, the micronutrient consumption in the elderly Dutch population was taken as a case study to inform about chemical sensitivity in the elderly.

2. THEORIES OF AGEING

At present, in spite of intensive interest and significant progress in aging research, there is not yet a universal agreement on one specific theory of aging. On the contrary, the number of theories is large, some more favored by investigators than others, although in the final analysis, many if not all of these theories may work together. Here we highlight specific changes that may affect chemical sensitivity and possibly micronutrient uptake in the elderly.

Aging is a continuous and complex multi-factorial process that involves the interaction of genetic, epigenetic and environmental factors. It is widely accepted that physiological aging is the result of two opposing processes: (i) processes that destabilize the organism and increase the probability of diseases and death, and (ii) longevity assurance mechanisms that prevent, repair, or contain damage. Processes of the first group are often chemical and physico-chemical in nature or related to endogenous metabolic processes, and are either inevitable or only under marginal biological control. In contrast, protective mechanisms are partially genetically determined and are subject age-related changes [12,13].

An important genetic event that occurs during ageing is the accumulation of gene mutations resulting in differences in gene expression and possible alterations in the abundance and post-translational modifications of proteins [14]. Changes in the regulation of gene expression may also occur through modifications in the organization of DNA and protein with increasing age. Studies have shown increases in histone $H1^{0}$ with concomitant decreases in histone $H1^{I}$ and $H1^{II}$ in the heart, brain and liver in different species of mice as a function of age [15]. An assessment of histone acetylase activity in rats showed a 30 to 70% decrease in acetylation of histones with increasing age. This alteration of histone-DNA interaction makes DNA more prone to toxicant-induced damage. Further, [^{3}H]-acetate incorporation showed an age-dependent trend where histone H4 was pivotal in rats that were 2 months of age, while histone H3 is more pivotal in older rats (12, 16 and 24 months of age). Rats that were 27 months old showed equal levels of histone H3 and H4 [16]. Hence, there is an indication of age-dependent variation in chromatin maintenance and modulation that may affect the sensitivity to toxicants in the elderly.

Repair systems in the elderly may be significantly reduced, leading to an accelerated induction of mutations [17]. Studies have demonstrated that DNA mismatch repair (MMR) is significantly reduced in senescent cells through a significant down-regulation of human MutS homologue 2 (hMSH2) and MSH6 proteins; both important proteins that form heterodimers (MutSα-complex) and are involved in the initiation step in the MMR pathway [18]. This senescence-dependent increase in genomic instability may account in part for the accumulation of genetic damage with increasing age. Other repair pathways have also been shown to be reduced with increasing age including base excision repair, nucleotide excision repair and double-strand break repair [19]. Repair enzyme levels responsible for the repair of alkylating mutagens such as 3-methyladenine-DNA *N*-glycosylase are significantly lower in older mice (15 to 17 months old), in comparison to young adult mice (7 to 8-week old) while the levels of O^6-methylguanine-DNA methyl transferase remain unaltered with increasing age [20]. The *in vitro* exposure of human peripheral lymphocytes from smokers to (±)-*anti*-benzo(*a*)pyrenediolepoxide resulted in an increase in lymphocytic DNA damage in older subjects presumably because of a decrease in lymphocytic excision repair capacity [21]. The increase in the frequency of mutations with age has been associated to a reduction of repair capacity towards reactive oxygen species-induced DNA damage [22], although recent studies have shown elevations in 8-oxoguanine DNA adducts from lymphocytes of individuals 63-70 years of age and 75-82 years of age, in comparison to individuals 20-35 years old [23]. In this study there was also an elevation in *in vitro* oxidative stress response with increasing age, as indicated by a pronounced decrease in the yield of breaks in lymphocyte DNA from anti-oxidant supplemented subjects treated with hydrogen peroxide [23]. Thus, repair enzymes activity changes with increasing age and may result in DNA hypersensitivity to toxicants.

The changes that occur during ageing are not limited to only changes in DNA alone but also in other macromolecules such as proteins and lipids. An increase in protein glycosylation resulting in a cascade of irreversible chemical reactions that lead to a decrease in the elasticity and permeability of extracellular compartments and impairment of the passage of nutrients and waste products out of the cell [24]. The glycosylation of DNA and RNA can also result in detrimental effects. Recent data indicate that toxicants, both environmental and generated endogenously by metabolism, are major contributors to macromolecular damage and physiological deregulation that contribute to aging. For instance, electrophilic carbonyl compounds derived from lipid peroxidation appear to be particularly important. As a consequence, detoxification mechanisms, including the removal of electrophiles by glutathione transferase-catalyzed conjugation, are major longevity assurance mechanisms [13]. In addition, elevated levels of oxidized proteins have been detected in tissues from patients with age-related diseases such as atherosclerosis, neurodegenerative disorders and cataracts, suggestive that their repair and complete catabolism to their amino acid components is inefficient [25]. Damaged proteins are degraded by the proteasome or through the collaboration of the proteasome with lysosomes. Incomplete degradation of heavily modified substrates may accumulate within the lysosomal compartment to form lipofuscin-like aggregates and the accumulation of these aggregates results in impaired turnover of proteasomes and mitochondria, destabilization of lysosomes, leakage of proteases into the cytosol and apoptosis [25].

Overall, age-dependent variation in chromatin maintenance and modulation, repair enzymes activity deficiencies and incomplete degradation of heavily modified substrates may result in hypersensitivity to toxicants in the elderly and may possibly also affect micronutrient uptake.

3. FACTORS WHICH MAY AFFECT CHEMICAL SENSITIVITY AND POSSIBLY MICRONUTRIENT UPTAKE

Although there are many physiological processes that are altered during ageing, focus will remain on age-related toxicokinetic changes that may have a significant impact in xenobiotic metabolism including absorption, distribution, metabolism and excretion in the elderly. In general, only some phase I enzymes e.g. esterases and some cytochrome P450s (CYP2C19) appear to be impaired in the elderly, while phase II reactions are not extensively affected, with the exception of a reduction in the glucuronidation capacity seen in the frailty-associated population.

3.1 Absorption

Absorption of chemicals and uptake into the blood occurs primarily via skin, lung and the gastrointestinal (GI) tract. Alterations in structure and function of these three organs clearly occur with age [12,26]. Several studies have indicated a decline in human skin permeability in elderly. The percutaneous absorption of 2,3,7,8-tetrachlorodibenzo-*p*-dioxin and 2,3,4,7,8-pentachlorodibenzofuran has been shown to be significantly decreased in older animals [27].

Little is known about effects of ageing on lung absorption, however, a decrease in the rate of alveolar-capillary gas exchange in elderly has been reported with greater body weight and lung volume changes [28]. A reduction in gastric acid secretion is commonly seen in the elderly [28]. The increase in pH alters the ionisation of compounds, enhancing or reducing their ability to diffuse passively across cellular membranes [28]. Decrease in gastric motility can prolong the transit time of chemicals in the gut thus enhancing their potential for absorption. However, GI absorption of xenobiotics is predominantly by passive diffusion and age-related changes in oral absorption of xenobiotics are not commonly seen. Active transport of molecules such as glucose appears to decrease with age, presumably as a result of a reduction in the number of sodium-linked glucose carriers [29,30]. The active transport of other small molecules such as calcium, phosphorus, galactose and iron has also been reported to decrease with age [12]. Overall, the limited available data suggests that subtle changes occur in absorption in the elderly.

3.2 Distribution

The distribution of chemicals throughout the body is governed by physiological changes that occur during ageing and the physico-chemical properties of the xenobiotic. The loss of body water and lean body mass with age results in a decrease in the volume of distribution for water-soluble compounds leading to enhanced toxicity, e.g. for ethanol [31] or ethylenediamine [32]. The higher percentage of body fat results in an increased accumulation of fat-soluble toxicants [33]. Lipophilic molecules rapidly pass across cellular membranes and accumulate in lipid rich tissues. Since adipose tissue volume increases but blood flow decreases with age, lipophilic compounds tend to show greater retention in the elderly (e.g. for polychlorinated biphenyls and halogenated solvents) [33]. Therefore, fat-soluble toxicants will have increased volume distribution and require a longer period of time in reaching steady state and be eliminated. Decreased cardiac output with age may also result in slower delivery of toxicants to the liver and kidney [33].

3.3 Metabolism and Excretion

There are several physiological changes that occur during ageing which can influence biotransformation reactions both in the liver and in extra-hepatic tissues (e.g. portals of entry, skin, respiratory tract, GI tract). However, few clear patterns exist concerning age-related changes in biotransformation because metabolic changes with ageing tend to be substrate, sex, strain, and species-dependent [12,26,34]. The decreased liver mass and up to 50% reduction in hepatic blood flow may result in a reduction in drug clearance while the reduction in albumin levels may alter the proportion of plasma protein binding and the volume of distribution of some xenobiotics [26,34].

Several age-related changes in cytochrome P450 levels have been shown particularly with CYP2C19, CYP2D6, and CYP3A4 [2]. The half-life of drugs metabolized by hepatic cytochrome P450s or through renal elimination has been shown to be between 50-75% longer in older than younger adults. Longer half-lives are also detected in liver or kidney disease [3]. For this reason, the elderly are at increased risk of drug interactions and side effects [35].

Generally, there is an age-related reduction in phase I reactions (e.g., oxidation, hydrolysis, reduction) and first-pass hepatic metabolism, while phase II reactions (e.g. glucuronidation, sulfation) are not usually affected by ageing [36]. The concentration of glutathione levels in the liver and kidney of aged rats (20 months of age) has been shown to be significantly lower (liver: 1.65-fold and kidney: 2.29-fold) than in young rats [37]. In addition, elderly rats had more DNA damage (liver: 1.84-fold and kidney: 1.86-fold increase in DNA migration in the comet assay) in comparison to young (9 weeks of age) rats [37]. These changes in the activity and/or concentration of drug metabolizing enzymes are important factors that account for differences in drug/metabolite exposure in the elderly population [38].

The livers of the elderly show a general decline in adaptive responsiveness and reduced reserve capacity, although clinical tests indicate that, in general, liver function is well maintained in this age group. Nevertheless, the changes in metabolic capacity may affect the kinetics of a chemical. For instance, the bioavailability, plasma concentration and half-life of a chemical may increase due to age-related changes in liver function and this may have consequences for the toxicodynamic effects of a chemical on target organs.

Finally, excretion of toxicants generally occurs in the kidney. Glomerular filtration decreases in the elderly as a result of decreased renal blood flow, loss of nephrons and decreased kidney size, and may result in prolonged exposure in the blood of xenobiotics that are eliminated by renal excretion [33].

Even though there are clear indications that the elderly may respond differently to chemical exposure compared to the average population because of toxicokinetic (drug/chemical disposition) and toxicodynamic (sensitivity) differences, no additional default UFs are recommended because the present UF is probably sufficient to protect the healthy elderly. It is recognised that in order to always cover the most sensitive person exposed to any chemical would require a very large default UF. That is of course not workable and it is usually assumed that a default UF of 10 is sufficient to protect the larger part of the population, including e.g. the elderly. For threshold effects, this factor of 10 is the standard procedure, as a default, when assessing exposure to the general population [39]. Additional UFs should nevertheless be applied on a case-by-case basis if the toxicological database indicates that the default UF of 10 for intra-species variability is insufficient to adequately safeguard the elderly population (or any other subpopulation). However, more research is needed to ensure that there is no additional risk, particularly in the not-so healthy elderly population. The effects of chronic exposure to high doses of toxicants in the elderly population also need to be investigated, especially if these are lipophilic or require inactivation by renal excretion.

4. DUTCH POLICY ON MICRONUTRIENT CONSUMPTION

Because data on chemical sensitivity in the elderly is lacking, evaluation as to whether the current uncertainty factor (UF) of 10 is sufficient to protect the elderly was investigated using the micronutrient consumption data in the elderly Dutch population. The Dutch policy was surveyed in order to investigate whether the increased trend of supplement intake in the elderly might lead to over-consumption. Dutch policy for micronutrient intake is geared towards preventing deficiency and over-consumption through the establishment of dietary reference values. These dietary reference values comprise of 1) an estimated average requirement (EAR) no data specific for elderly [43]. Dutch policy for micronutrient intake is geared towards preventing deficiency and over-consumption [41]. Existing which is the adequate requirement for 50% of the population; 2) a recommended daily allowance (RDA) which is the average daily intake level to meet nutritional requirements of 97.5% of healthy individuals (equivalent to the mean or EAR plus 2 SD); 3) a tolerable upper intake level (UL) which is the maximum level of total chronic daily intake of a nutrient (from all sources) judged to be unlikely to pose a risk of adverse health effects to humans; and 4) a LRNI which is the amount of a nutrient which is adequate for only 2.5% of the population group and represents those with the lowest need (Fig. 1).

Although for micronutrients the risk of adverse effects is associated with consuming too little (deficiency) or too much (over-consumption), this report will focus only on the latter. Dietary reference standards for evaluating and managing the risk of excessive intakes of vitamins and minerals have been established recently for a number of vitamins and minerals and are referred to as UL (sometimes also called Safe Upper Levels). The UL is the maximum level of total chronic daily intake of a nutrient from all sources including foods, water, nutrient supplements and medicines judged to be unlikely to pose a risk of adverse health effects to almost all individuals in the general population [41]. The term 'tolerable' implies a level of intake that can be physiologically tolerated by humans. ULs may be derived for populations in different life stages, for instance infants, children, adults, pregnant and lactating women by using relevant data or, in the absence of data for a particular life stage group, by extrapolating from the UL for other groups, e.g. for children on the basis of body weight. ULs have been derived for a number of vitamins and minerals, the most relevant now is from the EFSA [42]. The UL is not a recommended level of intake but rather is an estimate of the highest level of intake that carries no appreciable risk of adverse health effects. The UL is not meant to apply to individuals receiving the nutrient under medical supervision or to individuals with predisposing conditions which render them especially sensitive to one or more adverse effect(s) of the nutrient, e.g. those with genetic predisposition or certain metabolic disorders or disease states.

There is a trend of increased consumption of micronutrient supplements (i.e. vitamins and minerals) by the Dutch population, but there are European legislation for micronutrients in fortified foods (Regulation 1925/2006) and food supplements (Directive 2002/46) is now being translated into simultaneously setting maximum levels of micronutrients in foods and in supplements [44]. This setting of maximum levels in fortified foods and in food supplements involves the assessment of the margin between the amount of a micronutrient present from basic food intake (e.g. at 95-percentile) and the UL. The maximum amount of a micronutrient that may be added to fortified food and/or food supplements with little risk of adverse health effects in the population is therefore the UL minus the 95-percentile. Prior to setting maximum levels for fortified foods and food supplements a risk management decision needs to be taken whether to allow this maximum amount for fortified foods only, for food supplements only or for both (e.g. in a 50/50 ratio) (Fig. 2).

There are well-developed methods for setting maximum safe levels of food fortification, the latest therefore being developed by RIVM [45]. There is also methodology for setting maximum safe levels in food supplements [46], but this has thus far not really been scientifically challenged [11]. Maximum levels in fortified foods and food supplements can and need to be set for age- and gender-specific groups.

In the Netherlands, the healthy elderly population comprises of approximately 70-75% of 46-64 year-old individuals, 62-70% of 65-74 year-old individuals, and 50-65% of 75^+ year-old individuals [7]. For these individuals, no foreseeable risk is expected due to the consumption of micronutrients from supplements, dietary intake and other sources.

On the other hand, the not-so healthy elderly comprise of approximately 25-30% of 46-64 year-old individuals, 30-38% of 65-74 year-old individuals, and 35-50% of 75^+ year-old individuals [7]. These individuals may be taking daily prescription drugs to combat chronic diseases such as diabetes, circulatory or respiratory diseases. The effects of micronutrient consumption from supplements, dietary intake and other sources for the unhealthy elderly are not yet known. For this reason, dietary supplement intakes need to be continuously monitored using a combination of dietary recall methods with detailed questioning on health status, supplement and prescription drug use in the elderly population. In addition, the generation of an international and up-to-date database on the composition of available dietary supplements is needed to fill the current data gaps.

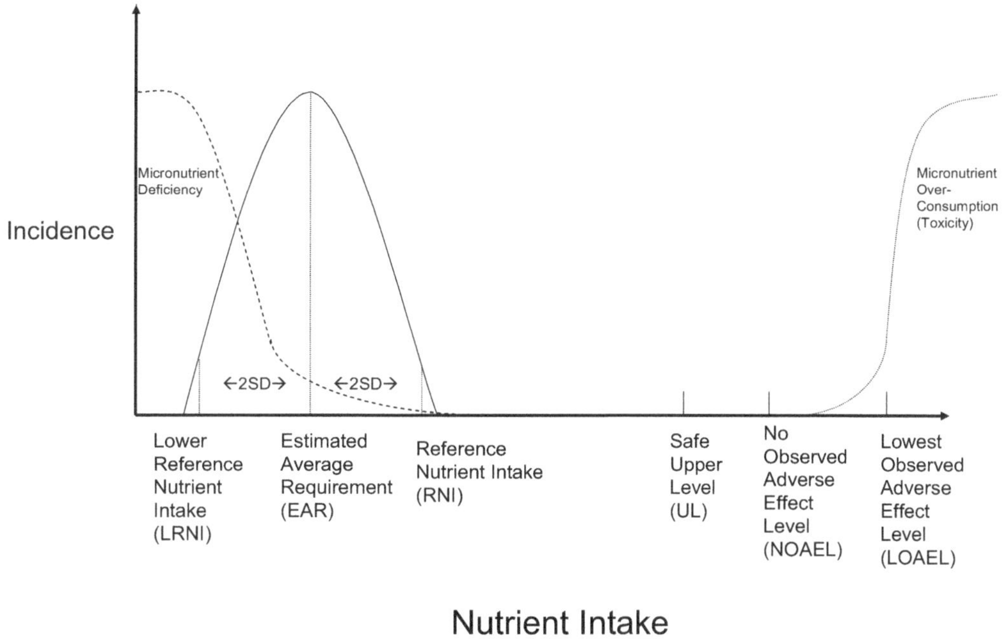

Fig. 1. Model for assessing dietary reference values for micronutrients
(Adapted from [40]; SD = Standard deviation)

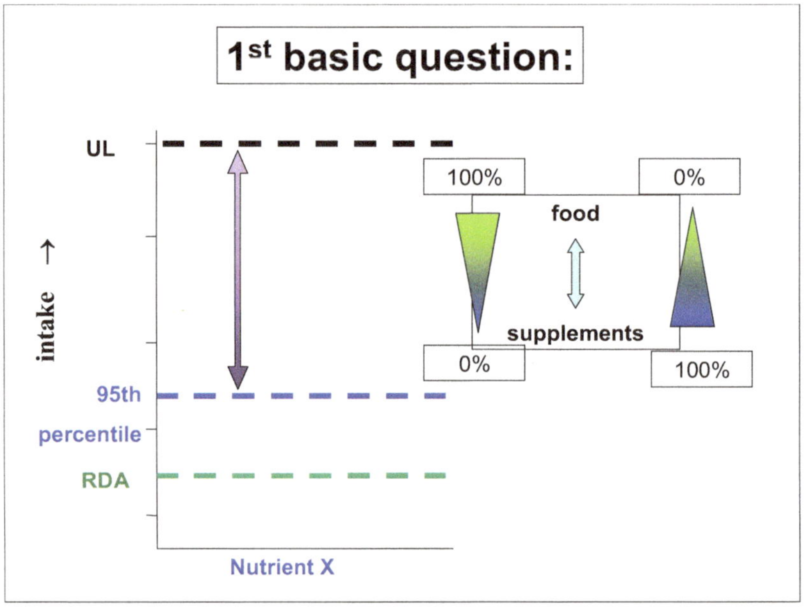

Fig. 2. Depiction of risk management scheme for the setting of maximum levels in fortified foods and food supplements

5. POTENTIALLY SENSITIVE SUB-POPULATIONS

Although this paper deals with the 'healthy elderly' populations, it is important to be aware of the existence of potential sensitive sub-populations among the elderly. The major causes of death in the Netherlands include cancer (41%), circulatory diseases (41%), respiratory diseases (14%), psychological disorders (6.4%), digestive disorders (5.4%), external causes of injury and poisonings (5.3%), and endocrine disorders as a result of dietary, metabolic diseases or immune disturbances (4.1%) [7]; all of which may be in one way or another exacerbated by chemical toxicity in the elderly. Of particular importance is diabetes because the prevalence increases exponentially after the age 40, with type II diabetes being the most common [47]. The prevalence of type II diabetes increases from 68 cases per 1000 per year in 45-64 year-old males to 171 cases per 1000 individuals per year in 65-74 year-old and in 75^{+} year-old males [47]. Similarly, the prevalence of type II diabetes increases from 48 cases per 1000 per year in 45-64 year-old females to 140 cases per 1000 individuals per year in 65-74 year-old females, and 165 cases per 1000 individuals per year in 75^{+} year-old females [47]. Immigrants with Surinam, Indonesian, Turkish or Moroccan origin are at greater risk of developing type II diabetes than the native Dutch population [48]. An early indication of diabetes is the metabolic syndrome where lipid status is a major influential component. There is a remarkable age-related difference in the serum concentration of some lipid parameters (triglycerides, free fatty acids) between men and women. At 20 years of age women have a relative healthy pattern (low concentration of lipids) but their lipid levels increase dramatically with age to levels comparable to those seen in men [49]. This rapid increase in lipid parameters may make women more prone to diabetes as they get older. Among other consequences, diabetes is known to impair renal function, and as a result may have a direct impact on chemical sensitivity. Therefore individuals with the above mentioned disorders, immigrants or sex may be potential sub-populations that need to be considered or at least kept in mind.

6. CONCLUSION

Although there is a trend of increased consumption of micronutrient supplements (i.e. vitamins and minerals) by the Dutch population, there are currently no elderly-specific data available. In the Netherlands, the healthy elderly population comprises of approximately 70-75% of 46-64 year-old individuals, 62-70% of 65-74 year-old individuals, and 50-65% of 75^{+} year-old individuals. For these individuals, no foreseeable risk is expected due to the consumption of micronutrients from supplements, dietary intake and other sources when existing European legislation for micronutrients in fortified foods (Regulation 1925/2006) and food supplements (Directive 2002/46) is simultaneously translated into setting maximum levels of micronutrients in foods and food supplements.

On the other hand, the effects of micronutrient consumption from supplements, dietary intake and other sources for the unhealthy elderly are not yet known. For this reason, dietary supplement intakes need to be continuously monitored using a combination of dietary recall methods with detailed questioning on health status, supplement and prescription drug use in the elderly population. In addition, the generation of an international and up-to-date database on the composition of available dietary supplements is needed to fill the current data gaps.

With regards to chemical sensitivity in the healthy elderly, even though there are clear indications that the elderly may respond differently to chemical exposure compared to the average population because of physiological differences, no additional default UFs are recommended because the present UF is probably sufficient to protect the healthy elderly. Additional UFs should nevertheless be applied on a case-by-case basis if the toxicological database indicates that the default UF of 10 for intra-species variability is insufficient to adequately safeguard the elderly population (or any other subpopulation). However, more research is needed to ensure that there is no additional risk, particularly in the not-so healthy elderly population. The effects of chronic exposure to high doses of toxicants in the elderly population also need to be investigated, especially if these are lipophilic or require inactivation by renal excretion.

ACKNOWLEDGEMENTS

We would like to thank Gerrit Wolterink and Martijn Dollé for their critical comments with regards to this publication. This investigation has been performed by order and for the account of the Food and Consumer Product Safety Authority (VWA), within the framework of project 340360, Knowledge Base Carcinogenicity, Mutagenicity and Reproduction Toxicology.

COMPETING INTERESTS

Authors have declared that no competing interests exist.

REFERENCES

1. WHO. World Health Organization. Assessing human health risk of chemicals: Derivation of guidance values for health-based exposure limits. World Health Organization, Geneva. Environmental Health Criteria. 1994;170.
2. Dorne JL. Human variability in hepatic and renal elimination: implications for risk assessment. J Appl Toxicol. 2007;27:411-420.
3. Ginsberg G, Hattis D, Russ A, Sonawane B. Pharmacokinetic and pharmacodynamic factors that can affect sensitivity to neurotoxic sequelae in elderly individuals. Environ Health Perspect. 2005;113:1243-1249.
4. Durnas C, Loi CM, Cusack BJ. Hepatic drug metabolism and aging. Clin Pharmacokinet. 1990;19:359-389.
5. Le Couteur DG, McLean AJ. The aging liver. Drug clearance and an oxygen diffusion barrier hypothesis. Clin Pharmacokinet. 1998;34:359-373.
6. WHO World Health Organization. Available:http://www.who.int/countries/nld/en/.
7. National-Kompas-Volksgezondheid; 2007. Available:http://wwwrivmnl/vtv/object_document/o1079n18744html.
8. Harbers MM, Van der Wilk EA, Kramers PGN, Kuunders MMAP, Verscuuren M, et al. Dare to compare! benchmarking dutch health with European Community Health Indicators (ECHI). RiVM Report 270051011; 2008.
9. EFSA. Tolerable upper intake levels for vitamins and minerals. European Food Safety Authority (EFSA); 2006.
10. Van Rossum CTM, Fransen HP, Verkaik-Kloosterman J, Buurma-Rethans EJM, Ocké MC. Dutch national food consumption survey 2007-2010. Diet of children and adults aged 7 to 69 years old. RIVM Report 350050006/2011; 2011.
11. Howlett J, Binns N. International Life Sciences Institute (ILSI) Micronutrient Landscape in Europe: Comparison of Intakes and Methodologies with Particular Regard to Higher Consumption; 2008. Available:http://wwwilsiorg/Europe/Publications/Micronutrient%20Report%202009pdf.
12. INCHEM. Envirnomental Health Criteria 144: Principles of Evaluating Chemical Effects on the Aged Population.

International Program on Chemical Safety. 1992;144.
13. Zimniak P. Detoxification reactions: relevance to aging. Ageing Res Rev. 2008;7:281-300.
14. Vijg J. DNA sequence changes in aging: how frequent, how important? Aging (Milano). 1990;2:105-123.
15. Niedzwiecki A, Lewis PN, Cinader B. Changes of histone H1 subtypes with aging in strains of mice that possess different immunological characteristics. J Gerontol. 1985;40:695-699.
16. O'Meara AR, Pochron SF. Age-related effects on the incorporation of acetate into rat liver histones. Biochim Biophys Acta. 1979;586:391-401.
17. Vijg J, Knook DL. DNA repair in relation to the aging process. J Am Geriatr Soc. 1987;35:532-541.
18. Chang IY, Jin M, Yoon SP, Youn CK, Yoon Y, et al. Senescence-dependent MutS alpha dysfunction attenuates mismatch repair. Mol Cancer Res. 2008;6:978-989.
19. Gorbunova V, Seluanov A, Mao Z, Hine C. Changes in DNA repair during aging. Nucleic Acids Res. 2007;35:7466-7474.
20. Washington WJ, Foote RS, Dunn WC, Generoso WM, Mitra S. Age-dependent modulation of tissue-specific repair activity for 3-methyladenine and O6-methylguanine in DNA in inbred mice. Mech Ageing Dev. 1989;48:43-52.
21. Stierum RH, Hageman GJ, van Herwijnen MH, van der Veer MS, Kleinjans JC. Age-related negative associations between parameters of cytogenetic damage and *ex vivo* (+/-)-anti-benzo(a)pyrene diolepoxide-induced unscheduled DNA synthesis in smoking humans. Cancer Epidemiol Biomarkers Prev. 1997;6:943-948.
22. Barnett YA, King CM. An investigation of antioxidant status, DNA repair capacity and mutation as a function of age in humans. Mutation Research/DNAging. 1995;338:115-128.
23. Humphreys V, Martin RM, Ratcliffe B, Duthie S, Wood S, et al. Age-related increases in DNA repair and antioxidant protection: a comparison of the Boyd Orr Cohort of elderly subjects with a younger population sample. Age Ageing. 2007;36:521-526.
24. Knight JA. The process and theories of aging. Ann Clin Lab Sci. 1995;25:1-12.
25. Dunlop RA, Brunk UT, Rodgers KJ. Oxidized proteins: mechanisms of removal and consequences of accumulation. IUBMB Life. 2009;61:522-527.
26. Crooks J, O'Malley K, Stevenson IH. Pharmacokinetics in the elderly. Clin Pharmacokinet. 1976;1:280-296.
27. Banks YB, Brewster DW, Birnbaum LS. Age-related changes in dermal absorption of 2, 3, 7, 8-tetrachlorodibenzo-p-dioxin and 2, 3, 4, 7, 8-pentachlorodibenzofuran. Fundam Appl Toxicol. 1990;15:163-173.
28. Stiles J, Tyler WS. Age-related morphometric differences in responses of rat lungs to ozone. Toxicol Appl Pharmacol. 1988;92:274-285.
29. Eastin WCJr, Birnbaum LS. Intestinal absorption of two glucose analogues in rats of different ages. Exp Gerontol. 1987;22:351-357.
30. Vincenzini MT, Iantomasi T, Stio M, Favilli F, Vanni P, et al. Glucose transport during ageing by human intestinal brush-border membrane vesicles. Mech Ageing Dev. 1989;48:33-41.
31. York JL. Body water content, ethanol pharmacokinetics, and the responsiveness to ethanol in young and old rats. Dev Pharmacol Ther. 1982;4:106-116.
32. Yang RS, Tallant MJ, McKelvey JA. Age-dependent pharmacokinetic changes of ethylenediamine in Fischer 344 rats parallel to a two-year chronic toxicity study. Fundam Appl Toxicol. 1984;4:663-670.
33. Dybing E, Soderlund EJ. Situations with enhanced chemical risks due to toxicokinetic and toxicodynamic factors. Regul Toxicol Pharmacol. 1999;30:27-30.
34. Crooks J. Aging and drug disposition-pharmacodynamics. J Chronic Dis. 1983;36:85-90.
35. Sjoqvist F, Alvan G. Aging and drug disposition--metabolism. J Chronic Dis. 1983;36:31-37.
36. O'Mahony MS, Schmucker DL. Liver disease in the elderly. Semin Gastrointest Dis. 1994;5:197-206.
37. Hashimoto K, Takasaki W, Yamoto T, Manabe S, Sato I, et al. Effect of glutathione (GSH) depletion on DNA damage and blood chemistry in aged and young rats. J Toxicol Sci. 2008;33:421-429.
38. Benedetti MS, Whomsley R, Canning M. Drug metabolism in the paediatric population and in the elderly. Drug Discov Today. 2007;12:599-610.
39. ECHA. Guidance on information requirements and chemical safety

assessment Chapter R.8: Characterisation of dose [concentration]-response for human health (version 2.1); 2012. Available:http://echa.europa.eu/documents/10162/13632/information_requirements_r10168_en.pdf.

40. Fletcher RJ, Bell IP, Lambert JP. Public health aspects of food fortification: a question of balance. Proc Nutr Soc. 2004;63:605-614.

41. SCF. Guidelines of the Scientific Committee on Food for the development of tolerable upper intake levels for vitamins and minerals. Scientific Committee on Food (SCF); 2000. Available: http://ec.europa.eu/food/fs/sc/scf/out80a_en.pdf.

42. EFSA. Tolerable Upper Intake Levels for Vitamins and Minerals; 2006. Available: http://wwwefsaeuropaeu/EFSA/Scientific_Document/upper_level_opinions_full-part33pdf.

43. Ocke MC, Buurma-Rethans EJM, Fransen HP. Dietary supplement use in the Netherlands: Current data and recommendations for future assessment. RIVM rapport 350100001; 2005.

44. EFSA. Discussion paper on setting of maximum and minimum amounts for vitamins and minerals in foodstuffs. European Food Safety Authority (EFSA) Health and Consumer Protection Directorate General; 2006. Available: http://ec.europa.eu/food/food/labellingnutrition/supplements/discus_paper_amount_vitamins.pdf

45. Kloosterman J, Fransen HP, De Stoppelaar J, Verhagen H, Rompelberg C. Safe addition of vitamins and minerals to foods: setting maximum levels for fortification in the Netherlands. Eur J Nutr. 2007;46:220-229.

46. ERNA-EHPM. Vitamin and mineral supplements: A risk management model. European Responsible Nutrition Alliance (ERNA) and the European Federation Association of Health Product Manufacturers (EHPM); 2004. Available: http://www.erna.org/UserFiles/Vitamin%20and%20Mineral%20Supplements.pdf.

47. National-Kompas-Volksgezondheid; 2007. Available:http://wwwrivmnl/vtv/object_document/o1259n17502html.

48. Diabetes-Fonds (2009). Available: http://wwwdiabetesfondsnl/indexphp?page=text&content=361.

49. Jansen EHJM, Kienhuis A. Supplements for the elderly. An inventory of vitamins and minerals available for elderly in The Netherlands. RIVM Letter Report 090143001; 2013. Available: http://www.rivm.nl/dsresource?objectid=rivmp:248182&type=org&disposition=inline&ns_nc=1

10

Incidence of Antimicrobial Residues in Meat Using a Broad Spectrum Screening Strategy

David Sanz[1*], Pedro Razquin[1], Santiago Condón[2], Teresa Juan[3], Benito Herraiz[4] and Luis Mata[1]

[1]Department of R&D, ZEULAB S.L., Polígono PLAZA, C/Bari, 25 dpdo, 50197, Zaragoza, Spain.
[2]Faculty of Veterinary Medicine, University of Zaragoza, C/ Miguel Servet, 177, 50013 Zaragoza, Spain.
[3]Agrifood Research and Technology Centre of Aragon (CITA), Montañana, 930, 50059 Zaragoza, Spain.
[4]S.A.T. Num 42 Los Gonzalez, C/ Extramuros, s/N, 44300 Monreal del Campo, Spain.

Authors' contributions

This work was carried out in collaboration between all authors. Author SC headed the project.

ABSTRACT

Aims: The aim of this paper was to assess the incidence of antimicrobial residues in market muscle samples from different animal species (bovine, ovine, poultry and porcine) using a new screening strategy.
Methodology: 4849 samples were evaluated with a methodology that combines a broad spectrum microbial test (Explorer) and a specific test for quinolones detection (Equinox). Supplementary tests were performed to achieve additional information about the nature of antimicrobials in positive samples.
Results: In a first step, 355 samples (7.3%) showed a positive result in Explorer and/or Equinox tests. The highest incidence of positive samples was obtained in poultry (9.7%) while the lowest rate was found in porcine samples (3.4%). Half of the positive screening samples (53%) showed also a positive result with supplementary tests indicating that tetracyclines, aminoglycosides sulphonamides and quinolones might be present in these samples. Aminoglycosides were the predominant residues in poultry while tetracyclines were more frequent in bovine and porcine samples. Sulphonamides were the main family of residues found in ovine.
Conclusion: Our results suggest that the current strategies used for control of antimicrobial

**Corresponding author: Email: dsanz@zeulab.com*

residues in muscle could not be adequate enough. In order to protect consumers from antibiotic exposition, it should be advisable to implement more efficient methods for the screening of antibiotic residues in muscle.

Keywords: Screening; antibiotic; antimicrobial residues; muscle; microbiological test.

1. INTRODUCTION

The use of pharmacological products in livestock –for treatment of animal diseases or prophylaxis purposes -might lead to the presence of antimicrobial residues in muscle and other organs from animals. As a consequence, several potential concerns may occur: Allergic reactions in sensitized individuals, technological problems in fermented products and toxicological effects [1-3]. Nevertheless, most experts consider that the selection of antimicrobial resistant pathogenic bacteria is the main risk derived from the use of antimicrobials in farm animals.

Maximum Residue Limits (MRLs) for veterinary drug residues in different animal tissues and milk were set by the European legislation [4]. Therefore, food of animal origin must be analyzed to ensure that residues do not exceed MRLs. Since antimicrobials do not share a common chemical structure, it could be necessary to use different analytical procedures to detect every family or even each single compound. Thus, in a first step, efficient control of residues would require screening tests, which are expected to be cheap, easy to perform, allow simultaneous analysis of large numbers of samples and give rapid results [5]. Screening tests are qualitative tools that can differentiate inhibitory samples (samples with antimicrobial residues concentration above detection limits of the test) from non-inhibitory samples. Microbial tests are generally used at screening level and they are based on a bacterial growth inhibition produced by residues contained in the samples [6]. A post-screening step (usually microbial multi-plate tests or immunological methods) may be performed after screening test to identify a family of antibiotics. Finally, positive screening samples must be confirmed, generally with chromatographic methods coupled to mass spectrometric detection [7,8].

Microbial agar diffusion tests have been widely used for detection of antibiotic residues in foods from animal origin. Among them, the most traditional methods are multiplate screening tests involving several bacterial strains; these methods were extensively reviewed by Pikkemaat [8]. Although they are relatively easy to perform, they are usually time-consuming, display a poor sensitivity for sulphonamides and could exhibit wide variations in the performance between laboratories [9]. Microbial inhibition tube tests using an indicator of bacterial growth have been also developed for the screening of antimicrobial residues in muscle. Generally, these tests use a medium seeded with *Geobacillus stearothermophilus* and a pH or redox indicator. The interpretation of results is based on the colour change of the medium caused by microbial growth when antibiotics are not present at inhibitory concentrations in the sample. Tube tests have several advantages since they are generally ready to use, easy to perform and can detect a broad range of antimicrobial residues. Furthermore, results can be obtained in a shorter time (< 3 h) and a photometric reading may be applied, avoiding variations due to visual interpretation made by different technicians or performed over different days [10,5].

However, screening tests based on *G. stearothermophilus* are not able to detect quinolones at MRL levels [11,12]. Consequently, these tests should be combined with specific methods able to detect quinolones in order to cover a wider range of antimicrobials at screening step [13].

The objective of the present study was to assess the incidence of antimicrobial residues in market muscle samples from different animal species (bovine, ovine, poultry and porcine). To achieve this goal, a large number of muscle samples were evaluated with a screening strategy that combines a broad spectrum microbial test (Explorer) with a specific test for quinolones detection (Equinox).

2. MATERIALS AND METHODS

2.1 Screening Tests

2.1.1 Explorer test

Explorer (ZEULAB, Zaragoza, Spain) is a qualitative test kit for the detection of inhibitory substances in raw meat and other matrices (kidney, liver, feed and eggs). The test is based on the inhibition of microbial growth of *G. stearothermophilus*. Each well contains an agar-based medium spread with the target bacteria and a pH indicator. When the test is incubated at 65°C, spores germinate and cells grow producing acid and changing the medium pH. Variations of pH will cause changes of the medium colour from blue to yellowish. When the sample contains inhibitors concentrations above detection limits (LOD) of the test, microorganisms will not grow and neither colour changes will be observed.

Samples were extracted by heating a piece of lean meat (3±0.5 g) without adipose or conjunctive tissue in a microwave ("defrost" setting for 1–2 min) and clarified by centrifugation (2000 g for 3 min). Sample meat fluid was added (0.1 ml) to each well with a micropipette. The wells were pre-incubated at room temperature for 20 min to allow the sample to diffuse through the well. Afterwards, the sample was eliminated by washing the wells with distilled water. Finally, the wells were sealed with an adhesive film and incubated at 65°±1C. The endpoint of the assay was reached when the negative control sample (antibiotic-free meat fluid) turned yellow. An objective interpretation of the results was made by performing photometric measurements. The plate was ready to be read when the result for the negative control sample (difference of absorbance at 590 nm and 650 nm) was between 0.15 and 0.25 OD (optical density units). A sample was declared positive when:

$$SA590\text{ nm} - SA650\text{ nm} \geq NA590\text{ nm} - NA650\text{ nm} + 0.15$$

where

NA is the negative control absorbance and SA the sample absorbance.

The performance characteristics of Explorer test for detection of antibiotic residues in muscle from different animal species have been described previously, including detection capabilities (CCβ), specificity, false-positive rate and robustness [10,14].

2.1.2 Equinox

Equinox (ZEULAB, Zaragoza, Spain) is a specific kit test for quinolones detection in several food matrices. The test is based on the inhibition of microbial growth of *Escherichia coli* ATCC 11303. The kit includes ampoules with a standardized number of freeze-dried bacteria and ampoules with a specific detection medium containing a redox indicator. During incubation time at 37°C bacterial cells will multiply and modify the redox potential of the medium. As a consequence, a colour change in the medium (from blue to brown/orange) will be observed. Samples containing concentrations of quinolone residues above the Equinox LODs will inhibit the growth of *E. coli* and will prevent the indicator colour change.

The extraction procedure of samples was performed as described previously (2.1.1). Prior to the analysis, the ampoule with *E. coli* was resuspended with the specific detection medium. The assay was carried out in microtiter plates, mixing gently 50 uL of sample or control with 200 uL of reconstituted *E. coli*. The wells were sealed with an adhesive sheet and incubated at 37±1°C.

The endpoint of the assay was reached when the negative control sample (antibiotic-free meat fluid) had turned brown-orange. Equinox results were evaluated by a photometric measurement for an objective interpretation. The assay ended when the result for the negative control sample (difference of absorbance at 590 nm and 650 nm) was between 0.2 and 0.5 OD (optical density units). A sample was declared positive when:

$$SA590\text{ nm} - SA650\text{ nm} \geq NA590\text{ nm} - NA650\text{ nm} + 0.4$$

where

NA is the negative control absorbance and SA the sample absorbance.

An evaluation of the performance of Equinox as a test for detection of quinolone residues in muscle was carried out by Sanz et al. [13].

2.1.3 Supplementary screening tests

To achieve additional information about the chemical nature of antimicrobials contained in screening positive samples, the following supplementary screening tests were performed:

- Test for tetracyclines: Plate with *Bacillus subtilis* (BGA), pH 6 [15].
- Test for aminoglycosides: plate with *B. subtilis* (BGA), pH 8 [15].

- Test for beta-lactams/macrolides: Plate with *Kocuria rhizophila* ATCC 9341, pH 8 [15].
- Test for sulphonamides: the plate seeded with *B. subtilis* (BGA) at pH 7.2 [15] was not suitable for our purpose since it cannot detect sulphonamides in muscle at MRLs [16]. As an alternative, an additional analysis was performed using 4-aminobenzoic acid (PABA, Sigma-Aldrich, Steinheim, Germany): a sample with PABA (50 µg/ml of muscle fluid) and without PABA were analyzed again with Explorer test at the same time. A loss of inhibitory effect was observed in Explorer test when PABA was added to samples spiked with sulphonamides (up to 2000 µg/Kg of sulfadiazine, sulfametoxipiridazine, sulfametazine or sulfathiazole) [17].
- Test for beta-lactams: since plate seeded with *K. rhizophila* ATCC 9341 at pH 8 was not able to discriminate if inhibitory samples contained beta-lactams or macrolides, an alternative was performed to detect specifically beta-lactams. Thus, an additional analysis with Explorer was performed using penicillinase (Sigma-Aldrich, Steinheim, Germany) in a similar way as described previously for sulphonamides: Explorer results were compared when analyzing at the same time a sample without and with penicillinase (100 µg/ml of muscle fluid) [18]. When penicillinase was added to a sample spiked with penicillins (up to 1000 µg/Kg of amoxicillin, bencilpenicillin, cephalexin or ampicillin) a loss of inhibition effect was observed [17].

2.2 Evaluation of LODs

Antimicrobial standards of known purity with certificates of analysis were purchased from Sigma-Aldrich (Steinheim, Germany). Antibiotic and sulphonamide stock solutions (1 mg/ml) were prepared and aliquots were kept at -20°C for no more than 2 months. For evaluation of detection limits, intermediate dilutions were obtained in water immediately before every assay and final testing concentrations were prepared in negative bovine muscle fluid.

LODs of Explorer and Equinox for different antimicrobials were published previously [10,13]. Sensitivity data from both studies are compiled in Table 1. Moreover, additional sensitivity data for some antimicrobials were determined in the present study.

2.3 Samples

A total of 4849 muscle samples were analyzed, comprising different animal species: bovine (1302), ovine (1283), poultry (1280) and porcine (984). Bovine, ovine and porcine samples were taken from thoracic diaphragm at different slaughterhouses in the north of Spain while poultry ones (chicken thighs) were purchased from different supermarkets. Samples were individually identified with a code related to animal specie and origin. Muscle samples were transported under refrigeration and screening tests (Explorer and Equinox) were performed in the laboratory upon arrival. Then, samples were frozen and kept at -20°C.

2.4 Sample Extraction

The extraction protocol recommended by the manufacturer of the tests was applied to the samples (see 2.1.1 and 2.1.2.). Briefly, meat fluid was extracted by heating in a microwave and was subsequently clarified by centrifugation [10,13].

2.5 Analysis of Samples

The antimicrobial screening strategy performed in this study is summarized in Fig. 1. Muscle samples were evaluated, in a first step, with Explorer and Equinox tests. Positive samples detected with Explorer were confirmed by a new assay with this test. If the inhibitory effect was observed also in the second analysis, the sample was further analyzed with specific screening tests for tetracyclines, aminoglycosides, beta-lactams, macrolides and sulphonamides. Besides, positive samples detected by Equinox were tested again with the same test to confirm the results. No more tests were performed to positive samples in first and second analysis, since Equinox is a highly specific test for quinolones.

3. RESULTS AND DISCUSSION

3.1 Results

3.1.1 Screening of meat samples

In the first step of the study, 4849 samples were analyzed for antibiotic residues with Explorer and

Equinox. As described previously, every positive result in Explorer or Equinox was confirmed with a second analysis in the same test. Only samples that showed inhibitory effect in both assays would be considered as positives. Every sample that showed a positive result at the first screening step remained as positive in a second analysis with the same test. As it is summarized in Table 2, 355 samples (7.3%) were identified as positive by Explorer and/or Equinox test. The highest incidence of positive samples was obtained in poultry (9.7%) while the lowest rate was found in porcine samples (3.4%).

Table 1. LODs (µg/Kg) of Equinox and Explorer for several antibiotics and sulphonamides in bovine muscle (from [10, 13])

LOD (ug/kg)	Equinox	Explorer	UE-MRL
Doxycycline	≤5000	200	100
Tetracycline	n.e.	500*	100
Oxytetracycline	n.e.	700*	100
Penicillin G	>5000	5*	50
Amoxicillin	n.e.	10	50
Cefalexin	n.e.	>500	200
Gentamycin	n.e.	400*	50
Neomycin	≤10000	300*	500
Tylosin	>10000	100	100
Erythromycin	n.e.	200*	200
Enrofloxacin	100	n.e.	100[a]
Norfloxacin	200	n.e.	
Sarafloxacin	100	n.e.	
Marbofloxacin	100	n.e.	150
Ciprofloxacin	25	n.e.	100[a]
Danofloxacin	200	n.e.	200
Difloxacin	300	n.e.	400
Flumequine	2000	n.e.	200
Sulfathiazole	>10000	200	100
Sulfamethoxypyridazine	n.e.	300*	100
Sulfadiazine	n.e.	200*	100

*LODs: Limits of detection, UE-MRL: maximum residue limits EEC (2377/90), [a]: sum of enrofloxacin and ciprofloxacin, n.e.: not evaluated, *: data provided by manufacturer*

Table 2. Screening results of meat samples with Explorer and Equinox

	Number of samples	Results	
		Negative (%)	Positive (%)
Bovine	1302	1210 (92.9)	92 (7.1)
Ovine	1283	1177 (91.7)	106 (8.3)
Poultry	1280	1156 (90.3)	124 (9.7)
Porcine	984	951 (96.6)	33 (3.4)
Total (%)	4849	4494 (92.7)	355 (7.3)

3.1.2 Supplementary screening test (classification of samples)

Supplementary analyses were performed to Explorer positive samples to confirm results and obtain additional information about the nature of these samples. Results obtained with these tests are summarized in Table 3. Half of the positive screening samples (3.9% of total analyzed samples) showed also a positive result with at least one of the supplementary tests. According to Table 3, these samples might contain residues of tetracyclines (22.3%), aminoglycosides (23.9%), sulphonamides (21.8%), and quinolones (15.4%). Furthermore, results suggest that aminoglycosides and tetracyclines could be combined in 31 samples (16.5%) since an inhibition was observed in two different specific tests at the same time. None of the evaluated samples was found positive at plate seeded with *K. rhizophila* or Explorer with penicillinase. In such a way, a low probability of finding residues of macrolides and beta-lactams would be expected in positive screening samples.

A different pattern was observed when comparing the results from different animal species. Thus, aminoglycosides might be present in 66% of poultry samples but were less common in other species (23-41%). By contrast, data suggest that tetracyclines might be the most frequent residues in bovine and porcine samples while sulphonamide residues were the main family of residues expected in ovine. Quinolone residues might be found in every species (16-26%) except for bovine samples.

3.2 Discussion

The screening strategy used in this work aimed to detect a large range of antimicrobials in muscle samples. The screening system combined 2 microbial methods: A broad spectrum test (Explorer) and a specific test for quinolones (Equinox). Gaudin et al. [10] reported that Explorer was able to detect compounds belonging to different antimicrobial families (penicillins, cephalosporins, tetracyclines, sulphonamides and macrolides) in muscle samples from different species (bovine, porcine, ovine and poultry). These authors observed that detection capabilities were around MRL levels for tested antimicrobials and concluded that Explorer might be used as a wide spectrum screening test for antimicrobials in muscle samples. However, a lack of sensitivity to quinolones is expected in microbiological tests based on *G. stearothermophilus* [11,12]. Equinox test was found to detect several quinolones (enrofloxacin, ciprofloxacin, marbofloxacin, norfloxacin, sarafloxacin, danofloxacin, difloxacin) in muscle samples of different animal species. The test showed an adequate sensitivity and LODs reported for most of evaluated quinolones were at or below the established MRLs. Furthermore, Equinox exhibited a much lower sensitivity for other groups of antimicrobials since concentrations of doxycycline, tylosin, neomycin, penicillin G and sulfathiazole up to ten times higher than MRL levels were required to inhibit the test. Therefore, it was concluded that Equinox was a suitable and very specific tool for the screening of quinolone residues in muscle samples [13].

Complementary sensitivity patterns reported for Explorer and Equinox suggest that the screening system used in this work could be an appropriate choice for detection of a wide spectrum of antimicrobial residues in muscle. Thus, a positive result in Equinox may be associated to a sample that contains quinolone residues above Equinox

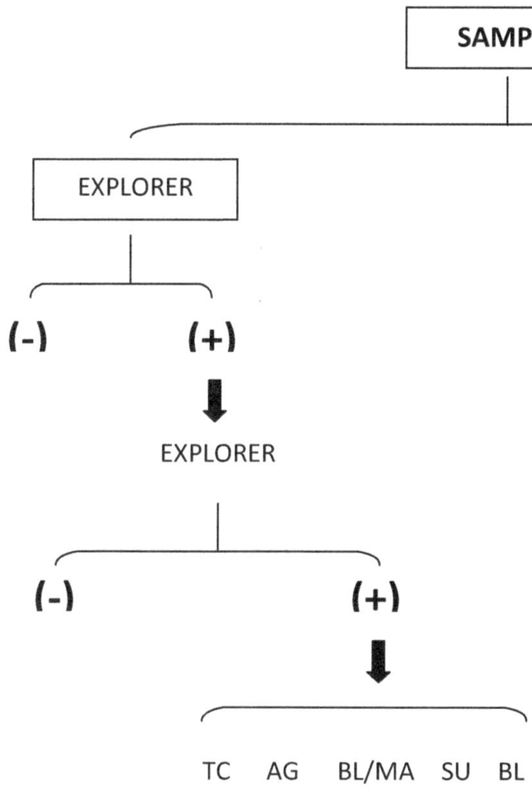

Fig. 1. Antimicrobial screening strategy

TC: test for tetracyclines, AG: test for aminoglycosides, BL/ML: test for beta-lactams/macrolides, SU: test for sulphonamides, BL: test for beta-lactams

Table 3. Classification of samples: Supplementary screening test results

	TC	AG	TC + AG	SU	Q	BL/MA	BL
Bovine	24	6	12	2	0	0	0
Ovine	8	4	12	33	11	0	0
Poultry	2	31	4	2	14	0	0
Porcine	8	4	3	4	4	0	0
Total	42	45	31	41	29	0	0
%	22,3	23,9	16,5	21,8	15,4	0	0

Results are expressed as the number of samples giving a positive result with each supplementary test, TC: test for tetracyclines, AG: test for aminoglycosides, SU: test for sulphonamides, Q: test for quinolones, BL/ML: test for beta-lactams/macrolides, BL: test for beta-lactams

LODs while a positive result in Explorer would indicate the presence in the sample of residues from other antimicrobial groups. A positive result observed simultaneously with both tests might be caused by a very high concentration of non-quinolone antimicrobial residues or by the presence of quinolones together with other antimicrobials [13].

It is generally agreed that samples should be screened in a first step with easy, quick and inexpensive methods. These methods are expected to distinguish positive samples (containing levels of antibiotic residues above MRL levels) from the great majority of negative samples [19]. Therefore, screening tests should have the capability for a high sample throughput. Laborious microbial methods involving several plates and bacterial strains could limit their applicability to analyze simultaneously a large number of samples. In this way, the use of Explorer and Equinox tests would allow to analyze a considerable number of muscle samples with a reasonable effort. Moreover, both tests provide a remarkable advantage over other screening tests since the results can be interpreted in a objective way through a photometric reading. Thus, variations in visual reading made by different technicians or performed over different days are avoided, ensuring computer recording and traceability of assays. In addition to that, these ready-to-use tests allow prolonged storage, enhancing operational flexibility in analysis laboratories.

Beta-lactams, tetracyclines, aminoglycosides, macrolides, sulphonamides and quinolones are considered the most widespread antibacterial drugs used in food producing animals [6]. Sensitivity patterns of supplementary specific tests applied to positive screening samples in this work would cover the most commonly used antibacterial drugs in food producing animals and could allow to display a first classification of positive samples according to the family of residues.

The incidence of positive samples observed in our study after analyzing 4849 muscle samples with Explorer and Equinox tests was 7.3%. Nevertheless, Explorer sensitivity data provided by the manufacturer could imply that Explorer could not be sensitive enough to detect tetracyclines at MRL levels (Table 1). Consequently, the real incidence rate of samples containing antimicrobial residues above MRL levels might be even higher than figures included in this work.

Above half of the positive samples showed a positive result with at least one of supplementary screening tests. Several reasons could explain disagreements between results of screening and additional tests. For example, higher LODs in supplementary screening tests than in Explorer for certain antimicrobials could justify this fact. Thus, the presence of macrolides or aminoglycosides residues at concentrations able to inhibit Explorer could not be sufficient to get a positive result in plates seeded with *K. rhizophila* and *B. subtilis* at pH 8 [10,20,16]. Moreover, degradation of antimicrobial residues in samples along freezing storage could be also involved in obtained results [21,22].

Several studies performed in different countries to evaluate the incidence of antimicrobial residues in meat show substantial variability. Regional animal husbandry, antimicrobial treatment patterns and slaughter practices could be involved. Moreover, differences in the number of tested samples or used methodologies could make difficult an evaluation of published data. As a consequence, comparisons between studies should be taken carefully. Pikemaat et al. [23] observed a 10.8% incidence rate of inhibitory samples after analyzing 591 slaughter animals samples from different species with several microbial tests (Nouws antibiotic test, STAR protocol and Premitest). Furthermore, a higher incidence rate was reported in other study [24] when evaluating 351 muscle samples with a test based on *B. subtilis*. On the other hand, results published by Okerman et al. [19] were more in accordance with our data. Beta-lactams, tetracyclines and fluoroquinolones were detected in 8.3% of chicken samples with a combination of three plates seeded with different bacterial strains.

Results from supplementary screening assays (Table 3) indicate that up to 39% of inhibitory samples may contain tetracycline residues. Our data are in agreement with results reported by Darwish et al. [25] in animal-derived foods and by Salama et al. [26] in chicken samples. These authors observed that 38-42% of samples contained tetracyclines residues. Moreover, high levels of tetracycline residues were found in other studies in different animal species [19,27-29]. However, lower tetracycline rates (4.3-18%) were observed by other authors [23,30].

Occurrence of beta lactams reported in published works showed considerable variations. While Okerman et al. [19] found beta-lactam antibiotics in less than 1% of chicken samples when performing confirmation methods other authors reported that 18% of animal-derived foods in Africa contained beta-lactam residues [25]. Results obtained in our work (Table 3) suggest that no beta lactam residues were present in analyzed samples.

Er et al. [31] and Pena et al. [32] observed a high incidence of residual quinolones in muscle samples (44-58%). On the contrary, quinolone residues were identified much less frequently by other authors [33-36]. Furthermore, Okerman et al. [19] found no quinolones in chicken samples using a multiplate microbial test. Overall occurrence of quinolone residues in different animal species observed in our study might range from 16 to 26%.

Published data on occurrence of antimicrobial residues in muscle from different countries suggest that strategies used for control of residues could not be adequate enough. As a consequence, muscle samples with antibiotic residues above MRL levels could reach consumers and toxicological effects or allergic reactions might occur [1,2]. Moreover, a high incidence of samples containing antimicrobial residues may be linked to an overuse or inappropriate use of antimicrobial drugs in food-producing animals [37]. Hence, the antimicrobial resistance problem might be driven by killing susceptible strains and selecting those that are resistant [38,39]. In this way, the implement of more suitable strategies in antimicrobial residues monitoring plans could help to minimize the incidence of residues concentrations above MRL levels in foods and might have a positive effect on controlling the development and spread of antimicrobial resistance mechanisms.

4. CONCLUSION

Monitoring large numbers of slaughter animals for the presence of antimicrobial residues requires cheap and easy qualitative screening methods. In practice this screening step is primarily performed using microbiological screening tests, because of their high cost-effectiveness compared to physical–chemical methods. The proposed screening strategy (combination of Explorer and Equinox tests) appears to be a useful tool since it would enable a broad screening of antimicrobials in muscle samples. The analysis of a large number of muscle samples from different animal species have been performed using the proposed strategy. Positive screening results were observed in 7.3% of analyzed samples. Tetracyclines, sulphonamides and aminoglycosides were the most frequent residues found in inhibitory samples.

ACKNOWLEDGEMENTS

This investigation was carried out as part of the project PM025/2007 supported by financial assistance from the Diputación General de Aragón (DGA, Spain).

COMPETING INTERESTS

Authors have declared that no competing interests exist.

REFERENCES

1. Cerny A. Side effects and consequences of frequently used antibiotics in clinical practice. Schweizerische medizinische Wochenschrift. 1996;126:528-534.
2. Demoly P, Romano A. Update on beta-lactam allergy diagnosis. Curr Allergy Asthma Rep. 2005;5(1):9-14.
3. Mäyrä-Mäkinen A. Technological significance of residues for the dairy industry. In: Residues of Antimicrobial Drugs and Other Inhibitors in Milk. IDF Special Issue N° 9505. International Dairy Federation. 1995;136-143.
4. European Commission Council Regulation (EEC) No.2377/90 of 26 June 1990: laying down a Community procedure for the establishment of maximum residue limits of veterinary medicinal products in foodstuffs of animal origin. Official Journal of the European communities. 1990;L224:1–8.
5. Okerman G, Croubels S, Cherlet M, De Wasch K, De Backer P, Van Hoof J. Evaluation and establishing the performance of different screening tests for tetracycline residues in animal tissues. Food Addit Contam. 2004;21:145-53.
6. Gaudin V, Maris P, Fuselier R, Ribouchon JL, Cadieu N, Rault A. Validation of a microbiological method: The STAR protocol, a five-plate test, for the screening of antibiotic residues in milk. Food Addit Contam. 2004;21(5):422–433.
7. Mc Glinchey TA, Rafter PA, Regan F, McMahon GP. A review of analytical

methods for the determination of aminoglycoside and macrolide residues in food matrices. Anal Chim Acta. 2008;624:1–15.

8. Pikkemaat MG. Microbial screening methods for detection of antibiotic residues in slaughter animals. Anal Bioanal Chem. 2009;395:893-905.
9. Stead S, Sharman M, Tarbin JA, Gibson E, Richmond S, Stark J, Geijp E. Meeting maximum residue limits: An improved screening technique for the rapid detection of antimicrobial residues in animal food products. Food Addit Contam. 2004; 21(3):216-21.
10. Gaudin V, Hedou C, Verdon E. (Validation of a wide-spectrum microbiological tube test, the EXPLORER® test, for the detection of antimicrobials in muscle from different animal species. Food Addit Contam. 2009;26:1162–1171.
11. Cullor JS. Test for identifying antibiotic residues in milk: How well do they work? Veterinary Medicine. 1992;87:1235-1241.
12. Pikkemaat MG, Mulder PP, Elferink JW, de Cocq A, Nielen MW, van Egmond, HJ. Improved microbial screening assay for the detection of quinolone residues in poultry and eggs. Food Addit Contam. 2007;24(8):842–850.
13. Sanz D, Mata L, Condón S, Sanz MA, Razquin P. Performance of a new microbial test for quinolone residues in muscle. Food Anal Methods. 2011;4(2):212-220.
14. Mata L, Sanz D, Razquin P. Validation of the Explorer® 2.0 test coupled to e-Reader® for the screening of antimicrobials in muscle from different animal species. Food Addit Contam: Part A. 2014;31(9):1496–1505.
15. Bogaerts R, Wolf F. A standardized method for detection of residues of antibacterial substances in fresh meat. Fleischwirtschaft.1980;60:672–674.
16. Okerman G, Van Hoof J, Debeuckelaere W. Evaluation of the European four-plate test as a tool for screening antibiotic residues in meat samples from retail outlets. J AOAC Int. 1998;81(1):51-56.
17. International Dairy Federation (IDF). Detection and confirmation of inhibitors in milk and milk products. IDF Bulletin. 1991;258.
18. Ferrini AM, Mannoni V, Aureli P. Combined Plate Microbial Assay (CPMA): A 6-plate-method for simultaneous first and second level screening of antibacterial residues in meat. Food Addit Contam. 2006;23(1):16-24.
19. Okerman G, Croubels S, De Baere S, Van Hoof J, De Backer P, De Brabander H. Inhibition tests for detection and presumptive identification of tetracyclines, beta-lactam antibiotics and quinolones in poultry meat. Food Addit Contam. 2001;18(5):385-393.
20. Korsrud GO, Boison JO, Nouws JF, MacNeil JD. Bacterial inhibition tests used to screen for antimicrobial veterinary drug residues in slaughtered animals. J AOAC Int. 1998;81(1):21-24
21. Alfredsson G, Ohlsson A. Stability of sulphonamide drugs in meat during storage. Food Addit Contam. 1998;15(3):302-306.
22. Boison JO, Korsrud GO, MacNeil JD, Yates WDG, Papich MG. Effect of cold-temperature storage on stability of benzylpenicillin residues in plasma and tissues of food-producing animals. J AOAC Int. 1992;75:974-978.
23. Pikkemaat MG, Rapallini M, Oostra S, Elferink JWA. Comparison of three microbial screening methods for antibiotics using routine monitoring samples. Anal Chim Acta. 2009;*637*:298-304.
24. Johnston RW, Reamer RH, Harris EW, Fugate HG, Schwab B. A new screening method for the detection of antibiotic residues in meat and poultry tissues. J Food Prot. 1981;44:828-831.
25. Darwish WS, Eldaly EA, El-Abbasy MT, Ikenaka Y, Nakayama S, Ishizuka M. Antibiotic residues in food: the African scenario. Jpn J Vet Res. 2013; 61(Supplement):S13-S22.
26. Salama NA, Abou-Raya SH, Shalaby AR, Emam WH, Mehaya FM. Incidence of tetracycline residues in chicken meat and liver retailed to consumers. Food Addit Contam Part B. 2011;4(2):88-93.
27. Adesokan HK, Agada CA, Adetunji VO, Akanbi IM. Oxytetracycline and penicillin-G residues in cattle slaughtered in south-western Nigeria: Implications for livestock disease management and public health. J S Afr Vet Assoc. 2013;84(1):E1-5.
28. Al-Ghamdi MS, Al-Mustafa ZH, El-Morsy F, Al-Faky A, Haider I, Essa H. Residues of tetracycline compounds in poultry products in the eastern province of Saudi Arabia. Public Health. 2000;114(4):300-304.

29. De Wasch K, Okerman L, De Brabander H, Van Hoof J, Croubels S, De Backer P. Detection of residues of tetracycline antibiotics in pork and chicken meat: correlation between results of screening and confirmatory tests. Analyst. 1998;123:2737-2741.
30. Cetinkaya F, Yibar A, Soyutemiz GE, Okutan B, Ozcan A, Karaca MY. Determination of tetracycline residues in chicken meat by liquid chromatography-tandem mass spectrometry. Food Addit Contam. 2012;5(1):45-49.
31. Er B, Onurdag FK, Demirhan B, Ozgacar SÖ, Oktem AB, Abbasoglu U. Screening of quinolone antibiotic residues in chicken meat and beef sold in the markets of Ankara, Turkey. Poult Sci. 2013;92(8):2212-2215.
32. Pena A, Silva LJG, Pereira A, Meisel L, Lino CM. Determination of fluoroquinolone residues in poultry muscle in Portugal. Anal Bioanal Chem. 2010;*397*:2615–2621.
33. Salehzadeh F, Salehzadeh A, Rokni N, Madani R, Golchinefar F. Enrofloxacin residue in chicken tissues from Tehran slaughterhouses in Iran. PJN. 2007;6:409–413.
34. Silfrany RO, Caba RE, Solís de Los Santos F, Hanning I. Detection of quinolones in poultry meat obtained from retail centers in Santiago Province, the Dominican Republic. J Food Prot. 2013;76(2):352-354.
35. Weiss C, Conte A, Milandri C, Scortichini G, Semprini P, Usberti R, Migliorati, G. Veterinary drugs residue monitoring in Italian poultry: Current strategies and possible developments. Food Control. 2007;18:1068–1076.
36. Zhao S, Li X, Ra Y, Li C, Jiang H, Li J, Qu Z, Zhang S, He F, Wan Y, Feng C, Zheng Z, Shen J. Developing and optimizing an immunoaffinity cleanup technique for determination of quinolones from chicken muscle. J Agric Food Chem. 2009;57:365–371.
37. Anonymous. Background Document for the Joint WHO/FAO/OIE Expert Workshop on non-human. antimicrobial usage and antimicrobial resistance: Scientific Assessment, Geneva, Switzerland; 2003. Available:http://www.who.int/gfn/links/en/background.pdf (Accessed 9 September 2014).
38. Levy SB, Marshall B. Antibacterial resistance worldwide: Causes, challenges and responses. Nat Med. 2004;10(12 Suppl):S122-9.
39. Chantziaras I, Boyen F, Callens B, Dewulf J. Correlation between veterinary antimicrobial use and antimicrobial resistance in food-producing animals: A report on seven countries. J Antimicrob Chemother. 2014;69(3):827-834.

11

Association of Supplemental Nutrition Assistance Program (SNAP) with Food Security and Nutrition Status among Persons Living with HIV

Irene Hatsu[1*], Fatma Huffman[2], Paulette Johnson[3], Marianna Baum[2], Barbara Thomlison[4] and Adriana Campa[2]

[1]Department of Human Science, Ohio State University, Columbus OH, USA.
[2]Department of Dietetics and Nutrition, Florida International University, Miami, FL, USA.
[3]Statistical Consulting, Florida International University, Miami, FL, USA.
[4]School of Social Work, Florida International University, Miami, FL, USA.

Authors' contributions

This work was carried out in collaboration between all authors. Author IH conceptualized and designed the study, collected data, performed the statistical analysis, and wrote the first draft of the manuscript. Author PJ performed statistical analyses and reviewed manuscript. Authors MB, FH and BT contributed to study conceptualization and reviewed manuscript. Author AC contributed to study conceptualization, manuscript write-up and revision of manuscript. All authors read and approved the final manuscript.

ABSTRACT

Aims: Nutritional status can be compromised by food insecurity which is common among HIV infected persons. Providing food assistance is expected to improve food insecurity and nutritional status among persons infected with HIV. This study aimed at examining the relationship of participating in the Supplemental Nutrition Assistance Program (SNAP), the largest food assistance program in the United States, with food security and nutritional status among HIV infected adults.
Study Design: A cross-sectional study design was used in this study.
Place and Duration of Study: This study was conducted in Miami, FL, USA, between April 2011 and August 2012.
Methodology: We included 159 HIV infected individuals in this study, 113 participants were SNAP recipients while 46 were not. All study participants were, however, eligible to participate in SNAP. Each participant completed demographic and food security surveys as well as dietary and nutrition status assessment. Statistical analyses were conducted using univariate and multivariate analyses.

**Corresponding author: Email: Hatsu.1@osu.edu*

Results: More than half (56%) of the sample experienced food insecurity and had inadequate intakes of several nutrients. There were no significant differences in food security level and nutritional status between SNAP participants and eligible non-participants, even after controlling for demographic and health characteristics. Individuals with very low food security had 4.7 times increased odds (95% CI: 1.29-17.38) of illicit drugs use, which was prevalent (38%) among HIV+ SNAP participants in Miami. Drug users were more than twice likely to have inadequate intakes of vitamins B_1, B_2, B_6, and zinc, compared to non-drug users.

- **Conclusion:** Our results do not support an association between SNAP participation and food security or nutritional status in this cohort of HIV infected individuals with prevalent substance abuse. However, it demonstrates that food insecurity and inadequate nutrient intake continues to be prevalent among HIV infected adults and it is related to drug abuse. Resources need to be identified and targeted at addressing both food insecurity and poor nutritional outcomes among populations of HIV infected adults.

Keywords: Supplemental Nutrition Assistance Program (SNAP); food security; HIV infection; nutrition status; nutrient intake.

1. INTRODUCTION

Poor nutritional status is a predictor of morbidity and mortality in the course of HIV infection [1,2], even during antiretroviral treatment (ART) [3,4]. When compared to non-infected individuals, HIV infected persons have lower serum concentration of nutrients and are more susceptible to nutritional deficiencies [5-8]. With the advent of ART, morbidity and mortality in HIV infected individuals has improved, however, poor nutritional status is still present among many populations in the United States and in developing countries [3,9].

Efforts to improve nutritional status during HIV infection, especially among those in limited resource settings, has included the provision of food assistance. These have been beneficial in improving nutritional status, however, they had no effect on immunologic outcomes [10-13]. A recent study conducted in Haiti found that HIV infected adults with low BMI, who lived in very poor socioeconomic conditions, were able to significantly increase their BMI when provided food assistance [14]. Similar findings, in addition to improvements in body composition, were reported from resource adequate settings, although the interventions only provided nutrient supplements with occasional counseling [15-18]. Nonetheless, these interventions produced significant changes in BMI and other measures of nutritional status [15,19], an indication that some forms of nutritional assistance are beneficial, especially for HIV infected persons with poor socioeconomic status.

Food insecurity "exists whenever the availability of nutritionally adequate, and safe foods or the ability to acquire acceptable foods in socially acceptable ways, is limited or uncertain" [20]. Food insecurity is prevalent among HIV infected individuals regardless of setting [21-28]. It is linked with HIV disease through poverty, poor outlook of life, diminished functional capacity and unemployment, creating a vicious cycle that affects the severity of each condition [29]. Moreover, food insecurity impacts HIV disease transmission through increases in risky behaviors [28,30-33]. Poor mental health, including drug and alcohol use have also been associated with food insecurity among persons living with HIV [33].

Food insecurity for the HIV infected person increases the risk of experiencing compromised nutrition [29,34]. In developing countries, the provision of food aid in the form of ready-to-eat therapeutic foods or food rations to persons living with HIV, has ameliorated food insecurity [13,14]. Although food insecurity rates are high among persons living with HIV in developed countries [21,24,26,28], the literature is scarce on potential benefits of providing food aid to this population. In the United States, the Supplemental Nutrition Assistance Program (SNAP), formerly the Food Stamp Program, is the largest food assistance program provided to alleviate problems of food insufficiency in low income households [35]. The program provides monthly cash benefits on an electronic benefit transfer card. This card can be used in a manner similar to a bankcard in purchasing eligible food items (excludes hot-food items, alcoholic beverages, vitamin supplements and cigarettes) with retailers authorized to participate in the program. Though not conclusive, evidence shows that SNAP is effective in ameliorating food insecurity in the general population, with demonstrated nutritional and health benefits from participation [36-40]. Due to its size, SNAP is most likely the food assistance program used by the majority of low income HIV infected adults. However, to date, there is no known study that has investigated the potential nutrition and food security benefits associated with participation in SNAP among HIV infected adults. The aim of this study is to investigate the relationship of participating in SNAP with food security and nutritional status of HIV infected adults in Miami, FL. The State of Florida has one of the highest number of HIV cases in the United States [41]. According to the Center for Disease Control and Prevention, Miami-Dade county (including the city of Miami) ranked first and third respectively, in HIV and AIDS reported case rates in 2013 [41]. Additionally, 16% of the residents of Miami-Dade County live below the federal poverty level [42]. With high incidence and prevalence rates of HIV/AIDS, and widespread poverty, Miami is an excellent site to study the relationship between HIV infection, participation in nutrition assistance programs, and nutritional health parameters. We hypothesized that participation in SNAP will be associated with significantly improved food security level and nutritional status. Parameters of nutritional status examined in this study include nutrient intake, anthropometric and body composition measurements, as well as biochemical indicators. The study was approved by the Institutional Review Board of the Florida International University (FIU).

2. MATERIALS AND METHODS

2.1 Study Design and Setting

This was a cross-sectional study conducted between April 2011 and August 2012. Its purpose was to compare nutritional status and food security levels among HIV infected adults who were participating in SNAP and eligible non-participants. Participants were recruited consecutively through the FIU-Borinquen HIV and Nutrition Research Clinic (HNRC), located in the Borinquen Health Care Center, Miami, Florida, and other local centers that provide HIV related services to low income infected individuals in Miami-Dade County, Florida. The HNRC conducts several large HIV and Nutrition related studies, and it is attended by several hundreds of persons living with HIV. Recruitment was conducted via flyers posted around these centers, referrals from case managers, and by word of mouth. Participants were considered eligible to participate in the study if they: (1) were 18 years or older, (2) were participating in or eligible to participate in SNAP, and (3) had medical documentation of HIV seropositive status.

Informed consent was obtained from eligible participants who were willing to participate. All who consented were scheduled to complete a study visit. Eligibility for SNAP was determined by the ACCESS Florida Pre-screening eligibility tool. This gathers information on income, assets, household size, and expenditure, and is used as a basic screening tool by the State of Florida for SNAP eligibility [43]. SNAP participation was defined as having received SNAP benefits for at least one month in the last 12 months. The visit also included completing a self-administered questionnaire and dietary, anthropometric, and body composition assessments. Complete blood chemistry (obtained within three months of the study visit) was also obtained from medical records. Assessments were completed by a professional nutritionist. A $10.00 incentive was given to participants for completing the study visit.

2.2 Survey Instruments

2.2.1 Demographic, socioeconomic and lifestyle information

Several questions were used to assess the participants' sociodemographic and economic status. Information collected included age, gender, race/ethnicity, country of birth, marital status, living with a child or not, education, employment, living conditions, monthly income, smoking status, recreational drug use, alcohol use, ART medication use, vitamin use, and the use of other food assistance programs. This questionnaire was field-tested prior to the initiation of the study among 21 HIV infected individuals. These individuals were recruited from the same sources as the study population and had similar characteristics as the study population. The questionnaire was revised based on the field-testing results.

2.2.2 Food security

Recent food security status was measured using the United States Department of Agriculture's (USDA) 6-Item Short Form US Household Food Security Survey Module [44]. This brief survey was selected in order to decrease participant's response burden. Although it is a shorter version of the original 18-item questionnaire, it is able to measure food security and distinguish it from food insecurity [44]. It has enough specificity, sensitivity, and minimal bias compared to the original modules [44]. Food security scores were coded and calculated using standard methods. Responses of "often true", "sometimes true", "almost every month", "some months but not every month" and "yes" were coded as affirmative (1). The total sum of affirmatives for all questions was considered the food security score [44]. The maximum score possible is 6 with higher scores indicating greater severity of food insecurity. Scores of 0-1 were classified as food secure, while 2-4 and 5-6 were classified as low food security and very low food security respectively. The latter two classifications are considered food insecurity [44].

2.2.3 Nutritional status assessment

2.2.3.1 Nutrient intake

Using a 24-hour dietary recall, a professional nutritionist assessed dietary and nutrient intake of study participants. They were each queried about foods and beverages consumed the day prior to the study visit. The brand of food, cooking method used, and the amount consumed were all recorded to ensure the comprehensiveness of information collected. Food models and measuring cups were used as props to accurately estimate the quantity of food consumed. The dietary information collected was analyzed using NutriBase Professional Nutrition Software [45] to determine caloric and micro-/macronutrient composition.

2.2.3.2 Body composition

Participants' height and weight were measured at the time of data collection. Height was measured without shoes using a wall mounted stadiometer. The results were recorded to the nearest 0.5 inch. Each measurement was done with participants' heels touching the base of the stadiometer. Participant's weight was obtained with them wearing light clothing and no shoes. Weight was measured to the nearest 0.1 lbs. using a standard calibrated scale. BMI was calculated using Weight (kg)/ Height (m^2). Body composition assessments were conducted using the bioelectrical impedance analysis (BIA) method [46]. Participants' waist and hip circumference were measured and used to calculate waist-to-hip ratio. Waist circumference (cm) was measured using a non-stretchable tape measured at the narrowest point of the waist; hip circumference (cm) was measured at the widest point of the hip [47].

2.2.3.3 Biochemical indicators

Laboratory measurements of serum albumin, hemoglobin and hematocrit concentration were also obtained to assess nutritional status. These were obtained from medical records, either provided by participants, or they signed a HIPAA authorization of medical release form with which we secured this information directly from the provider. The laboratory records were required to be less than 3 months old from the day of the study visit and needed to contain complete blood chemistry results.

2.3 Statistical Analysis

Statistical analyses were conducted using SPSS version 21.0 for Windows [48]. Descriptive statistics were used in characterizing the participants and were expressed as mean ± standard deviation, or percentages. Differences in characteristics by SNAP participation status were compared using student's t-test for continuous normally distributed variables, chi-square test for categorical variables, and Mann-Whitney U test for continuous variables that were not normally distributed.

Average nutrient intakes were adjusted for total calorie intake [49] and the Mann-Whitney U test assessed the differences in nutrient intakes between SNAP participants and eligible non-participants. Some of the body composition and blood chemistry variables were not normally distributed; as a result, both the Mann-Whitney U test and student's t-test were used to assess the differences in the variables between the groups. The results from both tests were similar, thus, only those from the student's t-test were reported. Adequacy of nutrient intake was determined using the Estimated Average Intake (EAR) requirements for each nutrient and Adequate Intake (AI) for nutrients whose EAR are yet to be determined [50]. Chi-square tests were then used to compare the adequacy of nutrient intake by SNAP participation status.

Chi-square tests were also used to compare the severity of food insecurity by SNAP participation status. The associations between SNAP participation and 1) food security status and 2) nutritional status (dietary intake), were examined using logistic regression. The association between SNAP participation and nutritional status (body composition and blood chemistries) was assessed using multiple linear regression. The regression models were built using SNAP as the independent variable and food security and nutrition status as the dependent variables. Variables controlled for in the logistic and multiple linear regression analyses were: age, gender, ethnicity, country of birth, living with a child or not, work status, household size, ART use, smoking status, alcohol use, drug use, vitamin use, and use of other food assistance programs. Participants with missing data (N=16) were excluded from the final analysis. The significance level for all analysis was set at $P<.05$.

3. RESULTS

3.1 Demographic Characteristics

After providing informed consent, 175 participants enrolled and participated in the study. However, due to missing data (such as SNAP participation status, food security status, etc.) only 159 were included in analyses. Participants were mostly male (67%), African-American (75%), with a mean age of 46.81±8.03 years. Over 90% were single, with nearly half (48%) without children. Over half (52%) reported having more than a high school education, with about 45% and 42% of participants claiming unemployment and disability respectively.

The mean duration for program participation among SNAP recipients was 10.52±2.90 months. Significantly more SNAP participants (49%) were on disability than non-participants (24%), and fewer (39.8%) were unemployed compared to the non-participants (57%), $P = .015$, $\Phi = 0.23$. As shown in Table 1, more SNAP participants (84%) were on ART compared to the non-participants (65%), $P = .001$, $\Phi = 0.37$. More SNAP participants used illicit drugs (38%) than non-participants (13.0%), $P = .002$, $\Phi = 0.25$. The distribution of alcohol use also differed significantly by SNAP status, $P = .03$. Forty four percent of SNAP participants drank alcohol at least 2-3 times/week compared to only twenty six percent of non-participants.

Table 1. Demographic characteristics by SNAP participation status among HIV infected adults

Variable	SNAP (n=113)	Non-SNAP (n=46)	*P*-value
Age[a]	46.81±7.52	46.80±8.76	.99
Male	77 (68.1)	30 (65.1)	.72
Ethnicity			.27
African American	88 (77.9)	31 (67.4)	
Hispanic	17 (15.0)	7 (15.2)	
White	4 (3.5)	4 (8.7)	
Other	4 (3.5)	4 (8.7)	
US born ***	103 (91.2)	30 (65.2)	<.001
Single	102 (90.3)	43 (93.5)	.52
No children	53 (46.9)	24 (48.8)	.35
More than High School	63 (55.8)	20 (43.5)	.16
Employment status*			.02
Unemployed	45 (39.8)	26 (56.5)	
Employed	13 (11.5)	9 (19.6)	
On disability	55 (48.7)	11 (23.9)	
Monthly income < $1000	95 (84.1)	34 (73.9)	.14
Living condition			.55
Alone	44 (38.9)	18 (39.1)	
With family	52 (46.0)	18 (39.1)	
Shelter	17 (15.0)	11 (21.7)	
Uses of other food assistance	16 (14.2)	7 (15.2)	.86
Smokes cigarettes	78 (69.0)	26 (56.5)	.13
Alcohol use	63 (55.8)	18 (39.1	.06
Drug use**	43 (38.1)	6 (13.0)	.002
On ART***	106 (93.8)	30 (65.2)	<.001
Takes vitamins	51 (45.1)	21 (45.1)	.95

*[a]Reported as mean±SD; All other variables are reported as n (%); ART: Antiretroviral therapy, *P<.05, **P<.01, ***P<.001*

3.2 Association between SNAP Participation and Food Security

About 32% of the sample experienced low food security while 24% experienced very low food security. There were no significant differences in the level of food security between groups. Nearly half (46%) of SNAP participants were food secure while 39% of non-SNAP participants experienced food security. Logistic regression examined the association between food security status and SNAP participation, controlling for confounding variables (Table 2). SNAP participation status was not significant in the logistic regression model.

The logistic regression also identified relationships between levels of food security and some demographic characteristics. Individuals experiencing very low food security relative to those food secure had 4.7 times greater odds (95% CI: 1.29-17.38) of using illicit drugs, and 8.3 times decreased odds [AOR= 0.12 (95% CI: 0.02-0.85)] of having children at home. Similarly, those experiencing low food security compared to those who were food secure had 3.6 times greater odds (95% CI: 1.10-11.48) of illicit drug use, and 4.2 times greater odds (95% CI: 1.21-14.37) of using other food assistance programs.

3.3 Association between SNAP Participation and Nutrition Status

No significant differences were seen in the nutrient and fiber intakes between SNAP participants and non-participants. Mean caloric intake for both groups was approximately 2000 kcals per day. Energy adjusted protein intake for both groups was below the recommendation for healthy adults while carbohydrate intake was above recommendation. Both groups also consumed below recommended amounts for fiber. For about half of the nutrients assessed, more than 50% of the sample had intakes below the recommended Dietary Reference Intakes (DRI), whether EAR or AI. More than 90% had intakes below DRI requirements for vitamin E, potassium and magnesium. Vitamin D intake was below EAR for 100% of the study sample. In those receiving SNAP, duration of participation as a continuous variable did not significantly change nutrient intake. There were no statistically significant differences in percentages of participants with intake below DRI for any of the nutrients by SNAP participation status (Table 3).

Table 2. Logistic regression of food security on SNAP participation status among HIV infected adults

Food security status	SNAP participation			
	Unadjusted		Adjusted	
	OR	95% CI	OR	95% CI
Food secure	----	----	----	----
Low food security	0.76	0.34-1.68	0.69	0.24-1.98
Very low food security	0.75	0.32-1.79	0.45	0.14-1.46

Model controlled for age, gender, ethnicity, country of birth, living with a child or not, work status, household size, ART use, smoking status, alcohol use, drug use, vitamin use, and use of other food assistance programs

Table 3. Percentages of HIV infected adults with nutrient intakes below Estimated Average Requirements (EAR) by SNAP participation status

Nutrient	SNAP (n=113)	Non-SNAP (n=46)	*P*-value
Vitamin A *μg*	92 (81.4)	38 (82.6)	.86
Thiamin *mg*	56 (49.6)	21 (45.7)	.66
Riboflavin *mg*	53 (46.9)	21 (45.7)	.87
Niacin *mg*	43 (38.1)	17 (37.0)	.89
Pantothenic acid *mg* [a]	91 (80.5)	38 (82.6)	.77
Vitamin B_6 *mg*	60 (53.1)	27 (58.7)	.52
Vitamin B_{12} *μg*	50 (44.2)	23 (50.0)	.51
Vitamin C *mg*	75 (66.4)	33 (71.7)	.51
Calcium *mg*	82 (72.6)	33 (71.7)	.92
Vitamin D *μg*	113 (100)	46 (100)	1.00

Table 3 continued……..

Vitamin E *mg*	111 (98.2)	46 (100)	.36
Folate *µg*	70 (61.9)	28 (60.9)	.89
Magnesium *mg*	102 (90.3)	44 (95.7)	.26
Phosphorus *mg*	41 (36.3)	15 (32.6)	.66
Potassium *mg* [a]	111 (98.2)	46 (100)	.36
Sodium *mg* [a]	10 (8.8)	6 (13.0)	.43
Iron *mg*	18 (15.9)	7 (15.2)	.91
Selenium *µg*	43 (38.1)	11 (23.9)	.08
Zinc *mg*	67 (59.3)	31 (67.4)	.34

[a]AI was used; EAR for nutrient not yet determined

Logistic regressions of SNAP participation status on adequacy of nutrient intakes showed no significant results. Several demographic characteristics were associated with the ability to achieve adequacy of nutrient intakes. Adequacy of calcium intake was associated with older age (P = .04), while adequacy of vitamin A intake was associated with younger age (P = .009) and not having children at home (P = .02). Adequacy of vitamin B_{12} and iron intake was associated with being male (P = .02, .007, respectively). Of the lifestyle related variables, drug and alcohol use were found to be associated with below EAR intake for some nutrients. After adjusting for control variables, the odds of drug users having below EAR intakes were 2.9 (95% CI: 1.09-7.49) for vitamin B_1, 3.3 (95% CI: 1.20-9.01) for vitamin B_2, 2.5 (95% CI: 0.95-6.45) for vitamin B_6, and 2.6 (95% CI: 1.00-6.56) for zinc, relative to non-users. Alcohol drinkers had 2.5 times greater odds (95% CI: 1.02-6.14) of below EAR intake for copper compared to non-drinkers. Use of ART was associated with above EAR intake for iron (AOR= 5.6, 95% CI: 1.17-26.40).

The average BMI for this population was 27.81±5.82 with an average waist-to-hip ratio of 0.9. There were no significant differences in mean values of body composition and biochemical indicator variables by SNAP participation status (Results not shown). Gender was significantly related to all variables except albumin. Being female was associated with having higher body mass index (β = .34, P<.001), higher fat mass (β = .47, P<.001), lower lean body mass (β = -.50, P<.001) and lower waist-to-hip ratio (β = -.27, P = .004). Multiple linear regression analyses were conducted to assess the association of SNAP participation status on body composition and biochemical variables. About 30% of the variability of most variables (with the exception of albumin and waist-to-hip ratio) was explained by the control variables. However, SNAP participation status explained less or equal to 1.4% additional variability of body composition and biochemical variables after controlling for these variables.

4. DISCUSSION AND CONCLUSION

The current study did not find an association between SNAP participation and food security, although some [51,52] but not all studies [53,54] have found that participation in SNAP alleviates food insecurity in the general population. More than half (56%) of the participants in this study experienced lower levels of food security. This rate is higher than both the national food insecurity rate of 14.9% and the state of Florida rate of 15.4% reported in 2011 [55]. The prevalence of very low food security in our study population was more than four times that reported for the general public in the same year [55]. High food insecurity rates have consistently been reported among persons living with HIV in North America. Eighty one percent of HIV infected adults in a study conducted in Miami, with a similar population, experienced food insecurity in addition to wasting [23]. Another study by Vogenthaler et al. [26] in Atlanta and Miami, reported a 34% food insecurity rate among those living with HIV. Studies conducted in Canada and California reported rates of food insecurity between 48-64% [21,23,24,28]. It is possible that the ability of SNAP to alleviate the effects of heightened food insecurity in HIV infected people is minimal compared to that observed in the general population, due to the increased nutritional demands, poor nutritional status and the severe socio-economic conditions surrounding this population in Miami.

Illicit drug use had a strong association with lower levels of food security in our population. Drug users had more than three times increased odds for experiencing food insecurity. Our findings are consistent with those noted in other HIV infected cohorts across the country [33,56,57], and suggests the need to include food insecurity intervention with drug intervention programs for persons living with HIV. The use of other food assistance programs was also associated with food insecurity and this finding was not surprising. Individuals having difficulty to attain food sufficiency are more likely to use several food assistance programs. Demographic characteristics like income, unemployment and instability in housing were not associated with food insecurity in this study sample.

The progression and management of HIV infection is affected by several factors including diet and nutritional status. Due to the collaborative interaction between nutrient status, infection and immunity, marginal or low intake of nutrients, especially those significantly associated with HIV disease is of particular concern [58,59]. Several studies have investigated the caloric and macronutrient intake of persons infected with HIV, with varying results [60-63]. Our study shows that calorie and macronutrient intake in this population is lower than what is recommended for healthy adults. Due to the demands of the disease, individuals infected with HIV have increased caloric and macronutrients needs [64]. Resting metabolic rates (RMR) is increased by about 10% for HIV infected adults, hence inadequate energy intake can contribute to poor nutritional outcomes [65]. Among individuals infected with HIV, low serum levels of Vitamins A, E and B_{12} have been linked to neurological abnormalities, increased oxidative stress and increased risk of mortality [66,67]. Our HIV infected cohort had inadequate intake of Vitamins B_{12} (46% of participants), as well as A and E (>80% of participants). These percentages are higher than those reported for another HIV cohort, where 21%-64% of the women had below DRI intakes for several vitamins including vitamins A and E [62]. Several studies report lower intakes for vitamins A and E among persons living with HIV; however, in these studies comparison of mean/median intakes with DRI's (specifically, RDA: Recommended Daily Allowance) was made without specifying proportions consumed below the DRIs [59,60,55,68].

Vitamin D plays an important role in immunity and has been associated with disease progression, anemia and mortality among HIV infected persons [69,70]. Vitamin D deficiency is common among HIV infected adults, and Blacks or Hispanics are at increased risk for deficiency [71]. In our study, 100% of participants had below DRI intake for vitamin D. This is of concern because more than half of the standard ART combinations used in Miami include tenofovir, which has been known to contribute to altered Vitamin D metabolism [72]. In addition, 75% of our participants are African Americans and 15% Hispanics; two ethnic groups at heightened risk for vitamin D deficiency. Vitamin D status of persons living with HIV needs to be a priority, especially for Nutrition Practitioners in South Florida. Further investigation to assess vitamin D status using biomarkers is warranted for this population, as deficiency may be a mediator of disease progression and other comorbidities. Although selenium status is associated with disease progression and risk of mortality [59,66,73], supplementation is not routinely recommended during HIV infection due to lack of sufficient evidence supporting associated benefits [74]. More than a third of our participants were found to consume below DRI requirements for selenium. Selenium deficiency, however, is only prevalent among HIV infected individuals with poor dietary intakes; not all HIV cohort studies have found deficiencies [59]. Again, we found nearly 62% of our study population to have inadequate zinc intake. Recently, Baum et al. [75] showed that adequacy of zinc intake achieved through supplementation leads to a fourfold decrease in the chances of developing immunological failures. Our findings show a continued need to emphasize the adequacy of zinc intake among persons living with HIV, as deficiency is still common.

Recent HIV studies report a trend of increasing weight among several cohorts [76-78] and this was evident in our study. About 32% of our population was overweight and 30% obese, with only 1.9% being underweight. This trend reflects the participants' access to the lifesaving ART, and perhaps, the nations' obesity epidemic. The cause of overweight and obesity observed among HIV infected persons are multifaceted, however, diet and specific ART agents are the main influencers of weight gain [79]. Overall, we did not find any significant association between SNAP participation and nutritional status. However, our results may have been influenced by significant drug use among SNAP participants, which decreases the generalizability of our findings to other HIV-positive populations. To the best of our knowledge, this is the first study examining the nutrition and food security benefits derived from food assistance among persons living with HIV in a developed country. Several other studies (from developing countries) have explored the use of food assistance programs to improve the nutritional status and food security of HIV infected adults [12-14]. This research differs from the others in two ways: 1) participants received public food assistance through financial allocations to spend on food; they were not provided food rations and 2) researchers had no control on the type of food consumed. Although SNAP is not designed specifically for persons living with HIV, findings from this study could serve as basis for other studies exploring the use of food assistance programs to help improve the food security and nutritional status of HIV infected persons in developed countries. Future studies are needed to help us better understand where to invest resources to improve the lives of persons living with HIV.

The limitations of this study include its cross-sectional nature, relying on self-reported data which may weaken links between SNAP participation and better food security and nutritional outcomes. Our study did not take into account the differences in the monetary amounts received as SNAP benefits, the types of food purchased and consumed by SNAP participants and the food security and nutritional status of SNAP participants prior to participation in the program. It also did not take into account disease status of participants. The external validity of the study is also decreased because we used a small, convenience sample of HIV infected persons from Miami. The generalization of study results to other HIV infected populations therefore needs to be executed with caution. Finally, the adequacy of nutrient intakes was based on a single 24-hr recall, and may not represent usual intake. Repeated 24-hour recalls may have been more appropriate dietary assessment method, especially for the determination of micronutrients intakes. Despite these limitations, this study shows that food insecurity and inadequate nutrient intake continues to exist among persons infected with HIV. It also demonstrates that in this population, food insecurity and inadequate nutrients intake are related to drug abuse. Food insecurity increases the risk for poor nutritional status [29,34]; a known predictor of morbidity and mortality during HIV infection [1,2]. As such, our findings have implications for HIV treatment and management. Resources need to be identified and directed towards addressing food insecurity and poor nutritional outcomes among HIV infected populations in developed countries. This is especially important in drug abusing subpopulations.

CONSENT

All authors declare that written informed consent was obtained from each participants before inclusion in this study.

ETHICAL APPROVAL

All authors hereby declare that all experiments have been examined and approved by the Florida International University Institutional Review Board and have therefore been performed in accordance with the ethical standards laid down in the 1964 declaration of Helsinki. IRB Approval # 041111-00.

ACKNOWLEDGEMENTS

We thank all our study participants and the research staff at Florida International University (FIU) HIV and Nutrition research clinic. We also acknowledge FIU's University Graduate School for providing the Data Evidence Acquisition Fellowship which supported data collection for this study. The infrastructure for this study was partially funded by NIH grants from NIDA "HIV and HIV/HCV-Infection, Disease Progression, Oxidative Stress and Antioxidants, Grant No. R01DA023405, and from NIAAA "Alcohol & Antiretrovirals in HIV Infection, Oxidative Stress and Liver Disease Grant No. R01AA018011. The funding agencies did not have any role in the design, conduct and analyses of this study.

COMPETING INTERESTS

Authors have declared that no competing interests exist.

REFERENCES

1. Chlebowski RT, Grosvenor MB, Bernhard NH, Morales LS, Bulcavage LM. Nutritional status, gastrointestinal dysfunction, and survival in patients with AIDS. Am J Gastroenterol. 1989;84(10):1288-93.
2. Melchior JC, Niyongabo T, Henzel D, Durack-Bown I, Henri SC, Boulier A. Malnutrition and wasting, immuno-depression, and chronic inflammation as independent predictors of survival in HIV-infected patients. Nutrition. 1999;15(11): 865-69.
3. Koethe JR, Heimburger DC. Nutritional aspects of HIV-associated wasting in sub-

Saharan Africa. Am J Clin Nutr. 2010; 91(4):1138S-42S.
4. Liu E, Spiegelman D, Semu H, Hawkins C, Chalamilla AA, Nyamsangia S, et al. Nutritional status and mortality among HIV-infected patients receiving antiretroviral therapy in Tanzania. J Infect Dis. 2011;204 (2):282-90.
5. Semba RD, Shah N, Strathdee SA, Vlahov D. High prevalence of iron deficiency and anemia among female injection drug users with and without HIV infection. J Acquir Immune Defic Syndr. 2002;29(2):142-44.
6. Periquet BA, Jammes NM, Lambert WE, Tricore J, Moussa MM, Garcia J, et al. Micronutrient levels in HIV-1-infected children. AIDS. 1995;9(8):887-93.
7. Beach RS, Mantero-Atienza E, Shor-Posner G, Javier JJ, Szapocznik J, Morgan R, et al. Specific nutrient abnormalities in asymptomatic HIV-1 infection. AIDS. 1992; 6(7):701-08.
8. Lacey CJ, Murphy ME, Sanderson MJ, Monteiro EF, Vail A, Schorah CJ. Antioxidant-micronutrients and HIV infection. Int J STD AIDS. 1996;7(7):485-89.
9. Wanke CA, Silva M, Knox TA, Forrester J, Speigelman D, Gorbach SL. Weight Loss and Wasting Remain Common Complications in Individuals Infected with Human Immunodeficiency Virus in the Era of Highly Active Antiretroviral Therapy. Clin Infect Dis. 200;31(3):803-05.
10. Kadiyala S, Gillespie S. Rethinking food aid to fight AIDS. Food Nutr Bull. 2004; 25(3):271-82.
11. Ndekha MJ, Van Oosterhout JJ, Zijlstra EE, Manary M, Saloojee H, Manary MJ. Supplementary feeding with either ready-to-use fortified spread or corn-soy blend in wasted adults starting antiretroviral therapy in Malawi: A randomized, investigator blinded, controlled trial. BMJ. 2009;338: b1867.
12. Rawat R, Kadiyala S, McNamara PE. The impact of food assistance on weight gain and disease progression among HIV-infected individuals accessing AIDS care and treatment services in Uganda. BMC Public Health. 2010;10:316.
13. Cantrell RA, Sinkala M, Megazinni K, Lawson-Marriot S, Washington S, Chi BH, Tambatamba-Chapula B, Levy J, Stringer EM, Mulenga L, Stringer JS. A Pilot Study of Food Supplementation to Improve Adherence to Antiretroviral Therapy Among Food-Insecure Adults in Lusaka, Zambia. J Acquir Immune Defic Syndr. 2008;49(2):190-95.
14. Ivers LC, Chang Y, Gregory Jerome J, Freedberg KA. Food assistance is associated with improved body mass index, food security and attendance at clinic in an HIV program in central Haiti: A prospective observational cohort study. AIDS Res Ther. 2010;7:33. (DOI: 10.1186/1742-6405-7-33)
15. Clark RH, Feleke G, Din M, Yasminb T, Singh G, Khan FA, et al. Nutritional treatment for acquired immunodeficiency virus-associated wasting using beta-hydroxy beta-methylbutyrate, glutamine, and arginine: A randomized, double-blind, placebo-controlled study. J Parenter Enteral Nutr. 2000;24(3):133-39.
16. McDermott AY, Shevitz A, Must A, Harris S, Roubenoff R, Gorbach S. Nutrition Treatment for HIV Wasting: A Prescription for Food as Medicine. Nutr Clin Pract. 2008;18(1):86-94.
17. Pichard C, Sudre P, Karsegard V, Yerly S, Slosman DO, Delly V, Perrin L, Hirschel B. A randomized double-blind controlled study of 6 months of oral nutritional supplementation with arginine and 3 fatty acids in HIV-infected patients. AIDS. 1998; 12(1):53-63.
18. Stack JA, Bell SJ, Burke PA, Forse RA. High-Energy, High-Protein, Oral, Liquid, Nutrition Supplementation in Patients with HIV Infection: Effect on Weight Status in Relation to Incidence of Secondary Infection. J Am Diet Assoc. 1996;96(4): 337-41.
19. Shabert JK, Winslow C, Lacey JM, Wilmore DW. Glutamine-antioxidant supplementation increases body cell mass in AIDS patients with weight loss: A randomized, double-blind controlled trial. Nutrition. 1999;15(11):860-64.
20. Andersen SA. Core indicators of nutritional state for difficult-to-sample populations. J Nutr. 1990;120(11):1559-600.
21. Normen L, Chan K, Braitstein P, Anema A, Bondy G, Montaner JS, et al. Food Insecurity and Hunger Are Prevalent among HIV-Positive Individuals in British Columbia, Canada. J. Nutr. 2005;135 (4):820-25.
22. Weiser SD, Tsai AC, Gupta R, Frongillo EA, Kawuma A, Senkungu J, et al. Food insecurity is associated with morbidity and patterns of healthcare utilization among

HIV-infected individuals in a resource-poor setting. AIDS. 2012;26(1):67-75.
23. Campa A, Zhifang Y, Lai S, X. HIV-Related wasting in HIV-infected drug users in the era of highly active antiretroviral therapy. Clin Infect Dis. 2005;41(8):1179-85.
24. Anema A, Kerr T, Weiser SD, Montaner JSG, Wood E. Prevalence and correlates of self-reported hunger among HIV+ and HIV- injection drug users in Canada 5th International AIDS Society Conference on HIV Pathogenesis and Treatment Cape Town, South Africa; 2009.
25. Mamlin J, Kimaiyo S, Lewis S, Tadayo H, Jerop FK, Gichunge C, et al. Integrating nutrition support for food-insecure patients and their dependents into an HIV care and treatment program in Western Kenya. Am J Public Health. 2009;99(2):215-21.
26. Vogenthaler NS, Hadley C, Lewis SJ, Rodriguez AE, Metsch LR, Del Rio C. Food insufficiency among HIV-infected crack-cocaine users in Atlanta and Miami. Public Health Nutr. 2010:13(9):1478-84.
27. Wang E, McGinnis K, Fiellin D, Goulet JL, Bryant K, Gibert CL, et al. Food Insecurity is Associated with Poor Virologic Response among HIV-Infected Patients Receiving Antiretroviral Medications. J Gen Intern Med. 2011;26(9):1012-18.
28. Weiser S, Frongillo E, Ragland K, Hogg R, Riley E, Bangsberg D. Food insecurity is associated with incomplete HIV RNA suppression among homeless and marginally housed HIV-infected individuals in san francisco. J Gen Intern Med. 2009; 24(1):14-20, 2009.
29. Anema A, Vogenthaler N, Frongillo E, Kadiyala S, Weiser S. Food insecurity and HIV/AIDS: Current knowledge, gaps, and research priorities. Curr HIV/AIDS Rep. 2009;6(4):224-31.
30. Dunkle KL, Jewkes RK, Brown HC, Gray GE, McIntryre JA, Harlow SD. Transactional sex among women in Soweto, South Africa: Prevalence, risk factors and association with HIV infection. Soc Sci Med. 2004;59(8):1581-92.
31. Oyefara JL. Food insecurity, HIV/AIDS pandemic and sexual behaviour of female commercial sex workers in Lagos metropolis, Nigeria. Sahara J. 2007;4(2): 626-35.
32. Weiser S, Bangsberg D, Kegeles S, Ragland K, Kushel M, Frongillo E. Food insecurity among homeless and marginally housed individuals living with HIV/AIDS in san francisco. AIDS Behav. 2009;13(5): 841-48.
33. Weiser SD, Young SL, Cohen CR, Kushel MB, Tsai AC, Tien PC, et al. Conceptual framework for understanding the bidirectional links between food insecurity and HIV/AIDS. Am J Clin Nutr. 2011;94(6): 1729S–39S.
34. Gillespie S, Kadiyala S, International Food Policy Research I. HIV /AIDS and food and nutrition security: from evidence to action. Washington, DC: IFPRI; 2005.
35. United States Department Agriculture. Building a Healthy America: A profile of the supplemental nutrition assistance program. UDSA Food and Nutrition Service; 2012.
36. Allen JE, Gadson KE. Nutrient Consumption Patterns of Low-Income Households. USDA, Economic Research Service; 1983.
37. Beebout H, Cavin E, Devaney B. Evaluation of the Nutrition assistance program in puerto rico: Volume II, effects on food expenditures and diet quality. Washington, DC: Mathematica Policy Research, Inc; 1985.
38. Devaney B, Moffitt R. Dietary Effects of the Food Stamp Program. Am J Agric Econ. 1991;73(1):202-11.
39. Fox M, Hamilton W, Lin BH. Effects of food assistance and Nutrition Programs on nutrition and health; Volume 4 executive summaryof the literature review: Food Assisstance and Nutrition Research. FANRR19-4; 2004.
40. Fraker TM, Martini AP, Ohls JC. The Evaluation of the Alabama Food Stamp Cash-out Demonstration: Recipient Impacts. USDA, Food and Nutrition Service; 1992;1.
41. Florida Health. HIV Disease: United States vs Florida Fact Sheet. Available:http://www.floridahealth.gov/diseases-and conditions/aids/surveillance/_documents/fact-sheet/2014/2014-us-vs-fl-fact-sheet.pdf. Assessed March 12, 2015.
42. Miami-Dade HIV/AIDS Partnership. Comprehensive Plan for HIV/AIDS. 2009-2011. Available:http://www.aidsnet.org/newmain/partnershipall/partnership/0911compplan.pdf (Accessed September 10, 2012)
43. Department of Children and Families (DCF). ACCESS Florida. Available:https://dcfaccess.dcf.state.fl.us/access2florida/prescreening/welcome.do?p

erformAction=init&mode=Inter (Accessed December 10, 2012)

44. Blumberg SJ, Bialostosky K, Hamilton WL, Briefel RR. The Effectiveness of a short form of the household food security scale. Am J Public Health. 1999;89(8): 1231-34.
45. Nutribase Professional Nutrition and Fitness Software (Version 9.0) [Computer Software]. Phoenix, AZ: Cybersoft.
46. Kyle UG, Bosaeus I, De Lorenzo AD, Deurenberg P, Elia M, Manuel Gómez J, et al. Bioelectrical impedance analysis-Part II: Utilization in clinical practice. Clinical Nutrition. 2004;23(6):1430-53.
47. Chan DC, Watts GF, Barrett PHR, Burke V. Waist circumference, waist-to-hip ratio and body mass index as predictors of adipose tissue compartments in men. QJM. 2003:96(6):441-47.
48. SPSS Statistics Software (Version 21.0) [Computer Software]. Armonk, NY: IBM.
49. Willett WC, Howe GR, Kushi LH. Adjustment for total energy intake in epidemiologic studies. Am J Clin Nutr. 1997;65(4):1220S-1228S.
50. Dietary Reference Intakes: Application for Dietary Assessment. National Academy of Science. Institute of Medicine; 2000.
51. Nord M. How much does the Supplemental Nutrition Assistance Program alleviate food insecurity? Evidence from recent programme leavers. Public Health Nutr. 2012;15(5):811-17.
52. Ratcliffe C, McKernan S-M, Zhang S. How much does the supplemental nutrition assistance program reduce food insecurity? Am J Agric Econ. 2011;93(4): 1082-1098, 2011.
53. Gundersen C, Oliveira V. The food stamp program and food insufficiency. Am J Agric Econ. 2001;83(4):875-87.
54. Huffman S, Jensen H. Do food assistance programs improve household food security? Recent Evidence from the United States: Center for Agricultural and Rural Development, Iowa State Univ. Working Paper 03-WP 335; 2003.
55. Coleman-Jensen A, Nord M, Andrews M, Carlson S. Household food security in the united states in 2011. USDA, Economic Research Services; 2011.
56. Kalichman SC, Cherry C, Amaral C, White D, Kalichman MO, Pope H, et al. Health and treatment implications of food insufficiency among people living with HIV/AIDS, Atlanta, Georgia. J Urban Health. 2010;87(4):631-41.
57. Hendricks KM, Erzen HD, Wanke CA, Tang AM. Nutrition Issues in the HIV-Infected Injection Drug User: Findings from the Nutrition for Healthy Living Cohort. J Am Coll Nutr. 2010;29(2):136-43.
58. Piwoz E, Preble E. HIV/AIDS and nutrition: a review of the literature and recommendations for nutritional care and support in sub-Saharan Africa Washington, DC: Academy for Educational Development, Support for Analysis and Research in Africa Project; 2000.
59. Bogden JD, Oleske JM. The essential trace minerals, immunity, and progression of HIV-1 infection. Nutr Res. 2007;27(2): 69-77.
60. Kim JH, Spiegelman D, Rimm E, Gorbach SL. The correlates of dietary intake among HIV-positive adults. Am J Clin Nutr. 2001; 74(6):852-61.
61. Onyango AC, Walingo MK, Mbagaya G, Kakai R. Assessing nutrient intake and nutrient status of HIV seropositive patients attending clinic at chulaimbo sub-district hospital, Kenya. J Nutr Metab; 2012. (DOI: 10.1155/2012/306530)
62. Vorster HH, Kruger A, Margetts BM, Venter CS, Kruger HS, Veldman FJ, et al. The nutritional status of asymptomatic HIV-infected Africans: Directions for dietary intervention? Public Health Nutr. 2004; 7(8):1055-64.
63. Woods MN, Spiegelman D, Knox TA, Forrester JE, Connors JL, Skinner SC, et al. Nutrient intake and body weight in a large HIV cohort that includes women and minorities. J Am Diet Assoc. 2002; 102(2):203-11.
64. FANTA. Food Assistance Programming in the Context of HIV. Food and Nutrition Technical Assistance (FANTA) Project and World Food Programme (WFP). Washington, DC: FANTA Project, Academy for Educational Development; 2007.
65. Hsu JWC, Pencharz PB, Macallan D & Tomkins A. Macronutrients and HIV/AIDS: A Review of Current Evidence. Geneva: WHO; 2005. Available:www.who.int/nutrition/topics/Paper_1_Macronutrients_bangkok.pdf
66. Baum MK, Shor-Posner G, Lai S, Zhang G, Lai H, Flecher MA, et al. High Risk of HIV-Related Mortality Is Associated With Selenium Deficiency. J Acquir Immune Defic Syndr. 1997;15(5):370-74.

67. Tang AM, Smit E. Selected vitamins in HIV infection: A review. AIDS Patient Care STDS. 1998;12(4):263-73.
68. Smit E, Graham NMH, Tang A, Flynn C, Solomon L, Vlahov D. Dietary intake of community-based HIV-1 seropositive and seronegative injecting drug users. Nutrition. 1996;12(7):496-501.
69. Villamor E. A potential role for vitamin D on HIV infection? Nutr Rev. 2006;64(5) 226-33.
70. Mehta S, Giovannucci E, Mugusi FM, Spiegelman D, Aboud S, Hertzmark E, Msamanga GI, Hunetr D, Fawzi WW. Vitamin D status of HIV-infected women and its association with HIV disease progression, Anemia, and Mortality. PLoS ONE. 2010;5(1):e8770. (DOI: 10.1371/journal.pone.0008770)
71. Dao CN, Patel P, Overton ET, Rhame F, Pals SL, Johnson C, et al. Low Vitamin D among HIV-Infected Adults: Prevalence of and Risk Factors for Low Vitamin D Levels in a Cohort of HIV-Infected Adults and Comparison to Prevalence among Adults in the US General Population. Clin Infect Dis. 2011;52(3):396-405.
72. Childs KE, Fishman SL, Constable C, Gutierrez JA, Wyatt CM, Dieterich DT et al. Short communication: Inadequate vitamin D exacerbates parathyroid hormone elevations in tenofovir users. AIDS Res Hum Retroviruses. 2010;26(8):855-59.
73. Singhal N, Austin J. A clinical review of micronutrients in HIV infection. J Int Assoc Physicians AIDS Care. 2002;1(2):63-75.
74. National Institute of Health. Dietary supplement fact sheet: Selenium office of dietary supplements. Accessed October 23, 2012. Available:http://ods.od.nih.gov/factsheets/Selenium-HealthProfessional
75. Baum MK, Lai S, Sales S, Page JB, Campa A. Randomized, controlled clinical trial of zinc supplementation to prevent immunological failure in HIV-infected adults. Clin Infect Dis. 2010;50(12):1653-60.
76. Crum-Cianflone N, Roediger MP, Eberly L, Headd M, Marconi V, Ganesan A, Weintrob R, Barthel V, Fraser S, Agan BK. Increasing Rates of Obesity among HIV-Infected Persons during the HIV Epidemic. PLoS ONE. 2010;5(4):e10106. (DOI: 10.1371/journal.pone.0010106)
77. Crum-Cianflone N, Tejidor R, Medina S, Barahona I, Ganesan A. Obesity among patients with HIV: the latest epidemic. AIDS Patient Care STDS. 2008;22(12): 925-30.
78. Hendricks KM, Willis K, Houser R, Jones CY. Obesity in HIV-Infection: Dietary Correlates. J Am Coll Nutr. 2006;25(4): 321-31.
79. Dorey-Stein Z, Amorosa VK, Kostman JR, Lo Re V 3rd, Shannon RP. Severe weight gain, lipodystrophy, dyslipidemia, and obstructive sleep apnea in a human immunodeficiency virus-infected patient following highly active antiretroviral therapy. J Cardiometab Syndr. 2008;3(2): 111-14.

12

Minimal Impact of Nutrition Education and Fruit and Vegetable Consumption on Biomarkers of Inflammation and Oxidative Stress

Meredith G. Wagner[1*], Yeong Rhee[2], Kerrie Hert-Honrath[2], Elizabeth H. Blodgett Salafia[3] and Donna Terbizan[2]

[1]Department of Nutrition and Dietetics, Concordia College, 901 8th Street South, Moorhead, MN 56562, USA.
[2]Department of Health, Nutrition and Exercise Sciences, North Dakota State University, Dept. 2620, P.O.Box 6050, Fargo, ND 58108, USA.
[3]Department of Human Development and Family Science, North Dakota State University, Dept. 2615, P.O.Box 6050, Fargo, ND 58108, USA.

Authors' contributions

This work was carried out in collaboration between all authors. Author MGW designed the study, wrote the protocol, conducted the research, performed the statistical analysis, wrote the first draft of the manuscript, and had primary responsibility for the final content. Author YR supervised the research and had primary responsibility for the final content. Author KHH conducted the research. Author EHBS provided assistance with statistical analysis and reviewed/edited the manuscript. Author DT reviewed/edited the manuscript. All authors read and approved the final manuscript.

ABSTRACT

Aims: To determine the effectiveness of a community-based fruit and vegetable education program and provision of fruits and vegetables on consumption of fruits, vegetables, antioxidants, and changes in biomarkers of inflammation and oxidative stress among overweight and obese adults.
Study Design: Randomized controlled design.
Place and Duration of Study: North Dakota State University, Fargo, North Dakota; 14 weeks.
Methodology: Forty-seven adults (31 women; 45.9 ± 11.8 y; body mass index 32.7 kg/m^2) were randomly assigned to one of three intervention groups. The control group received no intervention, the education group attended weekly nutrition education lessons, the fruit and vegetable group attended weekly nutrition education lessons and received one serving of fruits and two servings of vegetables per day for 10 weeks. Fasting blood was drawn and consumption of fruits, vegetables,

---

**Corresponding author: Email: wagner@cord.edu*

and antioxidants was assessed using three-day food records.
Results: Increased consumption of fruit from pre- to post-test was indicated among fruit and vegetable group participants, P = .01, and among education group participants, although this difference was not significant, P = .11. In contrast, a significant decrease in fruit servings consumed from pre- to post-test for control group participants was observed, P = .02. Vegetable consumption was the same for control group participants, decreased by 0.3 serving among education group participants, and increased by 0.4 serving among fruit and vegetable group participants. No significant differences in plasma TNF-α, TBARS, or CRP concentrations from pre- to post-test were indicated among the three groups, although the largest decrease was observed among fruit and vegetable group participants, P = .07.
Conclusion: Changes in fruit and vegetable consumption among participants were minimally associated with improvements in inflammation and oxidative stress biomarkers. Adequate and varied fruit and vegetable consumption is recommended to aid in the prevention and regulation of inflammation and oxidative stress.

Keywords: Antioxidant; overweight; obesity; tumor necrosis factor-alpha; C-reactive protein.

ABBREVIATIONS

*BMI (body mass index); CON (control group); CRP (C-reactive protein); EDUC (education group); FV (fruit and vegetable group); MDA (malondialdehyde); TBARS (thiobarbituric acid reactive substances); TNF-*α (tumor necrosis factor-alpha).

1. INTRODUCTION

In the U.S., more than 60% of adults are overweight or obese [1]. The rising rates of overweight and obesity are particularly concerning because of the numerous health conditions associated with excess weight, including cardiovascular disease, type 2 diabetes mellitus, and certain cancers [2,3]. An underlying mechanism believed to contribute to these detrimental health outcomes is the inflammatory process [4]. Excess adipose tissue contributes to disruption of metabolic homeostasis [5] and leads to an accumulation of free radicals [6]. The presence of free radicals perpetuates the secretion of proinflammatory cytokines, including tumor necrosis factor-alpha (TNF-α), which induce the release of C-reactive protein (CRP) from the liver [7-9]. The increased concentration of inflammatory cytokines and CRP result in a state of chronic, obesity-induced, low-grade inflammation, which is associated with increased disease risk [10]. Fortunately, the body is able to prevent accumulation of free radicals via antioxidants, which donate electrons in order to stabilize free radicals and prevent oxidative damage [11].

While some antioxidants are present in the body, others such as vitamins and minerals must be consumed through the diet [12,13]. Plant foods, including fruits and vegetables, have a substantially higher antioxidant content compared to meat, fish, and other animal-based foods [14]. Because of their excellent nutrient profiles, consumption of fruits and vegetables is widely encouraged [15-17]. However, according to the CDC [18], less than 15% of U.S. adults consume the recommended five or more daily servings of fruits and vegetables, with rates decreasing among some adult cohorts. Specifically, among overweight and obese adults, fruit and vegetable consumption was observed to significantly decrease from 22.6% to 21.5% over an eight-year period [19].

Previous studies have assessed the influence of educational interventions on fruit and vegetable consumption [20-22], antioxidant consumption [23,24], role of antioxidants in oxidative stress [14,25-27], and the correlation between fruit and vegetable consumption and inflammation in healthy individuals [16,17,28] and found favorable results. However, research examining the effects of both nutrition education and provision of fruits and vegetables on biomarkers of inflammation and oxidative stress is lacking. Therefore, this study was designed to determine the effectiveness of a community-based fruit and vegetable education program and provision of fruits and vegetables on consumption of fruits, vegetables, and antioxidants, and changes in biomarkers of inflammation and oxidative stress among overweight and obese adults. Goals were to increase participants' consumption of fruits and vegetables and evaluate subsequent changes in antioxidant consumption and inflammatory and oxidative stress biomarkers.

2. MATERIALS AND METHODS

2.1 Participants

A convenience sample of adults with a body mass index (BMI) of 25 kg/m^2 or greater was recruited via flyers and email notices from two neighboring communities with a combined population of approximately 135,000 residents. Exclusion criteria included being under 18 years of age, a current smoker, having a BMI lower than 25 kg/m^2, being pregnant or lactating, or having a history of bariatric surgery. Study protocol was approved by the North Dakota State University Institutional Review Board prior to implementation and all participants provided written informed consent.

2.2 Procedures

The study consisted of three phases over a 14-week period: pre-test (2 weeks), intervention (10 weeks), and post-test (2 weeks). Of the 67 participants originally enrolled in the study, 46 completed both pre- and post-test food records. Of these, 8 participants were in the control (CON) group, 15 were in the education (EDUC) group, and 23 were in the fruit and vegetable (FV) group. At pre-test, self-reported demographic information was collected. Participants were also asked to report medical history and use of prescription and over the counter medications, as inflammatory conditions or use of anti-inflammatory medications needed to be accounted for prior to data analysis. Information about habitual fruit and vegetable intake was also assessed at both pre- and post-test, using modified versions of two semi-quantitative food frequency questionnaires developed by nutrition experts at The University of Georgia [29].

Food and nutrient intake at pre- and post-test were assessed using a three-day food record. Validity of food records in assessment of energy and nutrient intake has been demonstrated in previous studies of adults [30-32]. Three-day food records require less participant burden, result in improved compliance, and provide accuracy comparable to seven-day food records [33]. Participants were instructed to record everything they ate or drank, as it was consumed, over the course of two weekdays and one weekend day. Data collected from the food records were entered by a registered dietitian and analyzed using The Food Processor® (ESHA Research, Salam, OR). To ensure validity, all data were entered by the same individual. Participant's age, gender, height, weight, and physical activity level were also entered into the program to allow for comparison with the U.S. Dietary Guidelines [34] for servings from each food group.

During pre- and post-test, anthropometric data were collected, including height, weight, waist circumference, and BMI. In addition, a total of 30 mL of venous whole blood sample was drawn after a 12-hour overnight fast. Blood samples at both pre- and post-test were collected from a total of 40 participants (6 in the CON group, 12 in the EDUC group, and 22 in the FV group). All blood draws were scheduled and performed by a phlebotomist at a local hospital. Within 30 minutes of blood being drawn, EDTA-plasma samples were separated from whole blood by laboratory staff at the hospital and frozen immediately at -80°C. Samples were transferred from the local hospital to the university in ice and stored in a freezer at -80°C.

Following pre-test, participants were randomly assigned to one of three intervention groups: CON, EDUC, or FV. Participants assigned to the CON group received no intervention but were given the option of receiving the fruit and vegetable information provided to other participants after the intervention was over. Participants in the EDUC group received weekly information about the antioxidant content of fruits and vegetables, the role of antioxidants in the inflammatory process, and the current recommendations for fruit and vegetable consumption. Participants assigned to the FV group received the same information as participants in the EDUC group. In addition, they received recommendations for incorporating fruits and vegetables into their current meal plan and samples of fruits and vegetables. The samples included fresh, frozen, or canned fruits and vegetables and were provided to participants at the weekly nutrition education sessions. The quantity provided to each FV group participant was the equivalent of one serving of fruit and two servings of vegetable per day for a week because on average, Americans fail to meet the recommended daily fruit and vegetable servings by these amounts [35]. Servings provided were consistent with the serving sizes listed in the 2010 U.S. Dietary Guidelines for Americans [34]. Participants in the FV group were encouraged to consume these fruits and vegetables as a supplement to their usual fruit and vegetable intake.

Nutrition lessons, approximately 30 minutes in duration, were provided weekly by licensed, registered dietitians at a local university. Curriculum materials, developed by nutrition experts at The University of Georgia [29,36], were used for the nutrition lessons. This curriculum was chosen because it was consistent with the purpose of the study, had been used for previous research [29,37,38], and was readily available to the public via the Internet [36]. Curriculum materials included scripts for each lesson, handouts describing how to purchase and store fruits and vegetables, lists of the nutrient content of various fruits and vegetables, and suggestions for incorporating fruits and vegetables into meals and snacks.

After conclusion of the 10-week intervention, post-test was initiated in order to evaluate changes in consumption of fruits, vegetables, and antioxidants as well as changes in biomarkers of inflammation and oxidative stress. Intake of fruits, vegetables, and antioxidants was again assessed using three-day food records.

2.3 Biochemical Analysis

Once all samples were collected, samples were thawed to room temperature and plasma CRP and TNF-α concentrations were measured according to manufacturer instructions using ELISA kits (CRP ELISA kit, Cayman Chemical Company, Ann Arbor, MI; TNF-α UltraSensitive ELISA Kit, Invitrogen Corporation, Camarillo, CA). Oxidative stress was measured using a TBARS kit (Cayman Chemical Company, Ann Arbor, MI). Microtiter plates were read by a microtiter plate reader (ELx 808 IU, Bio-Tek Instruments, Inc., Winooski, VT) at 550 nm. All standards and samples were run in duplicate for the CRP and TNF-α assays and in triplicate for the TBARS assay.

Results of plasma CRP concentrations were expressed as mg CRP/L and results of TNF-α were expressed as pg TNF/mL. Results of the oxidative stress analysis reflected the concentration of malondialdehyde (MDA), which is a byproduct of lipid peroxidation. Results were expressed as MDA equivalents (nmol MDA/mL plasma), with higher values indicating a higher degree of oxidative stress.

2.4 Statistical Analysis

All statistical analyses were performed using PASW version 18.0 (SPSS Inc., Chicago, IL). Chi square analyses were used to assess differences among intervention groups in regard to demographics. Descriptive statistics were generated for all participants completing both pre- and post-test. Paired sample t-tests were used to evaluate mean differences in weight, BMI, CRP, TNF-α, and TBARS concentrations from pre- to post-test for all participants and within the three groups. Pearson correlation analysis was used to evaluate relationships between changes in fruit and vegetable consumption and changes in biomarkers of inflammation and oxidative stress from pre- to post-test. Finally, one-way ANOVA was used to assess if nutrition education about antioxidant properties and the antioxidant content of fruits and vegetables, with and without provision of fruit and vegetable samples, had a differential effect on CRP, TNF-α, and TBARS concentrations from pre- to post-test compared to no intervention (CON group).

3. RESULTS

Participants were predominantly female (66.0%), white (91.5%), married or partnered (78.7%), and reported an annual household income in the range of $50,000-$99,999 (44.7%). Chi square tests for independence did not indicate differences in proportions in regard to gender, ethnicity (white, non-white), relationship status (married, other), or reported annual household income (≤$39,999, ≥$40,000) between intervention groups. The mean BMI among all participants at pre-test was 32.7 kg/m^2. Results of paired sample t-tests indicated no significant differences in weight and BMI from pre- to post-test (data not shown).

Changes in use of anti-inflammatory medications from pre- to post-test were accounted for and plasma concentrations beyond two standard deviations above or below the mean were eliminated prior to data analysis. One-way ANOVA indicated no significant differences in antioxidant consumption or CRP, TNF-α, or TBARS concentrations between intervention groups at pre-test.

Results of paired sample t-tests failed to indicate significant mean differences in inflammation and oxidative stress biomarkers from pre- to post-test (Table 1). Among all participants, mean TNF-α concentration decreased to an undetectable concentration at post-test. No significant differences in plasma TBARS concentrations from pre- to post-test were observed among CON, EDUC, or FV group participants. While differences in CRP concentrations between pre- to post-test were not significantly different for participants in any of the intervention groups, the largest decrease was observed among FV group participants, $t(19) = -1.96$, $P = .07$.

Results of paired sample t-tests indicated a significant increase in consumption of fruit from pre- to post-test among FV group participants, $t(22) = 2.69$, $P = .01$ (Table 2). An increase in fruit servings consumed at post-test compared to pre-test was also noted among EDUC group participants, but this increase did not represent a significant difference, $t(14) = 1.72$, $P = .11$. In contrast, results showed a significant mean decrease in the number of fruit servings consumed from pre- to post-test for CON group participants, $t(7) = -2.89$, $P = .02$. No significant differences in vegetable consumption existed from pre- to post-test for participants in any of the intervention groups. Vegetable consumption was the same for CON group participants, decreased by 0.3 serving among EDUC group participants, and increased by 0.4 serving among participants in the FV group.

Results of paired sample t-tests failed to indicate significant mean differences in antioxidant consumption from pre- to post-test among the total sample and the different intervention groups (Table 2). Despite absence of significant differences, a consistent increase in consumption of beta carotene from pre- to post-test was observed among all groups. In fact, increased consumption of beta carotene among FV group participants trended towards significance, $t(22) = 2.04$, $P = .05$. Changes in consumption of other antioxidants among the different intervention groups were inconsistent. Increases in vitamin C consumption were found among participants in the CON and FV group but not in the EDUC group. Consumption of vitamin E and selenium increased among CON group participants but decreased among EDUC and FV group participants.

Results of Pearson correlation analyses indicated moderate positive correlations between fruit consumption and consumption of vegetables ($r = 0.30$, $P = .03$) and beta carotene ($r = 0.37$, $P = .01$) at post-test. A strong positive correlation was indicated between fruit and vitamin C consumption ($r = 0.57$, $P<.01$). In addition, a moderate negative correlation existed between fruit consumption and CRP concentration ($r = -0.41$, $P = .01$). Significant positive correlations also existed between vegetable consumption and beta carotene ($r = 0.56$, $P<.01$) and selenium ($r = 0.30$, $P = .03$) consumption at post-test. Positive correlations were also indicated between the various antioxidants (Table 3). However, no significant relationships were found between post-test consumption of fruits, vegetables, and antioxidants, and TBARS concentration. TNF-α was not included in the correlation analysis due to undetectable concentrations at post-test.

Table 1. Mean concentrations of inflammation and oxidative stress biomarkers at pre- and post-test[a]

Variable	Control (n = 6)		Education (n = 12)		Fruit and vegetable (n = 20)	
	Pre	Post	Pre	Post	Pre	Post
CRP[b], *mg/L*	4.34 ± 1.16	4.31 ± 1.03	2.34 ± 0.64	2.11 ± 0.75	2.89 ± 0.84	1.52 ± 0.34
TNF-α, *pg/mL*	5.90 ± 1.13	n.d.	5.14 ± 0.91	n.d.	5.94 ± 0.66	n.d.
TBARS, *nmol/mL*[c]	13.25 ± 1.34	7.94 ± 2.06	16.54 ± 3.18	11.71 ± 2.58	13.63 ± 1.14	10.56 ± 1.41

[a]*Data are means ± standard error. n.d. = not detected; CRP = C-reactive protein; TNF-α = tumor necrosis factor-alpha, TBARS = thiobarbituric acid reactive substances*

[b]*Difference in total sample size due to two outliers being eliminated from fruit and vegetable group (N = 36).*

[c]*nmol/mL of malondialdehyde equivalents*

Table 2. Mean daily servings of fruits, vegetables, and antioxidants consumed at pre-test and post-test[a]

Variable	Control (*n* = 8)		Education (*n* = 15)		Fruit and vegetable (*n* = 23)	
	Pre	Post	Pre	Post	Pre	Post
Total fruits, *svgs*[b]	1.4 ± 0.3	0.8 ± 0.2*	1.2 ± 0.3	1.5 ± 0.3	1.5 ± 0.2	2.2 ± 0.2*
Total vegetables, *svgs*	1.4 ± 0.3	1.4 ± 0.2	1.8 ± 0.3	1.5 ± 0.2	1.6 ± 0.2	2.0 ± 0.2
Beta carotene, *mg*[c]	1.3 ± 0.6	2.2 ± 1.0	1.7 ± 0.7	2.3 ± 0.8	2.1 ± 0.4	3.7 ± 0.7
Vitamin C, *mg*	56.6 ± 11.7	80.8 ± 21.5	103.5 ± 24.1	90.7 ± 14.9	90.5 ± 13.9	115.0 ± 10.8
Vitamin E, *mg*	4.9 ± 1.6	5.1 ± 2.1	5.7 ± 1.3	4.7 ± 1.4	4.2 ± 0.7	3.9 ± 0.6
Selenium, *µg*	70.2 ± 10.4	80.7 ± 14.8	59.4 ± 5.9	53.8 ± 8.6	61.7 ± 6.8	54.6 ± 5.7

*[a]Data are means ± standard error. Test statistics are t statistics: *Different from pre, P<.05. svgs = servings; mg = milligrams; µg = micrograms.*
[b]Mean number of servings consumed derived from average daily consumption as recorded on three-day food records. Three-day food record data analyzed using The Food Processor® (ESHA Research, Salam, OR).
[c]Mean daily consumption of antioxidants was derived from average consumption as recorded on three-day food records. Three-day food record data were analyzed using The Food Processor® (ESHA Research, Salam, OR)

Table 3. Correlations among fruit and vegetable consumption, antioxidant consumption, and inflammation and oxidative stress biomarkers at post-test[a]

Variable	1	2	3	4	5	6	7	8
1. Total fruits	----	0.30*	0.37*	0.57*	0.22	0.05	-0.41*	-0.05
2. Total vegetables		----	0.56*	0.48	0.15	0.30*	-0.26	0.17
3. Beta carotene			----	0.46*	0.45*	0.13	-0.15	0.05
4. Vitamin C				----	0.47*	0.18	-0.07	-0.09
5. Vitamin E					----	0.06	-0.01	0.08
6. Selenium						----	0.20	0.29
7. CRP							----	-0.15
8. TBARS								----

*[a]Test statistics are Pearson correlation coefficients: *P<.05*
CRP = C-Reactive Protein; TBARS = Thiobarbituric Acid Reactive Substances

Finally, results of one-way ANOVA to evaluate differences in biomarkers of inflammation and oxidative stress from pre- to post-test revealed no significant differences between the three intervention groups regarding changes in CRP or TBARS concentrations over time.

4. DISCUSSION

This study investigated the effects of a 10-week nutrition education program and provision of fruits and vegetables on fruit and vegetable consumption and subsequent changes in biomarkers of inflammation and oxidative stress among overweight and obese adults. Results indicated that at pre-test, participants were obese, with a mean BMI of 32.7 kg/m^2 [39], and experiencing a moderate degree of inflammation and high degree of oxidative stress as evidenced by mean plasma concentrations (CRP = 2.91 mg/L; TNF-α = 5.70 pg/mL; TBARS = 14.50 nmol/mL) [40,41]. In addition, participants' mean weekly consumption of fruits and vegetables was below the current recommendations to consume five or more daily servings [18]. Inadequate consumption of fruits and vegetables is particularly concerning because research has demonstrated that fruit and vegetable consumption is associated with improved antioxidant consumption [23]. Adequate consumption of antioxidants, which include beta carotene, vitamin C, vitamin E, and selenium, has been associated with a reduction in both inflammation and oxidative stress [11,26,27,42].

Participants who received nutrition education and fruit and vegetable samples reported increased consumption of fruits, but not vegetables. Alternatively, a decrease in fruit consumption from pre- to post-test was noted among CON group participants. Such a decrease in consumption was due to seasonal changes, transitioning from fall to winter, and absence of an intervention effect of education on benefits of fruit consumption, which was not provided to CON group participants. In addition, small increases in vegetable intake were only noted among participants in the FV group. These findings suggest that provision of fruits and vegetables was effective at increasing fruit and vegetable consumption. This is the first study to examine the effects of providing fruits and vegetables on consumption patterns of overweight and obese adults. However, this does support previous findings, which suggest availability and exposure to fruits and vegetables are possible determinants of fruit and vegetable consumption [43,44].

Both fruit and vegetable consumption at post-test were related to intake of beta carotene at post-test. In addition, post-test fruit consumption was related to vitamin C intake and post-test vegetable consumption was related to selenium intake. These findings are in agreement with existing literature that identifies fruits and vegetables as key sources of antioxidants, including beta carotene and vitamin C, in the diet [23]. While the positive relationships between fruit, vegetable, and antioxidant consumption are encouraging, participants' overall mean consumption of fruits (1.7 servings) and vegetables (1.7 servings) following the intervention was below the recommended two or more daily servings of fruits and three or more daily servings of vegetables. Failure to consume an adequate quantity of fruits and vegetables likely hindered participants' consumption of beneficial antioxidants.

Despite minimal changes in fruit, vegetable, and antioxidant consumption from pre- to post-test, a negative relationship between fruit consumption and CRP concentration at post-test was indicated, which is consistent with previous findings in healthy individuals [16,17,28]. Similar relationships with TBARS concentrations were not found following correlation analyses. Changes in CRP, TNF-α, and TBARS concentrations over time and among different intervention groups were noted, despite absence of significant weight changes. TNF-α concentrations in all groups were within the normal high sensitivity reference range (1.2-15.3 pg/mL) [45] at pre-test, and were undetectable at post-test, meaning all concentrations were below the minimum detectable concentration (0.009 pg/mL) of the ELISA kit used. Such decreases in TNF-α concentration may be attributed to increases in fruit, vegetable, or antioxidant consumption from pre- to post-test among participants, although no one variable was consistently higher at post-test among all three groups. Decreases in CRP concentrations were greatest among participants in the FV group, which suggests a possible benefit of provision of fruits and vegetables on decreased inflammation. Previous research has suggested associations between dietary fiber [16], omega-3 fatty acids [26], and reductions in markers of inflammation. While intake of these dietary components was not a focus of the present study, the potential of these factors to impact markers of inflammation did exist.

Similar outcomes were observed in regard to oxidative stress. Trends toward a significant decrease in TBARS concentration were noted among FV and CON group participants more so than EDUC group participants. While a significant relationship was not observed, increased consumption of vitamin C among these two groups compared to decreased consumption in the EDUC group serves as a possible explanation for these results. As an antioxidant, vitamin C is especially effective at protecting against lipid peroxidation [11,46] and thus, helps to explain why decreased TBARS concentrations were observed among the groups with higher vitamin C consumption.

While changes in CRP, TNF-α, and TBARS concentrations were observed from pre- to post-test, these changes were not significantly different among participants who received the intervention compared to those who did not. These findings are contradictory to literature indicating associations between increased fruit and vegetable consumption and improvements in biomarkers of inflammation and oxidative stress [16,17,28]. A possible reason for lack of differences between participants in the EDUC and FV groups and those in the CON group may be due to the unequal sizes of the intervention groups and subsequent lack of statistical power.

The strength of this study was the randomized intervention design conducted in a sample of overweight and obese adults. A limitation was the unequal sizes of the intervention groups. More participants were purposefully included in the EDUC and FV groups compared to CON group in anticipation of attrition, but the final number of participants in the CON group fell short of the required sample size and may have prevented detection of significant differences. In regard to biochemical analysis of plasma samples, TNF-α concentrations at post-test were below the specificity of the ELISA kit used and therefore, undetectable. This prevented evaluation of changes from pre- to post-test and among the different intervention groups. A possible solution would be to measure TNF-α expression by immunoblotting techniques or gene expression instead of ELISA. In addition, alonger intervention may have allowed for detection of differences.

Results of the present study highlight a need for additional research, including long-term studies, evaluating the effects of nutrition education interventions with provision of fruits and vegetables on changes in consumption of fruits, vegetables, and antioxidants, and biomarkers of inflammation and oxidative stress in overweight and obese adults. Previous research suggests that improvements in fruit and vegetable consumption could potentially contribute to improved antioxidant intake [23,24], reduced inflammation and oxidative stress [16,17,25,26], and ultimately decreased chronic disease risk. Therefore, additional research that controls for confounding variables and includes long-term follow-up is warranted.

5. CONCLUSION

This intervention to provide nutrition education and fruit and vegetable samples to overweight and obese adults at risk for chronic disease resulted in increased fruit consumption, but inconsistent changes in vegetable and antioxidant consumption. In addition, minimal improvements in biomarkers of inflammation and oxidative stress were noted. Other factors such as changes in anthropometrics, physical activity, and macronutrient intake may have impacted these measures [47-49]. Despite the present findings, evidence suggests that a balanced diet, including antioxidant-rich fruits and vegetables, is beneficial in helping individuals maintain adequate antioxidant concentrations in order to aid in the prevention and regulation of inflammation and oxidative stress [23,50]. Therefore, adequate and varied consumption of fruits and vegetables is recommended.

CONSENT

All authors declare that written informed consent was obtained from all participants.

ETHICAL APPROVAL

All authors hereby declare that all experiments have been examined and approved by the appropriate ethics committee and have therefore been performed in accordance with the ethical standards laid down in the 1964 Declaration of Helsinki.

ACKNOWLEDGEMENTS

The authors thank L. Myers, J. Levine, and T. Heck for their assistance with teaching nutrition education classes.

This research was supported by scholarships and grants from the Department of Health, Nutrition, and Exercise Sciences and College of Human Development and Education at North Dakota State University, which had no roles in the study design, collection, analysis and interpretation of data, or in the writing of the manuscript.

COMPETING INTERESTS

Authors have declared that no competing interests exist.

REFERENCES

1. Centers for Disease Control and Prevention. Obesity and overweight. U.S. Department of Health and Human Services; 2014. Accessed 17 September 2014. Available:http://www.cdc.gov/nchs/fastats/obesity-overweight.htm
2. Centers for Disease Control and Prevention. Overweight and obesity: Health consequences. U.S. Department of Health and Human Services. 2012. Accessed 17 September 2014. Available:http://www.cdc.gov/obesity/adult/causes/index.html
3. National Heart, Lung, and Blood Institute. What are the health risks of overweight and obesity? U.S. Department of Health and Human Services; 2011. Accessed 5 October 2014. Available:http://www.nhlbi.nih.gov/health/dci/Diseases/obe/obe_risks.html
4. Dowd JB, Zajacova A, Aiello AE. Predictors of inflammation in U.S. children aged 3-16 years. Am J Prev Med. 2010; 39(4):314-20. DOI: 10.1016/j.amepre.2010.05.014.
5. Nguyen X-MT, Lane J, Smith BR, Nguyen NT. Changes in inflammatory biomarkers across weight classes in a representative U.S. population: a link between obesity and inflammation. J Gastrointest Surg. 2009;13:1205-12. DOI: 10.1007/s11605-009-0904-9.

6. Vieira VJ, Valentine R J, Wilund KR, Antao N, Baynard T, Woods JA. Effects of exercise and low-fat diet on adipose tissue inflammation and metabolic complications in obese mice. Am J Physiol Endocrinol Metab. 2009;296:E1164-71. DOI: 10.1152/ajpendo.00054.2009.
7. Camhi SM, Stefanick ML, Ridker PM, Young DR. Changes in c-reactive protein from low-fat diet and/or physical activity in men and women with and without metabolic syndrome. Metabolism. 2010;59: 54-61. DOI:10.1016/j.metabol.2009.07.008
8. Federico A, D'Aiuto E, Borriello F, Barra G, Gravina AG, Romano M, et al. Fat: a matter of disturbance for the immune system. World J Gastroenterol. 2010;16:4762-72. DOI: 10.3748/wjg.v16.i38.4762.
9. Marinou K, Tousoulis D, Antonopoulos AS, Stefanadi E, Stefanadis C. Obesity and cardiovascular disease: from pathophysiology to risk stratification. Int J Cardiol. 2010;138:3-8. DOI: 10.1016/j.ijcard.2009.03.135.
10. Stienstra R, Duval C, Muller M, Kersten S. PPARs, obesity, and inflammation. PPAR Res. 2007;1-10. DOI: 10.1155/2007/95974
11. Jones DP, DeLong MJ. Detoxification and protective functions of nutrients. In: Stipanuk MH. Biochemical and physiological aspects of human nutrition. Philadelphia: W.B. Saunders Company; 2000.
12. Cilla A, De Palma G, Lagarda MJ, Barbera R, Farre R, Clemente G, et al. Impact of fruit beverage consumption on the antioxidant status in healthy women. Ann Nutr Metab. 2009;54:35-42. DOI: 10.1159/000205318.
13. Mayes PA, Botham KM. Lipids of physiologic significance. In: Murray RK, Granner DK, Mayes PA, Rodwell VW, editors. Harper's illustrated biochemistry. New York: McGraw-Hill Companies, Inc.; 2003.
14. Carlsen MH, Halvorsen BL, Holte K, Bohn SK, Dragland S, Sampson L, et al. The total antioxidant content of more than 3100 foods, beverages, spices, herbs and supplements used worldwide. Nutrition Journal. 2010;9:3-13. DOI: 10.1186/1475-2891-9-3.
15. Hansen L, Dragsted LO, Olsen A, Christensen J, TjØnneland A, Schmidt EB, et al. Fruit and vegetable intake and risk of acute coronary syndrome. Br J Nutr. 2010;104:248-55. DOI: 10.1017/S0007114510000462.
16. Hermsdorff HHM, Zulet MA, Puchau B, Martinez JA. Fruit and vegetable consumption and proinflammatory gene expression from peripheral blood mononuclear cells in young adults: a translational study. Nutr Metab. 2010;7:42. DOI: 10.1186/1743-7075-7-42.
17. Polidori MC, Pratico D, Mangialasche F, Mariani E, Aust O, Anlasik T, et al. High fruit and vegetable intake is positively correlated with antioxidant status and cognitive performance in healthy subjects. J Alzheimers Dis. 2009;17:921-7. DOI: 10.3233/JAD-2009-1114.
18. Centers for Disease Control and Prevention. State indicator report on fruits and vegetables, 2009. U.S. Department of Health and Human Services; 2010. Accessed 12 September 2014. Available:http://www.fruitsandveggiesmatter.gov/health_professionals/data_behavioral.html
19. Andreyeva T, Long MW, Henderson KE, Grode GM. Trying to lose weight: Diet strategies among Americans with overweight or obesity in 1996 and 2003. J Am Diet Assoc. 2010;110:535-42. DOI: 10.1016/j.jada.2009.12.029.
20. Chapman K. Can people make healthy changes to their diet and maintain them in the long term? A review of the evidence. Appetite. 2010;54:433-41. DOI: 10.1016/j.appet.2010.01.017.
21. Guillaumie L, Godin G, Vezina-Im L. Psychosocial determinants of fruit and vegetable intake in adult population: a systematic review. Int J Behav Nutr Phys Act. 2010;7:12-23. DOI: 10.1186/1479-5868-7-12.
22. Ransley JK, Taylor EF, Radwan Y, Kitchen MS, Greenwood DC, Cade JE. Does nutrition education in primary schools make a difference to children's fruit and vegetable consumption? Public Health Nutr. 2010;13:1898-1904. DOI: 10.1017/S136898001000095.
23. Chun OK, Floegel A, Chung S, Chung CE, Song WO, Koo SI. Estimation of antioxidant intakes from diet and supplements in U.S. adults. J Nutr. 2010;140:317-24. DOI: 10.3945/jn.109.114413.
24. Kusano C, Ferrari B. Total antioxidant capacity: a biomarker in biomedical and

nutritional studies. Journal of Cell and Molecular Biology. 2008;7:1-15.

25. Dilis V, Trichopoulou A. Antioxidant intakes and food sources in Greek adults. J Nutr. 2010;140:1274-9. DOI: 10.3945/jn.110.121848.
26. McAnulty SR, Nieman DC, Fox-Rabinovich M, Duran V, McAnulty LS, Henson DA, et al. Effect of n-3 fatty acids and antioxidants on oxidative stress after exercise. Med Sci Sports Exerc. 2010;42:1704-11. DOI: 10.1249/MSS.0bb13e3181d85bd1.
27. Vertuani S, Angusti A, Manfredini S. The antioxidants and pro-antioxidants network: an overview. Curr Pharm Des. 2004;10: 1677-94.
28. Talegawkar SA, Beretta G, Yeum K-J, Johnson EJ, Carithers TC, Taylor Jr. HA, et al. Total antioxidant performance is associated with diet and serum antioxidants in participants of the diet and physical activity substudy of the Jackson Heart Study. J Nutr. 2009;139:1964-71. DOI: 10.3945/jn.109.107870.
29. Garcia JM. A fruit and vegetable education intervention in Georgia's Older Americans Act Nutrition Program improves intake, knowledge, and barriers related to consumption. Master's thesis. Athens: The University of Georgia; 2005.
30. Comrie F, Masson LF, McNeill G. A novel online food recall checklist for use in an undergraduate study population: a comparison with diet diaries. Nutrition Journal. 2009;8:13. DOI: 10.1186/1475-2891-8-13.
31. Drapeau V, King N, Hetherington M, Doucet E, Blundell J, Tremblay A. Appetite sensations and satiety quotient: predictors of energy intake and weight loss. Appetite. 2007;48:159-66.
32. Shuaibi AM, Sevenhuysen GP, House JD. Validation of a food choice map with a 3-day food record and serum values to assess folate and vitamin B-12 intake in college-aged women. J Am Diet Assoc. 2008;108:2041-50.
33. Larkin FA, Metzner HL, Guire KE. Comparison of three consecutive-day and three random-day records of dietary intake. J Am Diet Assoc. 1991;91:1538-42.
34. US Department of Health and Human Services, US Department of Agriculture. Dietary Guidelines for Americans; 2010. Accessed 10 September 2014. Available: http://www.dietaryguidelines.gov
35. Centers for Disease Control and Prevention. State-specific trends in fruit and vegetable consumption among adults – United States, 2000-2009. Accessed 10 October 2014. Available:http://www.cdc.gov/mmwr/pdf/wk/mm5935.pdf
36. Johnson MA, Fischer JG, Andress E, Burnett SM, Crawley C, Cress ME, et al. NOAH net curriculum materials. The University of Georgia; 2003. Accessed 10 October 2014. Available:http://noahnet.myweb.uga.edu/plansfv.html
37. Hendrix SJ, Fischer JG, Reddy S, Lommel TS, Speer EM, Stephens H, et al. Fruit and vegetable intake and knowledge increased following a community-based intervention in older adults in Georgia senior centers. J Nutr Gerontol Geriatr. 2008;27:155-78. DOI: 10.1080/01639360802060249.
38. Wade JC. A fruit and vegetable education intervention in Northeast Georgia Older Americans act nutrition programs improves intake, knowledge, and barriers related to consumption. Master's thesis. Athens: The University of Georgia; 2003.
39. World Health Organization. BMI classification. World Health Organization, 2006 [updated 2014 October 2; cited 2013 November 17]. Available:http://apps.who.int/bmi/index.jsp?introPage=intro_3.html
40. Pearson T, Mensah G, Alexander R, Anderson J, Cannon III R, Criqui M, et al. Markers of inflammation and cardiovascular disease: application to clinical and public health practice: a statement from healthcare professionals from the Centers for Disease Control and Prevention and the American Heart Association. Circulation. 2003;107:499-511. DOI:10.1161/01.CIR.0000052939.59093.45.
41. Yagi K. Simple assay for the level of total lipid peroxides in serum or plasma. Methods Mol Biol. 1998;108:101-6.
42. Sureda A, Tauler P, Aguilo A, Cases N, Llompart I, Tur JA, et al. Influence of an antioxidant vitamin-enriched drink on pre- and post-exercise lymphocyte antioxidant system. Ann Nutr Metab. 2008;52:233-40. DOI: 10.1159/000140515.
43. Kidd T, Peters PK. Decisional balance for health and weight is associated with whole-fruit intake in low-income young

adults. Nutrition Research. 2010;30:477-82. DOI: 10.1016/j.nutres.2010.07.003.

44. Powell LH, Kazlauskaite R, Shima C, Appelhans BM. Lifestyle in France and the United States: An American perspective. J Am Diet Assoc. 2010;110:845-7. DOI: 10.1016/j.jada.2010.03.029.
45. TNF-α Reference Range. Personal communication: Sanford Health Lab, Fargo, ND; 2015.
46. Bender DA, Mayes PA. Vitamins & minerals. In: Murray RK, Granner DK, Mayes PA, Rodwell VW, editors. Harper's illustrated biochemistry. New York, NY: McGraw-Hill Companies, Inc.; 2003.
47. Desroches S, Archer WR, Paradis M-E, Deriaz O, Couture P, Bergeron J, et al. Baseline plasma c-reactive protein concentrations influence lipid and lipoprotein responses to low-fat and high monounsaturated fatty acids diets in healthy men. J Nutr. 2006;136:1005-11.
48. Rankin JW, Turpyn AD. Low carbohydrate, high fat diet increases c-reactive protein during weight loss. J Am Coll Nutr. 2007;26:163-9.
49. Van Dijk SJ, Feskens EJM, Bos MB, Hoelen DWM, Heijligenberg R, Bromhaar MG, et al. A saturated fatty acid-rich diet induces an obesity-linked proinflammatory gene expression profile in adipose tissue of subjects at risk of metabolic syndrome. Am J Clin Nutr. 2009;90:1656-64. DOI: 10.3945/ajcn.2009.27792.
50. Quadir T, Akhtar-Danesh N. Fruit and vegetable intake in Canadian ethnic populations. Can J Diet Pract Res. 2010;71:11-6. DOI: 10.3148/71.1.2010.11.

13

Short Review of Sulphites as Food Additives

Alvaro R. Garcia-Fuentes[1], Sabrina Wirtz[1], Ellen Vos[1] and Hans Verhagen[2,3*]

[1]Maastricht University, P.O.Box 616, 6200 MD Maastricht, The Netherlands.
[2]National Institute for Public Health and the Environment (RIVM); P.O.Box 1, 3720 BA Bilthoven, The Netherlands.
[3]University of Ulster, Northern Ireland Centre for Food and Health (NICHE), Cromore Road, Coleraine, BT52 1SA, Northern Ireland.

Authors' contributions

This work was carried out in collaboration between all authors. All authors read and approved the final manuscript.

ABSTRACT

Sulphites or sulphiting agents refer to sodium hydrogen sulphite, sodium metabisulphite, potassium metabisulphite, calcium sulphite, calcium hydrogen sulphite, and potassium hydrogen sulphite. As food additives, they are widely used by the food industry with a variety of commercial uses in food and beverages. Sulphites are effective bleaching agents, antimicrobials, oxygen scavengers, reducing agents, and enzyme inhibitors. Wine, beer, dehydrated fruits and vegetables, jam, juice, sugar, processed potatoes, seafood, meat and baked products are some of the food categories in which sulphites are added. Sulphites have been implicated in various health related issues. Asthmatic reactions and some antinutritional consequences such the degradation of thiamine (vitamin B_1) are adverse reactions associated with sulphites. In many countries, sulphites have been regulated. Sulphites are generally recognized as safe in the USA with some exceptions when using in raw fruits and vegetables. In the European Union sulphites are also controlled, and the permitted amount varies according to the food product.

Keywords: Food additives; sulphites; sulphiting agents; legal status; toxicity.

**Corresponding author: Email: hans.verhagen@rivm.nl*

1. INTRODUCTION

Sulphites or sulphiting agents are a group of sulphur-based compounds that are capable to release sulphur dioxide SO_2, which is the active component that helps to preserve food [1,2]. Sulphites have been added to foods as preservative agents and other purposes for centuries [2]. Sulphites were used by many ancient cultures such as Romans and Greeks to reduce or prevent spoilage and discoloration of foods, to sanitize wine vessels or as a purifier and disinfectant [1,3]. Furthermore, naturally occurring sulphites exist especially in fermented products, such as beer and wine. They are produced by yeasts during the fermentation process [1].

Sulphites have many functional uses in food. They are effective bleaching agents, antimicrobials, oxygen scavengers, reducing agents, and enzyme inhibitors. Their primary function is to act as a preservative or antioxidant to prevent or reduce/retard spoilage and browning during the preparation, storage, and distribution of many foods [2,4]. Additionally, sulphites are utilized to stabilize product colour and inhibit discoloration, as well as to improve flavour of many foods [5]. Sulphites are also used in the pharmaceutical and cosmetics industries as preservatives of the active ingredient. Thus, in addition to being cheap and convenient, sulphites are extremely versatile, and in many products, serve more than one purpose [3].

Despite all uses of sulphites in the food industry, sulphites have been related to asthmatic reactions in a small subset of sensitive individuals of the asthmatic population [2]. They have also been related with some anti-nutritional consequences such as degradation of thiamine (vitamin B_1) [5].

This short review presents the available data published regarding sulphites, their uses, related health problems, and the actual legislation around these food additives.

2. CHEMISTRY OF SULPHITES

Sulphites are a group of compounds characterized by the presence of the sulphite ion SO_3^{2-} [1]. The term sulphites is commonly used to describe the oxo species of sulphur oxidation state that include the ionic and non-ionic forms: SO_2, HSO_3^-, SO_3^{2-}, and $S_2O_5^{2-}$ [6]. The concentration of free sulphites in foods correlates with the preservative activity, with sulphite levels in foods and drinks usually expressed as theoretical yields or 'equivalents' of SO_2 [3]. The term S(IV) is used to refer to any of the four oxo species of sulphur oxidation (Fig. 1) [6,7]. The most common sulphites used by food industry are described in Table 1.

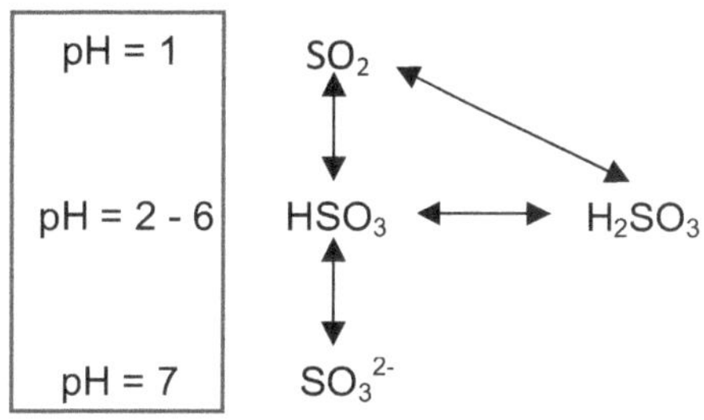

Fig. 1. Sulphites chemistry equilibrium, modified from Vally et al. [3]

A distinction is often made between free and bound S(IV) which forms reversible interaction products. The former refers to all the species of the additive which may rapidly and quantitatively be converted to SO_2, thereby acidifying a treated food. The latter represents hydroxysulphonate adducts formed by reaction of carbonyl groups with HSO_3^-. In fact, according to the nature of food, the type and extent of the technological processes, and the conditions and time of storage, sulphites react with reducing sugars, aldehydes, ketones, proteins to form combined compounds [6-8].

Table 1. Most common sulphites used by food industry

Name	E-number	Chemical formula
Sulphur dioxide	E-220	SO_2
Sodium sulphite	E-221	Na_2SO_3
Sodium hydrogen sulphite (Sodium bisulphite)	E-222	$NaHSO_3$
Sodium metabisulphite	E-223	$Na_2O_5S_2$
Potassium metabisulphite	E-224	$K_2O_5S_2$
Calcium sulphite	E-226	$CaSO_3$
Calcium hydrogen sulphite	E-227	$Ca(HSO_3)_2$
Potassium hydrogen sulphite	E-228	$KHSO_3$

In the digestive tract, the free and reversibly bound S(IV) species can be liberated [9]. The useful properties of sulphiting agents are generally due to the nucleophilicity of the sulphite ion that may react by addition to carbonyl groups, carbon-carbon double bonds, quinones, heterocyclic nitrogen compounds [6-8]. The most common reactions of S(IV) are the result of oxidation by molecular oxygen catalysed by transition metal ions [10].

3. SULPHITES AS FOOD ADDITIVE

3.1 Uses and Reactions in Food

Sulphites are used in a wide variety of food products such as beverages, dried fruits, jams, seafood, meat, sugar, and others (Table 2) [3]. In wine, the aim of adding sulphite is to prevent microbial growth and oxidation of volatile odorous compounds which may develop during fermentation [9].

Table 2. Major type of foods that may contain sulphiting agents [11,12]

Food category	Examples
Drinks	Bottled soft drinks and fruit juice, cordials, cider, beer, wine (including sparkling wine)
Processed fruits	Dried apricots, fruit bars, maraschino cherries, canned, bottled, or frozen fruit juices, commercial preparations of citrus fruit beverage mixes.
Processed vegetables	Canned vegetables (including potatoes), pickled vegetables (including sauerkraut), dried vegetables, instant mashed potatoes, frozen potatoes, potato salad
Fish and crustaceans	Canned clams; fresh, frozen, canned, or dried shrimp; frozen lobster; scallops; dried cod
Other commercial foods	Gravies, sauces and fruit toppings, maple syrup, jams, jellies, biscuits, bread, pie and pizza dough, gelatine, coconut, vinegar, beef stew.

Sulphites are the most effective and cheapest antibrowning agents [13,14]. Sulphites reduce quinones produced by catalysis of polyphenol oxidase to less reactive, and colourless compounds, thereby preventing pigmentation [15]. The quantity of sulphites necessary to prevent enzymatic browning varies according to the concentration and the nature of the substrate, the level of activity exhibited by the polyphenoloxidase, the desired period of control and the presence of other inhibitors [1].

Besides anti-browning and anti-microbial functions, sulphites are used as dough conditioners in bakery industry, especially in frozen pizza dough and pie crusts. They are also used in biscuits, tortilla shells, cookies and crackers. The level of sulphites required for dough conditioners are relatively low. Sulphites act by breaking the cysteine disulphide bonds that are present in the gluten fraction of the dough [1]. Cleavage of thiamine (vitamin B_1) in stored food [4], and destruction of cyanocobalamin (vitamin B_{12}) are undesired reactions of sulphites in foods [16].

3.2 Legal Status

In the European Union sulphiting agents are regulated by Regulation 1333/2008 of the European Parliament and of the Council on food additives. According with Annex II of this Regulation, sulphiting agents are considered food additives other than sweeteners or colorants. The name of the most commonly used authorized compounds and their E-number are shown in Table 1 [17].

This Regulation 1333/2008 allows the use of sulphiting agents in both food products and with other food additives, enzymes, flavourings and nutrients [18]. The permitted amounts vary depending of the food as listed in the Annex II [19] or the form of the ingredient in which they are added as listed in the Annex III [18]. The range goes from 10 mg/kg (e.g. table grapes) to 10,000 mg/kg in papain in solid form [17-19].

With introduction of the legislative Package on Food Improvement Agents in December 2008, the regulation of food additives in the European Union was reformed. Regulation 1333/2008 of the European Parliament and the Council, introduced one regime for the use of food additives in the Union, food colours sweeteners and the remaining food additives that until then were regulated in separate Directives (respectively Directive 94/36/EC, Directive 94/35/EC and Directive 95/2/EC).Apart from consolidating the regulation of food additives, Regulation 1333/2008 also started a process of re-evaluation of all food additives authorized for use in the Union prior to 20 January 2009. Based on the re-evaluation carried out by EFSA, the Commission will revise the list of food additives and the conditions of use specified therein. Hereby sulphites will be revised within the food additives other than colours and sweeteners category on 31 December 2018 [20].

In the USA, since 1959 sulphiting agents have been listed as generally recognised as safe (GRAS) when used in accordance with good manufacturing practice, except in meats or in foods recognised as a source of thiamine (vitamin B_1) [8]. However, in 1986 the practice of adding sulphites to fruits and vegetables that are to be sold or served raw to the public was banned by revocation of the GRAS status [2,5,21]. Also, any food containing more than 10 ppm of a sulphiting agent must have a label declaration [22].

4. BIOLOGICAL DATA

4.1 Absorption, Distribution, Metabolism and Excretion

Sulphites are readily absorbed from the digestive tract [23]. Normally, human serum has a sulphite concentration in the range 0–10 µM [24]. Sulphite is oxidized *in vivo* to sulphate by sulphite oxidase, a molybdenum containing enzyme, located in the intra-membrane space of mitochondria [25]. Sulphite oxidase is present in mammalian tissues and organs, mainly in liver, heart and kidney and very low amounts may be found in spleen, brain, skeletal muscle, and blood [26,27]. In humans, sulphite oxidase activity varies from slightly to substantially less than in other mammals and some individuals present levels well below the normal range [4]. During the oxidation of sulphite to sulphate in mitochondria, cytochrome c transfers the electrons produced to the electron transport chain, allowing generation of ATP in oxidative phosphorylation [24].

Metabolism of sulphites by sulphate oxidase has been shown to be rapid. Therefore, sulphite does not accumulate in the tissues on chronic administration. Sulphate, which represents the normal non-toxic sulphite metabolite, is excreted rapidly through the urine [31]. In several species, less than 10% of the dose administered appeared in the urine as sulphite [16]. A proportion of the sulphite absorbed is converted to thiosulfate. When elevated levels of thiosulfate in body fluids are observed it may be due to sulphite oxidase deficiency [28].

4.2 Adverse Reactions

Sulphites have been implicated as initiators of asthmatic reactions in a small subset of sensitive individuals of the asthmatic population [2]. Adverse reactions, including anaphylactic reactions, dermatitis, urticaria, flushing, hypotension, abdominal pain and diarrhoea, have been reported in those individuals [29]. The mechanism of action of sulphites in asthmatic reactions is not established, and various studies have proposed different possible mechanisms.

One study suggests that the reaction of nitric oxide and its carriers with sulphites may be a major source of sulphites' biological toxicity by affecting the pathway of the drugs used to treat asthma [30]. Another study provided evidence that this response may be mediated through stimulation of an oro-bronchial reflex by sulphur dioxide vapour released from foods, rather than by an immunological mechanism [31]. A third study says that partial sulphite oxidase deficiency is a possible mechanism involved in sulphite sensitivity due to alterations in drug metabolizing enzymes that can contribute to the varying susceptibility and response of sulphite-sensitive individuals to different drugs and/or therapeutics used for treatments.

Sensitivity to sulphites can develop at any time during a person's lifespan, with some initial reactions not showing up until a person has reached their forties or fifties [11]. Dose-related respiratory hyper-reactivity has also been documented in a small number of individuals following consumption of potassium bisulphite-treated red wine and wine which contained smaller amounts of sulphites formed naturally as products of fermentation [32]. The most sensitive patients have reacted to a dose of 5 mg sodium bisulphite [33].

On the other hand, gastric lesions had been induced in rats by sodium metabisulphite [34]. Furthermore, gastro-intestinal reactions are also known to occur in man; with very high doses of sulphites, abdominal pain and vomiting have been observed in human volunteers [32].

5. TOXICOLOGICAL DATA

5.1 Acute, Sub-chronic and Chronic Toxicity

The acute toxicity of free sulphites is low with a LD50 of 1000-2000 mg/kg bw. The acute toxicity of bound sulphites may even be lower with a LD50 that exceeds 5000 mg/kg bw [35].

Til et al. studied the toxicity of sulphite in newly weaned Wistar rats. 120 males and 120 females, were divided into six groups each of 20 males and 20 females. The groups were maintained on the stock diet containing 0.0 (control), 0.125, 0.25, 0.5, 1.0 or 2.0% sodium metabisulphite for periods up to 2 years and over three generations. A level of 2% sulphite caused slight growth retardation in two generation of rats, however it was no significant. Blood was present in the faeces in groups given 1% sulphite or more. Pathological examination revealed hyperplastic changes in both the fore- and glandular stomach with levels of 1 and 2% sulphite in each of the three generations. Also, it was perfomed a series of short-term studies in 10 male and 10 female rats on high sulphite levels (0–8%) for 10–56 days. Diets containing 6% sulphite caused considerable growth depression and reduced food intake and food efficiency. Anaemia occurred at 2% and above. Increase in weight of the spleen occurred with levels of 4% and above. The most sensitive criteria of sulphite damage in the present studies was the presence of blood in the faeces and changes in gastric morphology. The dietary level of sulphite that did not show any significant adverse effect was 0.25% in the diet, which is equivalent to 72 mg SO_2/kg of body weight after conversion and correction for sulphite losses [36].

The same research group examined sub-chronic and chronic oral toxicity of sodium metabisulphite. 20 male and 20 female pigs were fed with diets at levels of 0.0, 0.125, 0.25, 0.5, 1.0 or 2.0% of $Na_2S_2O_5$. After 15 weeks, 14 males and 14 females from each group were killed. The six males and six females remaining in each group were kept on the same diets for up to 48 weeks. The concentrations of sulphite still present in the various experimental diets when they were consumed by the pigs were calculated to be on average 0.06, 0.16, 0.35, 0.83 and 1.72% respectively. Organ-to-body weight ratios of the liver, kidneys, heart and spleen were significantly increased with dietary levels of 1.72%, after both 15 and 48 weeks; and with dietary levels of 0.83% only after 48 weeks. The no-effect level established in this study was 0.35% $Na_2S_2O_5$ in the diet of pigs for 48 weeks [37].

More recently, Kadi et al. determined the sodium metabisulphite sub-chronic toxicity on 24 female Wistar rats divided in four groups treated with 0.0, 0.25, 1 and 4% of sodium metabisulphite in their drinking water for 90 days. Administration of $Na_2S_2O_5$ at 1% and 4% had significant effects on body weight, food and water consumed. There was an increase in biochemical parameters (calcium, urea, creatinine, uric acid, transaminases) and decrease immunoglobulin levels. The hematology revealed a decrease in red blood cells and hemoglobin, as well as leucocytosis. They concluded that sub-chronic intake of $Na_2S_2O_5$ 1% and 4% seems to alter immune function, biochemical, hematological and physiological parameters in Wistar rats [38].

5.2 Genotoxicity and Carcinogenicity

Sulphites are capable of inducing mutations in yeast, *E. coli* and *Vicia faba*, but only at high concentrations (0.2 – 3 M) and pH values in the range of 5 – 5.6 [39]. Hayatsu and Miura described the mutagenic activity of sodium bisulphite on *E. coli* K12 after a treatment with a solution of 3 M $NaHSO_3$and pH of 5.6 during 1.5 hours. Sodium bisulphite caused mutations by modification of cytosine [40]. Similarly, Mukai et al. showed that a concentration of 1 M sodium bisulphite solution at pH 5.2 in *E. coli*K12is specific for cytosine-guanine mutants [41].

In a study *in vitro*, sodium bisulphite at low concentrations (0.05, 0.1, 0.5, 1.0, and $2.0x10^{-3}$ M) induced chromatid-type aberrations, but not chromosome-type aberrations in human blood lymphocytes; high concentrations induced both chromatid- and chromosome-type aberrations. The authors concluded that sulphur dioxide is a clastogenic and genotoxic agent [42].

Til et al. however, did not find any carcinogenic effect in Wistar rats on the stock diet containing 0.0 (control), 0.125, 0.25, 0.5, 1.0 or 2.0% sodium metabisulphite for periods up to 2 years and over three generations [36].

Tanaka et al. examined the carcinogenicity of potassium metabisulphite in ICR/JCL mice. Three groups of 50 male and 50 female mice each received 0, 1, and 2% potassium metabisulphite solution in distilled water for 24 months. There was no significant difference in tumor incidence between the two experimental groups and the control group, suggesting that potassium metabisulphite is not carcinogenic in mice [43].

In summary, sulphites has been shown genotoxicity *in vitro* but not *in vivo* [44]. Furthermore, sulphites has a number of reactions with nucleic acids, and cellular molecules, but no harm has been yet been demonstrated in laboratory animals, generally rats or mice, to sulphite levels of 50 or 100 times in excess of the usual maximum human environmental exposure [45].

5.3 Acceptable Daily Intakes (ADI's)

The Joint FAO/WHO Expert Committee on Food Additives (JECFA) established an ADI of 0 –0.7 mg/kg-bw in 1974 [46], which was maintained in its latest evaluation in the 51st report about evaluation of certain food additives. This ADI is for use with all the sulphiting agents presented in Table 1.The ADI was based on long-term studies in rats with a NOEL of 2.5 g/kg (0.25%) sodium metabisulphite in the diet, equivalent to 70 mg/kg of body weight per day of sulphur dioxide equivalents [44].

In the EU, the Scientific Committee for Food (SCF) in its 35th report established an ADI of 0 – 0.7 mg/kg of body weight in accordance to the established by the JECFA [32].

Surveys on sulphite content in food and intake, carried out in different countries, showed that the acceptable daily intake (ADI) for this additive is likely to be exceeded [4]. Nonetheless, a study performed in Belgium found that the intake of sulphites in the Belgian adult population is likely to be below the acceptable daily intake (ADI) with the possible exception for some high consumers of wines with high sulphite levels. Further, it was shown that for several food groups, actual use levels of sulphites are much lower than the maximum levels set in legislation [9].

6. CONCLUSIONS

Sulphites are well known food additives that constitute an important and versatile category within the food additives. They are widely used in food industry and they have some applications in other industries. Besides their functions, sulphites constitute a very cheap food additive.

Sulphites are associated with some adverse reactions such as asthmatic and anaphylactic episodes in some sensitive individuals. However, there is no data that suggests the present Acceptable Dietary Intake levels have any adverse reactions in the normal population. Only caution should be taken in cases with sulphite oxidase deficiency, which is the main enzyme that degrades sulphites into sulphate.

Overall, the available data suggests that sulphites can be used safely as a food additive. Therefore, EFSA's re-evaluation on food additives is opportune and its Opinion will be relevant to future uses of sulphites within the food industry.

ACKNOWLEDGEMENTS

H. V. and S. W. designed the work. A.R. G-F. conducted the literature research, analysed the data and wrote the first version. H. V., S. W. and E. V. were responsible for subsequent reviewing and scientific editing, while A.G-F. was the primary responsible for final content. All authors have read and approved the final manuscript. The authors declare no conflicts of interest and this work received no grant from any funding agency.

COMPETING INTERESTS

Authors have declared that no competing interests exist.

REFERENCES

1. Vasile M. Sulfites and foods, toxicity and hypersensitivity. J Eco Agri Tourism. 2010;6(4):63-6.
2. Fazio T, Warner CR. A review of sulphites in foods: Analytical methodology and reported findings. Food Addit Contam. 1990;7(4):433-54.
3. Vally H, Misso NLA, Madan V. Clinical effects of sulphite additives. Clin Exp Allergy. 2009;39(11):1643-51.
4. Quattrucci E, Masci V. Nutritional aspects of food preservatives. Food Addit Contam. 1992;9(5):515-25.
5. Ruiz-Capillas C, Jiménez-Colmenero F. Application of flow injection analysis for determining sulphites in food and

beverages: A review. Food Chem. 2009;15;112(2):487-93.
6. Wedzicha BL. Chemistry of sulphiting agents in food. Food Addit Contam. 1992;9(5):449–59.
7. Scotter MJ, Castle L. Chemical interactions between additives in foodstuffs: a review. Food Addit Contam. 2004;21(2):93-124.
8. Pizzoferrato L, Di Lullo G, Quattrucci E. Determination of free, bound and total sulphites in foods by indirect photometry-HPLC. Food Chem. 1998;63(2):275–9.
9. Vandevijvere S, Temme E, Andjelkovic M, De Wil M, Vinkx C, Goeyens L, et al. Estimate of intake of sulfites in the Belgian adult population. Food Addit Contam Part A. 2010;27(8):1072–83.
10. Yang SF. Reactions of oxidation intermediates of sulphite species with some cellular components of plants. Food Chem. 1984;15(2):113–24.
11. Grotheer P, Marshall M, Simonne A. Sulfites: Separating Fact from Fiction [Internet]. 2011 [cited 2013 Oct 11]. Available: https://edis.ifas.ufl.edu/fy731
12. Sulfite Allergy [Internet]. Australasian Society of Clinical Immunology and Allergy. 2010 [cited 2013 Dec 7]. Available:http://www.allergy.org.au/patients/product-allergy/sulfite-allergy
13. Mogol BA, Yıldırım A, Gökmen V. Inhibition of enzymatic browning in actual food systems by the Maillard reaction products. J Sci Food Agric. 2010;90(15):2556–62.
14. Nicolas JJ, Richard-Forget FC, Goupy PM, Amiot M, Aubert SY. Enzymatic browning reactions in apple and apple products. Crit Rev Food Sci Nutr. 1994;34(2):109–57.
15. McEvily AJ, Iyengar R, Otwell WS. Inhibition of enzymatic browning in foods and beverages. Crit Rev Food Sci Nutr. 1992;32(3):253-73.
16. Gunnison AF, Palmes ED. Species variability in plasma S-sulfonate levels during and following sulfite administration. Chem Biol Interact. 1978;21(2–3):315–29.
17. European Commission. Regulation (EC) No 1333/2008 of the European Parlament and of the Council on food additives. Off J Eur Union [Internet]. 2008;(L354). Available:http://eur-lex.europa.eu/LexUriServ/LexUriServ.do?uri=OJ:L:2011:295:0178:0204:EN:PDF
18. European Commission. Regulation (EU) No 1130/2011 amending Annex III to Regulation (EC) No 1333/2008 of the European Parliament and of the Council on food additives by establishing a Union list of food additives approved for use in food additives, food enzymes, food flavourings and nutrients. Off J Eur Union. 2011;(L295/180).
19. European Commission. Commision regulation (EU) No 1129/2011 amending Annex II to Regulation (EC) No 1333/2008 of the European Parliament and of the Council by establishing a Union list of food additives. Off J Eur Union. 2011;(L295/1).
20. European Food Safety Authority (EFSA). Call for food additives usage level and/or concentration data in food and beverages intended for human consumption [Internet]. 2013. Available:http://www.efsa.europa.eu/en/data/call/130327.htm
21. U.S. Food and Drug Administration. CFR - Code of Federal Regulations Title 21. Part 182 [Internet]. 1994 [cited 2013 Oct 11]. Available:http://www.accessdata.fda.gov/scripts/cdrh/cfdocs/cfcfr/CFRSearch.cfm?fr=182.3616
22. U.S. Food and Drug Administration. CFR - Code of Federal Regulations Title 21. Part 130 [Internet]. 1998 [cited 2013 Oct 11]. Available:http://www.accessdata.fda.gov/scripts/cdrh/cfdocs/cfcfr/CFRSearch.cfm?fr=130.6
23. Gibson WB, Strong FM. Metabolism and elimination of sulphite by rats, mice and monkeys. Food Cosmet Toxicol. 1973;11(2):185–98.
24. Pundir CS, Rawal R. Determination of sulfite with emphasis on biosensing methods: a review. Anal Bioanal Chem. 2013;405(10):3049–62.
25. Johnson JL, Waud WR, Rajagopalan KV, Duran M, Beemer FA, Wadman SK. Inborn errors of molybdenum metabolism: combined deficiencies of sulfite oxidase and xanthine dehydrogenase in a patient lacking the molybdenum cofactor. Proc Natl Acad Sci. 1980;77(6):3715–9.
26. Gunnison AF, Dulak L, Chiang G, Zaccardi J, Farruggella TJ. A sulphite-oxidase-deficient rat model: Subchronic toxicology. Food Cosmet Toxicol. 1981;19:221-32.
27. Cohen HJ, Fridovich I. Hepatic Sulfite Oxidase THE NATURE AND FUNCTION OF THE HEME PROSTHETIC GROUPS. J Biol Chem. 1971;246(2):367–73.
28. Joint FAO/WHO Expert Committee on Food Additives. Sulfur dioxide and sulfites. World Health Organization; 1976.

Available:http://www.inchem.org/documents/jecfa/jecmono/v21je15.htm
29. Tutuncu B, Kuçukatay V, Arslan S, Sahin B, Semiz A, Sen A. Alteration of drug metabolizing enzymes in sulphite oxidase deficiency. J Clin Biochem Nutr. 2012;51(1):50–4.
30. Harvey SB, Nelsestuen GL. Reaction of nitric oxide and its derivatives with sulfites: A possible role in sulfite toxicity. Biochim Biophys Acta BBA - Mol Cell Res. 1995;1267(1):41–4.
31. Gunnison AF, Jacobsen DW, Schwartz HJ. Sulfite Hypersensitivity. A Critical Review. Crit Rev Toxicol. 1987;17(3):185–214.
32. European Commission, Scientific Committee for Food. Reports of the Scientific Committee for Food. European Commission; 1996.
33. Walker R. Sulphiting agents in foods: Some risk/benefit considerations. Food Addit Contam. 1985;2(1):5–24.
34. Beems RB, Spit BJ, Koëter HBWM, Feron VJ. Nature and histogenesis of sulfite-induced gastric lesions in rats. Exp Mol Pathol. 1982;36(3):316–25.
35. Til HP, Feron VJ. Toxicology of sulphiting agents I: Animal studies. Food Addit Contam. 1992;9(5):587–95.
36. Til HP, Feron VJ, de Groot AP. The toxicity of sulphite. I. Long-term feeding and multigeneration studies in rats. Food Cosmet Toxicol. 1972;10(3):291-310.
37. Til HP, Feron VJ, de Groot AP, van der Wal P. The toxicity of sulphite. II. Short- and long-term feeding studies in pigs. Food Cosmet Toxicol. 1972;10(4):463–73.
38. Kadi FZE, Bénali AI, Bénali M, Belbraouet S. Effect of Sodium Metabisulphite on Blood Metabolic Status of Wistar Rats. Food Nutr Sci. 2014;05(15):1529–37.
39. Gunnison AF. Sulphite toxicity: A critical review of in vitro and in vivo data. Food Cosmet Toxicol. 1981;19:667–82.
40. Hayatsu H, Miura A. The mutagenic action of sodium bisulfite. Biochem Biophys Res Commun. 1970;39(1):156–60.
41. Mukai F, Hawryluk I, Shapiro R. The mutagenic specificity of sodium bisulfite. Biochem Biophys Res Commun. 1970;39(5):983–8.
42. Meng Z, Zhang L. Cytogenetic damage induced in human lymphocytes by sodium bisulfite. Mutat Res Toxicol. 1992;298(2):63–9.
43. Tanaka T, Fujii M, Mori H, Hirono I. Carcinogenicity test of potassium metabisulfite in mice. Ecotoxicol Environ Saf. 1979;3(4):451–3.
44. Joint FAO/WHO Expert Committee on Food Additives. Evaluation of certain food additives: fifty-first report of the Joint FAO/WHO Expert Committee on Food Additives. World Health Organization; 2000.
45. Shapiro R. Genetic effects of bisulfite (sulfur dioxide). Mutat Res Genet Toxicol. 1977;39(2):149-75.
46. Joint FAO/WHO Expert Committee on Food Additives. Toxicological evaluation of some food additives including anticaking agents, antimicrobials, antioxidants, emulsifiers and thickening agents [Internet]. Geneva: World Health Organization. 1974;520. Available:http://www.inchem.org/documents/jecfa/jecmono/v05je01.htm

14

The Application of Magnetic Resonance Spectroscopy to Investigate the Effect of a Commercial Energy Drink

Tracy L. Burrows[1*], Kirrilly M. Pursey[1] and Peter Stanwell[1]

[1]School of Health Sciences, Faculty of Health and Medicine, University of Newcastle, Callaghan, NSW, Australia.

Authors' contributions

This work was carried out in collaboration between all authors. Authors TLB and PS designed the study. Author KMP collected data. All authors read and approved the final manuscript.

ABSTRACT

Aims: Proton magnetic resonance spectroscopy (MRS) is an imaging technique that provides non-invasive, *in-vivo* measurement of brain biochemistry. The objective of this study was to explore the effects of a commercial energy drink (ED) on brain metabolites in adult males using MRS.

Methods: Ten males (18 to 30 years), of mean body mass index 23.8±2.2 kg/m^2 [range 21.4-27.8 kg/m^2] were recruited and randomized to either the intervention group (n=5) or control group (n=5). The intervention group were asked to consume two 250ml cans of Red Bull and the control group consumed no drink. One dimensional MRS on a 1.5T Achieva MRI scanner was obtained of bi-occipital gray matter. Subjects were placed in the scanner and a spectrum acquired every 5.75 minutes for 60 minutes. Linear mixed models were generated to determine the main effects due to differences between groups and times and the interaction.

Results: There was a significant group by time effect for N-acetylaspartate (NAA) (P= .02, d= 0.7) and a statistical trend for macromolecule 20 (MM20) (P= .06 d= 0.1) with the control group having higher concentrations than that of the intervention. Mean differences from baseline to 45 minutes for NAA in the intervention group were -0.8 mM (-12%) compared with 0.15 mM (<1%) for control and for MM20 -0.12 mM (<1%) in the intervention group compared with 0.03 mM (<1%) for control.

Conclusions: Consumption of two cans of ED was found to produce few changes in brain metabolites with moderate effect sizes. Changes detected in this study demonstrate the sensitivity

**Corresponding author: Email: tracy.burrows@newcastle.edu.au*

of the MRS method to explore nutrition related variables with 1.5T clinical MRI unit. The study provides some insight into an understudied, topical area of research.

Keywords: Energy drink; magnetic resonance spectroscopy; MRS.

1. INTRODUCTION

Proton magnetic resonance spectroscopy (MRS) is a technique to investigate brain metabolism in-vivo and to study endogenous biochemistry in the human brain non-invasively. It is capable of assessing the neurochemical profile of *in-vivo* brain tissue and thus provides biomarkers of neurological and metabolic disorders [1,2,3,4]. Disregarding the use in clinical studies, this methodology has not been widely used in the field of experimental nutrition, primarily due to its high cost, the need for specialized laboratory facilities and trained staff [5]. A small amount of previous diet related research using MRS has aimed to explore the pharmacologic effects of intravenous nutrient supplementation on the brain *in-situ* [6,7] and has been used for rapid assessment of food ingredient monitoring and surveillance [8]. However, the effects of whole foods or beverages on human health which are consumed as part of an individual's regular dietary intake has not yet been explored using MRS. The non-invasive quality of this technique makes it suitable not only for diagnostic purposes but also exploratory studies. This includes examining the health effects of new age foods and beverages such as energy drinks (ED), where the effects of their short and long term consumption are not yet known and their health claims highly debated.

Energy drinks are new age beverages and their market size and consumption rates are growing at exponential rates, with males aged 18 to 34 years being the highest consumers [9]. These beverages are characterized by ingredients purported to enhance energy metabolism and to improve physical and cognitive performance [9-12]. An increasing number of publications are emerging from health authorities [13,14] and case reports [15-20] regarding the health implications associated with consumption of ED's often in combination with alcohol [21-25] and their unique combination of ingredients. The safety of ED's is not well substantiated in published literature and regulation is extremely variable worldwide [26].

A systematic review identified that few studies (n=15) have investigated the implications associated with the consumption of the whole ED drink, with very few investigating effects of short and long term consumption [26]. In a second review it was reported that apart from one of the major ingredients, caffeine, there is weak evidence to substantiate nutritional health claims associated with ED [27]. Most studies have investigated individual ingredients, with caffeine the most commonly studied. Other common ED ingredients which vary widely across the spectrum of ED's include taurine, B-vitamins, glucose and herbal supplements including guarana and are often added in quantities that exceed recommended dietary intakes [9,28]. Interestingly, some of these ED ingredients such as taurine are present in human tissue and are able to be directly detected in the brain using *in-vivo* MRS. Due to homeostatic mechanisms, the basal levels of major brain metabolites are not expected to increase after ED consumption. However, a small pilot study [29] which investigated brain chemistry using MRS reported significant increases in taurine following ED consumption. These results are yet to be substantiated.

The objective of this exploratory study was to explore the effects of the commercial energy drink on the concentration of the metabolite taurine in the brain as well as other metabolites surviving specified statistical thresholds. A secondary aim was to demonstrate the application of novel imaging techniques such as MRS to the field of nutrition.

2. METHODS

Ten males aged 18 to 30 years were recruited from June to July, 2011. Exclusion criteria was consistent with MRS scanning procedures which included metal implants, pre-existing medical conditions and body mass >150 kg. Young adult male participants were recruited to limit potential confounding factors such as gender differences in brain metabolites which have been previously reported in brain MRS including males having less N-acetylaspartate (NAA) when compared to females [30]. Eligible participants were invited to participate and voluntary, written informed consent was obtained. Participants were randomized to either the intervention group (n=5) or control group (n=5). The intervention group were asked to consume two 250 ml cans of a commercially available ED, Red Bull (160 mg caffeine, 2000 mg taurine, 1200 mg glucuronolactone), and the control group consumed no ED drink. The study was approved by the University of Newcastle Human Research Ethics Committee. The dosage of Red Bull was informed by a pilot MRS study conducted by research staff (data not shown) and falls within the Australian safe consumption guidelines for caffeine [31] and therefore unlikely to cause any adverse events to participants. To ensure a consistent washout period, participants were asked to fast for two hours before their scanning session (apart from water), and refrain from caffeine intake (tea, coffee, cola drinks, chocolate and ED) for ten hours before scanning [32]. This is in line with previous studies and standard imaging protocols [33].

Prior to scanning, participants completed a 15-item questionnaire regarding their usual consumption of ED adapted from a previously conducted study [34]. Basic demographic information (age, socio-economic status, ethnicity and occupation), frequency of consumption and types of EDs consumed were assessed. Socioeconomic status was calculated using the Socio Economic Indexes for Areas (SEIFA) deciles whereby postal codes receive a score of 1-10 with 1 being the lowest socioeconomic decile and 10 the highest [35]. ED consumption patterns with alcohol and the incidence of subjective adverse effects following the consumption of ED's were also assessed.

2.1 MRS Protocol

Participants attended a single scan session between the hours of 14:00 to 17:00 hours due to scanner availability and to reduce any possible influence associated with Circadian rhythms as previously described [36]. Depending on their group allocation, participants consumed either two cans of Red Bull (500 ml total) or consumed no drink, followed by MRI scan. Compliance of the consumption of the Red Bull was monitored. Participants remained *in a rested state* in the supine position for standardized ten minutes prior to scanning to ensure participants were assessed in the rested state.

One dimensional (1D) MRS on a 1.5T Achieva magnetic resonance imaging (MRI) scanner (Phillips, The Netherlands) using a phase-array head coil was obtained of bi-occipital gray matter. Subjects were placed in the scanner and a spectrum acquired every 5.75 minutes for 60 minutes. MR spectra were acquired with a point-resolved spectroscopy (PRESS), short echo time (TE) sequence (TR 1.5 s, TE 35 ms, 192 signal averages) to maximize visibility of neurochemicals [37]. Ten spectra were acquired sequentially (i.e. temporal resolution of 5 minutes, 45 seconds) for a total examination time of approximately 60 minutes. This time point was chosen based on a previous published study [29] and our own preliminary data to confirm feasibility (data not shown).

A baseline spectrum was obtained for the intervention group (i.e. before consuming ED) followed by nine further sequential spectroscopy acquisitions. In the case of non-ED drinkers, 10 sequential spectroscopy acquisitions were acquired. Following each examination, raw spectroscopy data was exported offline and analysed with LCModel [38] using an appropriate metabolite basis set for the chosen examination parameters. LCModel analysis allows for the incorporation of metabolite *a-priori* knowledge to maximise quality of spectral fit. Prior knowledge methods in LCModel take into account the type of imager, magnetic field strength, echo time, and repetition time to maximise accuracy of quantification through incorporation of known signal parameters, such as relative frequencies, amplitude ratios, scalar coupling, and phases of resonances, which are characteristics of individual metabolites.

Spectroscopy processing incorporated water scaling using a non-water suppressed reference spectrum for signal quantification. LCModel initially reported on 70 single or combined metabolite resonances, however the majority of these were reported with Cramer-Rao lower bounds (CRLB) >20% and were thus excluded from further analysis. The concentrations in this study are reported as absolute values in mmol/kg.

The CRLB values report on the variance of unbiased estimators and thus provides an assessment of noise-related errors on estimated parameters and, are thus indispensable in assessing reliability of the metabolite concentration estimates and providing a metric of quantification precision [8]. Quantification errors are caused by measurement noise and inadequate modelling of the overlapping background and other nuisance signals (residual water, lineshape distortions, ignoring metabolites with very small concentrations, etc). Quantitative results for signals attributable to glutamate (Glu), choline-containing compounds (tCho), N-acetylaspartate (NAA), Taurine, myo-inositol (ml), and glutamate+glutamine (Glx) were only reported if the corresponding Cramer-Rao-Lower-Bounds (CRLB values), calculated by LCModel, for spectral fit were <15%, while quantitative results for several macromolecules (MM09 and MM20) were only reported if the CRLB values were ≤20%.In this study separate values for Glu, tCho, NAA, ml and Glx (<15%) and MM09 and MM20 (≤20%) were chosen due the short T2 relaxation time (and hence broader signal) expected in these macromolecules.

2.2 Statistical Analysis

Metabolite concentrations were exported to SPSS (IBM, Version 19) and analyzed using descriptive statistics with *P*<.05 considered statistically significant. Linear mixed models were generated to determine the main effects due to differences between groups and times and also the group by time interaction. No imputation of missing values was carried out for subjects; the study and subjects were analyzed in their allocated, randomization group. The effects were time (0-60 minutes), group (intervention or control). Cohens*d* was used to compare effect sizes across different metabolites [39] and allows for a more direct comparison of effects on each outcome variable. These were calculated using the mean difference and the pooled standard deviation of the group ($d= M_1- M_2/ \sigma_{pooled}$).

3. RESULTS

Ten males were recruited and completed the study. The mean age of the participants was 23 years (range 22-25 years). Mean body mass index (BMI) of the sample was 23.8±2.2 kg/m^2 [21.4-27.8 kg/m^2] with the majority of participants (n=7) classified as of healthy weight BMI category using World Health Organization BMI cut points [40]. All participants were Australian citizens and none identified as indigenous descent. The majority of participants (n=7) were students and, the mean SEIFA decile was calculated to be 6.5 (range 2-9) and classified as moderate socio-economic status. No adverse reactions were recorded.

3.1 ED Consumption

Of the ten participants, six reported being regular consumers of ED's with an average consumption of one drink per week. The remaining four reported to be non-regular consumers of ED. There were no statistically significant differences in ED consumption patterns or anthropometric variables between the intervention and control groups at baseline. All consumers reported the use of ED's with alcohol and increased consumption during periods when in combination with alcohol. Of the non-ED consumers group, the primary reason prioritized for non-consumption was the concern about the high caloric density of the beverage (n=3) followed by taste (n=1). Red Bull was the most commonly consumed ED of choice (n=7), followed by V-Drink (Frucor Beverages, n=2), and Mother (Coca Cola Amatil, n=1). All consumers of ED reported the use of original varieties (full sugar) of ED's with only one participant reporting drinking sugar-free ED options. Reasons prioritized for the consumption of ED's included to improve physical energy and stamina (n=3), prolong wakefulness (n=2), and to improve mental acuity/concentration (n=1). Insomnia and heart palpitations were the most frequently reported subjective adverse effects (n=3 each).

3.2 MRS Outcomes

Brain metabolite concentrations at each time point are shown in Table 1, example spectra shown in Fig. 1. There were no significant differences between groups at baseline for measured brain metabolites. Taurine was not reliably detected in this study as evidenced by CRLB >20%.

The acquired set of metabolites in this study for both the intervention group and our own control data were within similar ranges to other published studies using a 1.5T magnet [41,42].

Linear mixed models showed there was a significant group by time effect for the metabolite NAA (*P*= .02, *d*= 0.7) (Fig. 2a) and a statistical trend for MM20 (*P*=.06, *d*= *0.1*) (Fig. 2b) with the control group having higher concentrations than that of the intervention. Mean differences baseline to 60 minutes for NAA in the intervention group were -0.8 mM (-12%) compared with 0.15 mM (<1%) for control and for MM20 -0.12 mM (<1%) for the intervention group compared with 0.03 mM (<1%) for control.

Table 1. Brain metabolite concentrations (mM / kg) by group and time

Time (Mins)	Glu		Ins		NAA		tCho		tCr		GluGln (Glx)		MM09		MM20	
	Control	Int	Control	Int	Control	Int	Control	Int	Control	Int	Control	Int	Control	Int	Control	Int
0	6.1±0.5	5.6±0.8	3.1±0.2	3.2±0.2	6.5±0.4	6.4±0.3	0.8±0.0	0.7±0.1	4.3±0.3	4.4±0.3	8.4±0.9	7.7±1.1	4.1±0.3	3.3±1.5	9.3±1.1	7.5±3.8
5	5.7±0.6	5.8±0.9	3.2±0.1	3.0±0.3	6.4±0.4	6.3±1.0	0.8±0.1	0.7±0.1	4.5±0.2	4.6±0.4	7.6±1.0	7.7±1.6	3.7±0.3	3.2±1.3	8.4±0.9	6.8±3.8
10	6.3±0.6	6.6±0.6	3.0±0.2	2.8±0.4	6.5±0.3	6.3±1.0	0.8±0.1	0.7±0.1	4.5±0.3	4.5±0.2	8.8±0.9	9.2±0.2	4.1±0.3	2.9±1.0	9.8±0.4	5.5±2.5
15	6.0±0.6	6.0±0.6	3.2±0.3	2.9±0.4	6.4±0.6	6.4±0.8	0.8±0.1	0.7±0.1	4.5±0.2	4.6±0.4	8.4±0.6	9.0±0.9	4.0±0.3	3.3±1.3	9.1±0.8	7.2±2.9
20	6.6±0.9	5.8±0.8	3.1±0.2	2.9±0.3	6.4±0.4	6.5±0.8	0.8±0.1	0.7±0.1	4.4±0.2	4.5±0.2	9.1±1.2	8.2±0.9	3.9±0.4	3.1±1.2	8.8±0.9	6.3±2.7
25	6.0±0.2	6.2±0.8	3.1±0.2	3.3±0.4	6.4±0.5	6.1±1.3	0.8±0.0	0.7±0.1	4.5±0.2	4.6±0.1	7.8±0.4	8.4±1.2	3.5±0.6	2.9±1.0	8.3±1.0	5.9±3.2
30	6.1±0.9	5.7±0.5	3.1±0.2	2.9±0.6	6.5±0.4	6.2±1.0	0.8±0.1	0.6±0.2	4.4±0.3	4.3±0.2	8.2±1.0	8.1±0.5	3.8±0.5	3.0±1.3	8.6±1.0	6.5±3.1
35	6.5±0.6	5.5±0.6	3.1±0.1	3.0±0.3	6.2±0.4	6.6±0.6	0.8±0.0	0.7±0.1	4.4±0.2	4.5±0.3	8.8±0.6	7.8±0.7	3.5±0.3	2.9±1.0	8.0±1.0	6.5±2.3
40	6.1±0.5	5.9±0.5	3.1±0.2	3.0±0.4	6.4±0.2	6.4±0.6	0.8±0.1	0.7±0.1	4.4±0.1	4.6±0.3	8.6±0.5	8.3±0.7	3.8±0.4	2.8±1.2	9.3±1.0	6.1±3.0
45	6.2±0.4	6.3±0.7	3.1±0.1	2.7±0.4	6.7±0.5	5.6±1.3	0.8±0.1	0.6±0.1	4.4±0.3	4.1±0.9	8.9±0.6	8.5±1.4	3.8±0.4	3.0±1.4	9.3±1.7	7.4±3.4

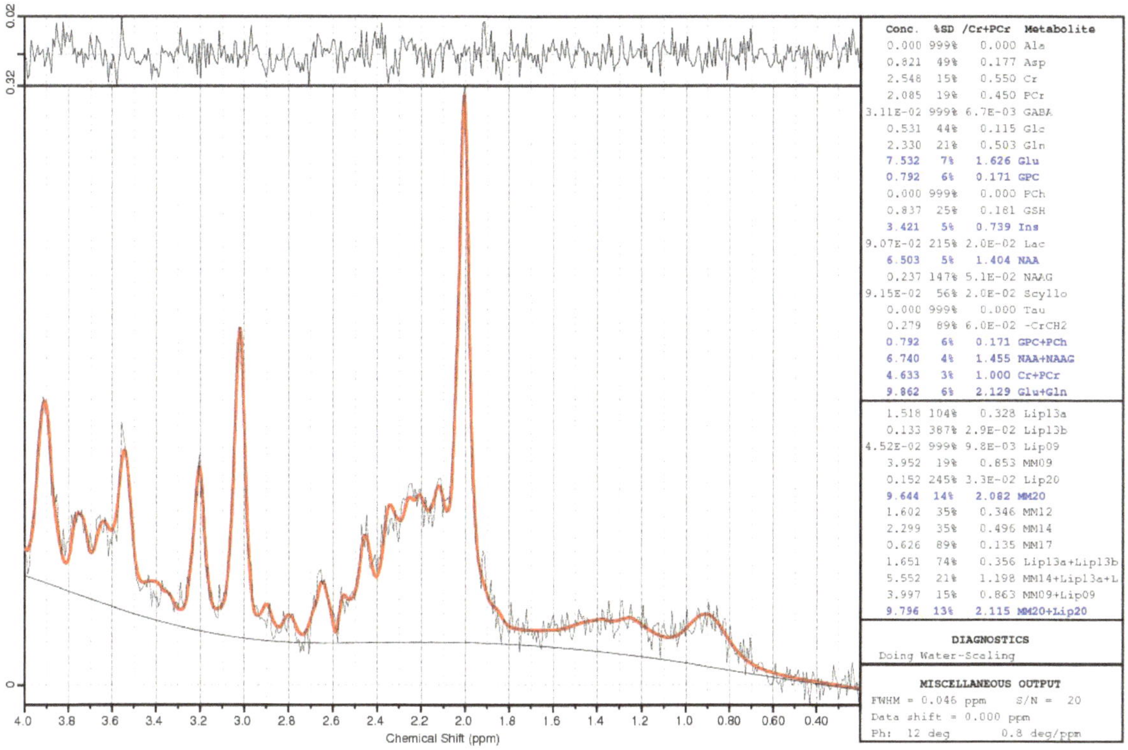

Fig. 1. Example of LCModel analysis of a proton magnetic resonance (1H-MR) spectrum acquired from bi-occiptalgray matter from a control participant. Analysis was undertaken using an appropriate simulated basis set

Although there was a significant time effect for Glx (P=0.04) (Figs. 2c and 2d), no group by time effect was found (P>.05). There was larger variability as evidenced by the large standard deviation in concentrations for all metabolites for the intervention group) compared to the control group.

4. DISCUSSION

This is the first study to explore changes in brain metabolites using MRS following the consumption of a commercial ED and compared with a no drink control. This study demonstrated that the consumption of ED within Australian guidelines for maximum safe levels of caffeine had differing effects on the brain metabolites NAA and MM20. Over a 60 minute time frame absolute concentrations of NAA and MM20 decreased after ED consumption. The changes detected in this study demonstrate the sensitivity of the MRS method even when measured with 1.5T magnet. Consumption of two cans of ED were found to produce few changes with a moderate effect size for NAA (Cohens*d*). The study provides insight into an understudied, topical area of research whereby nutritional claims for energy drinks are often highly debated in the literature. The use of MRS in this study provides an objective measure to assess short term changes in brain chemistry following ED consumption. Additionally, MRS did not appear to pose a recruitment issue for this small exploratory study.

There are increasing concerns as to the potential effects of combining alcohol with ED's [43,44], as was reported by the majority of ED consumers in this study. It has been previously reported that stimulant ingredients such as caffeine and guarana in ED's dull the depressant effects of alcohol [23]. This includes giving the consumer a false sense of physical and mental competence, decreasing awareness of alcohol-related impairment [21-23,43], increasing amounts of alcohol consumed [45,46] and increasing the likelihood of engaging in risk taking behaviours [47-51].

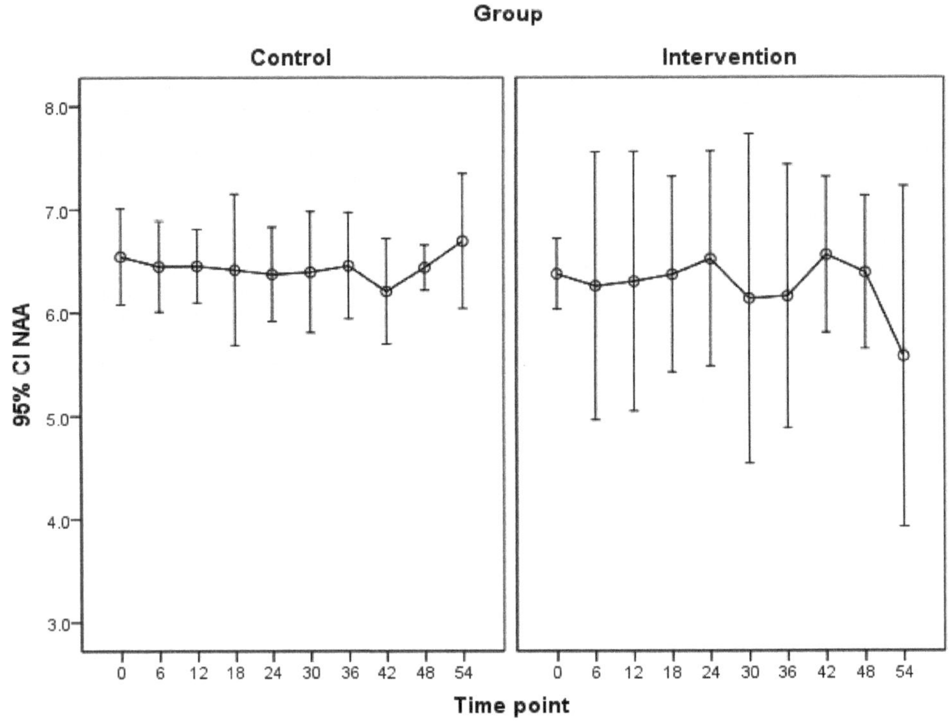

Fig. 2a. Mean concentrations of N-acetylaspartate (NAA) by group over time. Values shown are mmol/kg. Error bars represent the 95% confidence interval

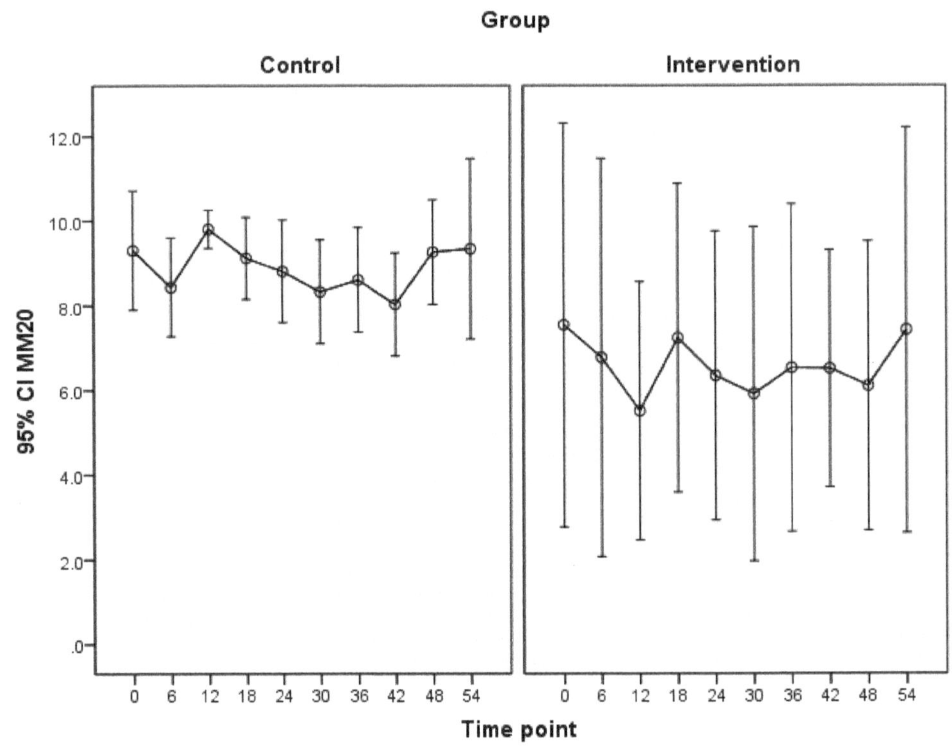

Fig. 2b. Mean concentrations of Macromolecules (MM20) by group over time. Values shown are mmol/kg. Error bars represent the 95% confidence interval

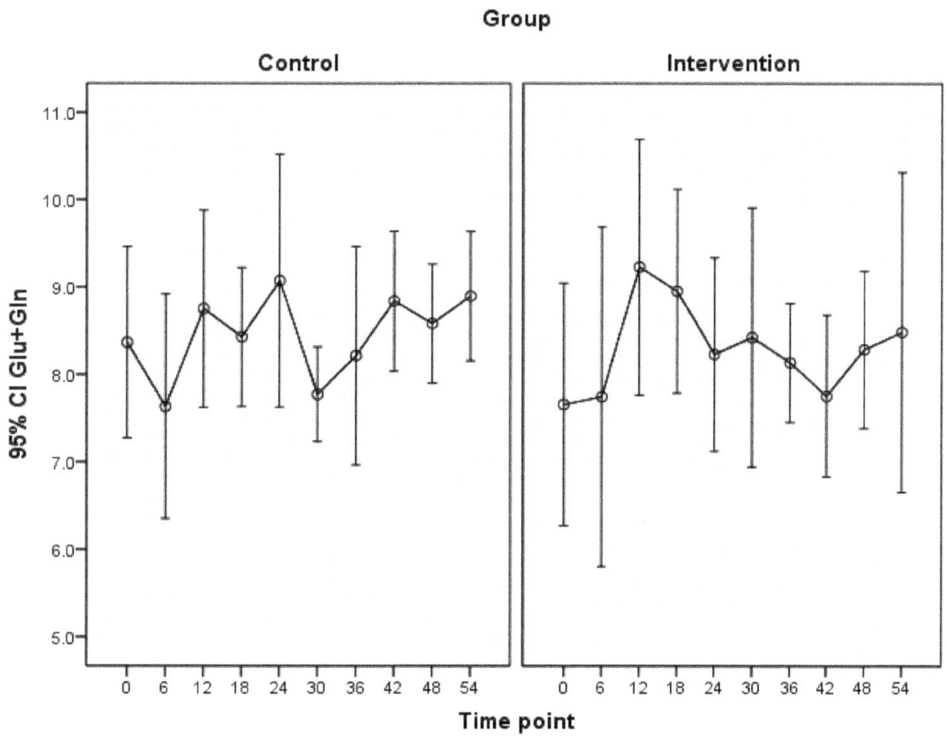

Fig. 2c. Mean concentrations of Glutamate+Glutamine (Glx) by group over time. Values shown are mmol/kg. Error bars represent the 95% confidence interval

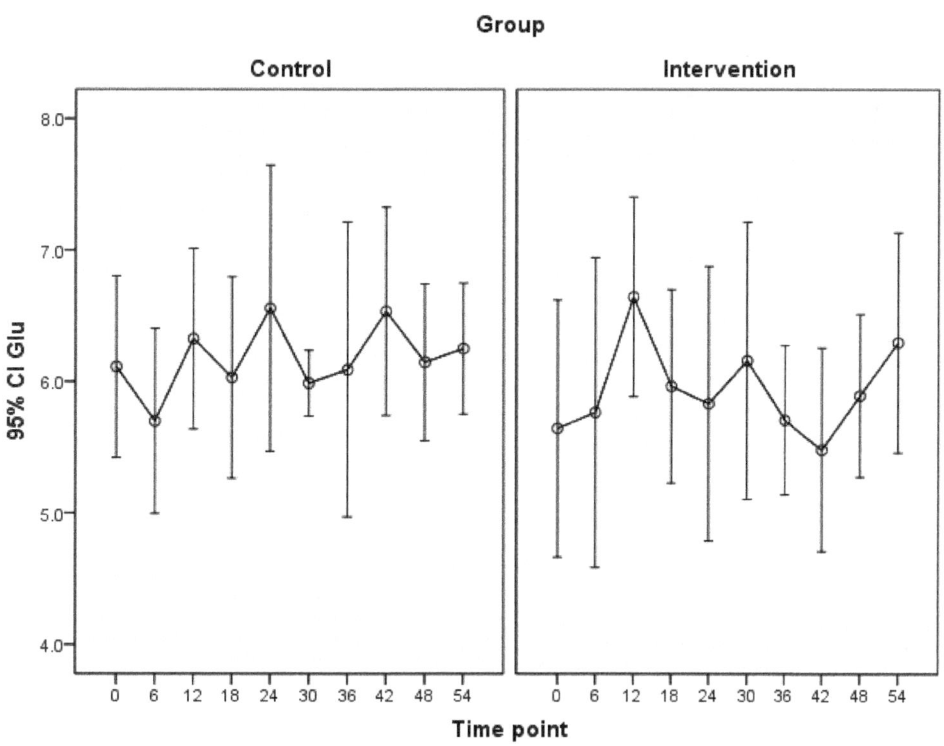

Fig. 2d. Mean concentrations of Glutamate (Glu) by group over time. Values shown are mmol/kg. Error bars represent the 95% confidence interval

N-acetylaspartate (NAA) is generally considered a putative neuronal marker with its almost exclusive localisation in neurons, making it in a reliable diagnostic marker. However, NAA is additionally believed to play a role in osmoregulation and can be released to the interstitial space upon osmotic challenge [37] N-acetylaspartate was found to decrease in this study after ED consumption. While statistically significant, the absolute change was still within reference values. It is also possible the changes in NAA observed here are due primarily to changes in N-acetylaspartylglutamate (NAAG), which is difficult to separate from NAA at the magnetic field strength employed in this study. Glutamate is the dominant excitatory amino acid in the human brain and thus is thought to have a role linking excitatory neurotransmission to energy production and usage [37]. Glutamine is synthesized from glutamate by glutamine synthetase in glia and converted back to glutamate by glutaminase in neurons, completing the glutamate–glutamine cycle. Both glutamate and glutamine are engaged in a number of important metabolic pathways like energy provision, synthesis of important chemicals like N-acetylaspartate (NAA), N-acetylaspartylglutamate (NAAG), γ-aminobutyrate (GABA) and glutathione (GSH), and protein synthesis [52]. Glutamine is commonly assessed in combination with glutamate (glutamate + glutamine) and referred to as Glx. Myo-inositol is an isomer of inositol and has been associated with osmolarity control in the human brain [37]. Choline-containing compounds are essential for membrane lipid synthesis and act as precursors for the biosynthesis of the neurotransmitter acetylcholine [53].

Direct comparison with other studies is difficult due to the lack of studies in this area, differences in measurement methodologies such as differing strengths of magnets, gradient performance and field homogeneity and also the reporting of brain metabolites as different units such as ratios with total creatinine (tCr). The decrease in MM20 in participants who consumed the ED was surprising. These resonances are believed to be attributable to proteins and MM present in brain tissue [54], and be primarily assigned to methyl and methylene resonances of protein amino acids such as leucine [55]. It therefore possible that the decrease observed in this study could be explained by the presence of Red Bull ingredients that co-resonate with these macromolecules and therefore added destructively to attenuate the MM20 signal.

This study articulates the MRS sequencing procedures that could be applied to assess the impact of other foods and beverages. With a plethora of functional foods and beverages appearing on supermarket shelves internationally, the application of new objective methodologies such as MRS to the field of nutrition could offer insights into their effects after short and long term consumption [5].

Taurine was not reliably detected in this study as evidenced by CRLB >20% using LCModel and metabolite basis set of 35ms. This does not support previous findings where a large elevation in taurine (n=7) was observed 25 minutes following consumption of 750mL of Red Bull[29]. Changes of the magnitude reported by Roser et al (1998) could have a number of potential implications for the normal functioning of the brain and osmoregulation [56]. Roser et al noted that the cerebellum was more sensitive for detecting taurine elevation rather than the bi-occipital gray matter measured in this study. This may have limited the ability to detect a difference in taurine levels to the magnitude of the previous study.

The main limitation of this study are the small sample size. However, the number of participants in the study are in line with other exploratory studies [5,57]. As a random convenience sample of young adult males was recruited, it is likely that this limited any age or gender differences which have been observed for some brain metabolites [5]. This reduces the generalizability of the study findings to other population groups and ethnicities. A limitation of the study was that the control group consumed no drink. This unbalanced intake of water may have led to differences in the concentration of brain metabolites. In future studies using MRS to identify differences in brain metabolites following intake of beverages should use a control beverage matched for fluid composition. Full demographic details of the participants and their educational ability were not collected as part of the study.

Although this study assessed usual consumption of ED, habitual caffeine consumption including commonly consumed products such as tea, coffee, cola and chocolate was not. Chronic caffeine consumption often leads to an increase in adenosine receptor density and metabolism [58,59]. Adenosine receptors function in the brain to regulate neurotransmitters including dopamine and glutamate [60,61]. Therefore it is possible that metabolite concentrations identified in this study could have been affected by individual differences in long term consumption patterns of other caffeinated beverages more broadly. Future studies using MRS to investigate the effects of food and beverages on the brain should assess habitual caffeine consumption of study subjects more comprehensively. Additionally, smoking status of subjects was not assessed in this pilot study. It is well documented that the CYP1A2 enzyme catalyses the metabolism of numerous compounds including caffeine [62,63]. CYP1A2 expression is affected by a number of dietary and environmental factors including smoking, which enhances the metabolism and clearance of caffeine [64,65]. Future MRS studies should assess smoking status when investigating caffeinated foods and beverages as this could influence the interpretation of the study findings.

The strengths of the current study are that it assessed ED consumption at baseline, included a control group and that it used a strong statistical approach to assess any changes that may have occurred in brain metabolites after ED consumption. Variables that influence metabolite concentration include echo time (TE), repetition time (TR), population variables (age, gender, brain location) so acquiring our own control data with the same acquisitionvariables in a similar population group is unique. Future studies should investigate the effects of ED's when combined with alcohol, utilize larger sample sizes, and a more culturally diverse population as well as using higher strength magnets that improve spectral resolution and thus metabolite identification and quantification [37]. Additionally, participants received a pre-scan nutrition protocol including fasting for two hours prior to scanning and avoidance of caffeine and alcohol products ten hours prior to scanning. Future studies may consider providing more extensive nutritional standardisation for the days leading into scanning to further control variables related to dietary intake and effects on brain metabolites.

The findings of the study including adaptation of the methodology may be used to generate meaningful power size calculations to detect possible changes and inform future research into the effects of other foods and beverages could have on brain metabolites. This is important given a recent review has highlighted the benefits of brain imaging procedures such as MRS in nutrition intervention studies [5].

5. CONCLUSION

This exploratory study found that the consumption of two cans of ED produced few changes in specific brain metabolites including taurine, NAA, Glx, and MM20 with moderate effect sizes. Changes detected in this study demonstrate the sensitivity of the MRS method to explore nutrition related variables with 1.5T clinical MRI unit. The non-invasive quality of MRS make it a highly valuable technique and investigation of the effects of ED on brain metabolites is only one of many paradigms that could be investigated. Other protocols conducted over a longer duration or using higher dosages or in combination with alcohol may produce different results and is worthy of future study. Given the increasing consumption of ED, this study provides some insight into a topical area of research.

CONSENT

All authors declare that 'written informed consent was obtained from the patient (or other approved parties) for publication of this case report and accompanying images.

ETHICAL APPROVAL

All authors hereby declare that all experiments have been examined and approved by the appropriate ethics committee and have therefore been performed in accordance with the ethical standards laid down in the 1964 Declaration of Helsinki. The study was approved by the University of Newcastle Human Research Ethics Committee, Approval No. H- [20110034].

ACKNOWLEDGEMENTS

This research was funded by an ECR grant for T Burrows University of Newcastle. The authors thank Dr Saad Ramadan his technical assistance in the acquisition of data.

COMPETING INTERESTS

There are no competing interests to declare in this study.

REFERENCES

1. Duarte JMN, et al. The neurochemical profile quantified by in vivo 1h nmr spectroscopy. Neuroimage. 2012;61(2):342-362.

2. Mountford CE. et al. Neurospectroscopy: The past, present and future. Chemical Reviews. 2010;110(5):3060-3086.
3. Kreis R. et al. reproducibility of cerebral phenylalanine levels in patients with phenylketonuria determined by 1h-mr spectroscopy. Magnetic Resonance in Medicine. 2009;62(1):11-16.
4. Mangia S, et al. Neurochemical profile of patients with type 1 diabetes measured by 1h-mrs at 4t. Journal of Cerebral Blood Flow & Metabolism. 2013;33(5):754-759.
5. Sizonenko S, et al. brain imaging and human nutrition: Which measures to use in intervention studies. British journal of nutrition. 2013;110(s1).
6. Terpstra M, et al. noninvasive quantification of human brain antioxidant concentrations after an intravenous bolus of vitamin c. Nmr in biomedicine. 2011;24(5):521-528.
7. Smith FE, et al. Quantitative lithium magnetic resonance spectroscopy in the normal human brain on a 3 t clinical scanner. Magn Reson Med. 2011;66(4):945-949.
8. Triebel S, et al. Rapid analysis of taurine in energy drinks using amino acid analyzer and fourier transform infrared (ftir) spectroscopy as basis for toxicological evaluation. Amino Acids. 2007;33(3):451-457.
9. HeckmanMA KSherry, de mejia eg. Energy drinks: An assessment of their market size, consumer demographics, ingredient profile, functionality, and regulations in the united states. Compre Rev Food Sci F. 2010;9(3):303-317.
10. Baum M, Weiss M. The influence of a taurine containing drink on cardiac parameters before and after exercise measured by echocardiography. Amino Acids. 2001;20(1):75-82.
11. Geiβ KR, et al. The effect of a taurine-containing drink on performance in 10 endurance-athletes. Amino Acids. 1994;7(1):45-56.
12. Ivy JL, et al. Improved cycling time-trial performance after ingestion of a caffeine energy drink. International Journal of Sport Nutrition and Exercise Metabolism. 2009;19(1):61-78.
13. Althea F. FDA warns caffeinated alcohol energy drinks are unsafe. National Journal; 2010.
14. Australia and new zealand food regulation ministerial council communique. [website]; 2011. Availablefrom:http://www.foodstandards.gov.au/scienceandeducation/newsroom/mediareleases/mediareleases2011/australiaandnewzeala5154.cfm
15. Iyadurai SJP, Chung SS. New-onset seizures in adults: Possible association with consumption of popular energy drinks. Epilepsy Behav. 2007;10(3):504-508.
16. Röggla G, et al. Seizures in a night club. The lancet. 2007;370(9583):220.
17. Machado-vieira R,. Viale CI, Kapczinski F. Mania associated with an energy drink: The possible role of caffeine, Taurine and Inositol Can Journ Psychiat. 2001;46(5):454.
18. Nagajothi N, et al. Energy drink-related supraventricular tachycardia. American Journal of Medicine. 2008;121(4):E3-4.
19. Terlizzi R, et al. Reversible postural tachycardia syndrome due to inadvertent overuse of red bull I. Clinical Autonomic Research. 2008;18(4):221-223.
20. Greene E, Oman K, Lefler M. Energy drink-induced acute kidney injury. Ann Pharmacother. 2014;48(10):1366-70.
21. Ferreira SE, et al. Effects of energy drink ingestion on alcohol intoxication. Alcohol Clin Exp Res. 2006;30(4):598-605.
22. Schoffl I, et al. Vodka Enegy: Too much for the adolescent nephron? Pediatrics. 2011;128(1):E227-e231.
23. Weldy DL. Risks of alcoholic energy drinks for youth. Journal of the American Board of Family Medicine. 2010;23(4):555-558.
24. Striley CW, khan SR. Review of the energy drink literature from 2013: Findings continue to support most risk from mixing with alcohol. Curr Opin Psychiatry. 2014;27(4):263-8.
25. Peacock A, et al. 'H'gh' risk? A systematic review of the acute outcomes of mixing alcohol with energy drinks. Addiction. 2014;109(10):1612-33.
26. Burrows T, et al. What are the health implications associated with the consumption of energy drinks? A systematic Review. Nutrition Reviews. 2013;71(3):135-148.
27. Mclellant, Lieberman H. Do energy drinks contain active components other than caffeine? Nutrition Reviews. 2012;70(12):730-744.
28. Shao A, Hathcock JN. Risk assessment for the amino acids taurine, l-glutamine and l-

arginine. Regulatory Toxicology & Pharmacology. 2008;50(3):376-99.

29. Roser W, et al. Dangerous increase of taurine in the human brain after consumption of 'n 'energy dr'nk'? In proceedings to the 16th scientific annual meeting of the International Society for Magnetic Resonance in Medicine; 1998.
30. Maudsley A, et al. Mapping of brain metabolite distributions by volumetric proton mr spectroscopic imaging (mrsi). Magn Reson Med. 2009;61(3):548-559.
31. Smith PF, et al. Safety aspects of dietary caffeine - report from the expert working group. Australia New Zealand Food Authority; 2000.
32. Seng KY, et al. Population pharmacokinetics of caffeine in healthy male adults using mixed-effects models. Journal of Clinical Pharmacy and Therapeutics. 2009;34(1):103-114.
33. Pursey K, et al. Neural responses to visual food cues according to weight status: A systematic review of functional magnetic resonance imaging studies. Frontiers Nutrition; 2014.
34. Malinauskas BM, et al. A survey of energy drink consumption patterns among college students. Nutrition Journal. 2007;6:35.
35. Census of population and housing: Socio-economic indexes for areas (Seifa). Australian bureau of statistics: Australia. P. Cat; 2011. No. 2033.0.55.001.
36. Kakeda ., et al. Influence of work shift on glutamic acid and gamma-aminobutyric acid (gaba): Evaluation with proton magnetic resonance spectroscopy at 3t. Psychiatry Research: Neuroimaging. 2011;192(1):55-59.
37. Duarte J, et al. The neurochemical profile quantified by in vivo 1h nmr spectroscopy. Neuroimage. 2012;61(2):342-62.
38. Provencher SW. automatic quantitation of localized in vivo 1h spectra with lcmodel. Nmr in Biomedicine. 2001;14(4):260-264.
39. Cohen J. Statistical power analysis behavioural sciences' 2nd ed, ed. N. Hillsdale. Lawrence Earlbaum and Associates; 1988.
40. Cole T, et al. Establishing a standard definition for child overweight and obesity worldwide: International Survey. BMJ. 2000;320:1240.
41. Parks M, et al. Longituduinal brain metabolic characterization of chronic alcoholics with proton magentic resonance spectroscopy. Alcohol Clin Exp Res. 2002;26(9):1368-1380.
42. Govindaraju V, Young K, Maudsley A. Proton nmr chemical shifts and coupling constants for brain metabolites. Nmr Biomedicine. 2000;13:129-153.
43. Curry K, Stasio MJ. The effects of energy drinks alone and with alcohol on neuropsychological functioning. Human psychopharmacology, 2009;24(6):473-481.
44. Finnegan D. The health effects of stimulant drinks. Nutrition Bulletin. 2003;28(2):147-155.
45. Arria AM, et al. Energy drink consumption and increased risk for alcohol dependence. Alcoholism: Clinical and Experimental Research. 2011;35(2):365-375.
46. Velazquez CE, et al. Associations between energy drink consumption and alcohol use behaviors among college students. Drug and Alcohol Dependence. 2012;123(1–3):167-172.
47. Higgins JP, tuttle TD, Higgins CL. Energy beverages: Content and safety. Mayo Clinic Proceedings. 2010;85(11):1033-1041.
48. Pennay A, lubman DI, Miller P. Combining energy drinks and alcohol: A recipe for trouble? Australian Family Physician. 2011;40(3):104-107.
49. Bigard AX. Risks of energy drinks in youths. Archives de Pediatrie. 2010;17(11):1625-1631.
50. Brache K, Stockwell T. Drinking patterns and risk behaviors associated with combined alcohol and energy drink consumption in college drinkers. Addictive Behaviors. 2011;36(12):1133-1140.
51. Snipes DJ, Benotsch EG. High-risk cocktails and high-risk sex: Examining the relation between alcohol mixed with energy drink consumption, sexual behavior, and drug use in college students. Addictive Behaviors. 2013;38(1):1418-1423.
52. Ramadan S, Lin A, Stanwell P. Glutamate and glutamine: A review of in vivo mrs in the human brain. Nmr in Biomedicine. 2013;26(12):1630-1646.
53. Mountford CE, et al. Neurospectroscopy: The past, present and future. Chem Rev. 2010;110:3060-3086.
54. Hoffman L, et al. Characterisation of the macromolecule localised h-mr spevtra of human brain. Magn Reson Med. 2001;46(5):855-863.

55. Behar K, et al. Analysis of macromolecule resonances in 1h nmr spectra of human brain. Magn reson med. 1994;32(3):294-302.
56. Pasantes-morales H, Schousboe A. Role of taurine in osmoregulation in brain cells: Mechanisms and functional implications. Amino Acids. 1997;12(3-4):281-292.
57. Babb SM, et al. Oral choline increases choline metabolites in human brain. Psychiatry Resear–h - Neuroimaging. 2004.;130(1):1-9.
58. Ribeiro JA, sebastlão AM. Caffeine and adenosine. Journal of Alzhei'er's Disease. 2010;20(0):3-15.
59. Daly JW. Caffeine analogs: Biomedical impact. Cellular and molecular life sciences. 2007;64(16):2153-2169.
60. Ciruela F, et al. Presynaptic control of striatal glutamatergic neurotransmission by adenosine a1-a2a receptor heteromers. J Neurosci. 2006;26(7):2080-7.
61. Fredholm BB, et al. Actions of caffeine in the brain with special reference to factors that contribute to its widespread use. Pharmacol Rev. 1999;51(1):83-133.
62. Kalow W, Tang BK. Use of caffeine metabolite ratios to explore cyp1a2 and xanthine oxidase activities. Clin Pharmacol Ther. 1991;50(5 pt 1):508-19.
63. Butler MA, et al. Determination of cyp1a2 and nat2 phenotypes in human populations by analysis of caffeine urinary metabolites. Pharmacogenetics. 1992;2(3):116-27.
64. Kalow W, Tang BK. Caffeine as a metabolic probe: Exploration of the enzyme-inducing effect of cigarette smoking. Clin Pharmacol Ther. 1991;49(1):44-8.
65. Sachse C, et al. Polymorphisms in the cytochrome p450 cyp1a2 gene (cyp1a2) in colorectal cancer patients and controls: Allele frequencies, linkage disequilibrium and influence on caffeine metabolism. British Journal of Clinical Pharmacology. 2003;55(1):68-76.

15

Short Review of Extracts of Rosemary as a Food Additive

Paul de Raadt[1], Sabrina Wirtz[1], Ellen Vos[1] and Hans Verhagen[2,3*]

[1]Maastricht University, Faculty of Law, 6200MD Maastricht, The Netherlands.
[2]National Institute for Public Health and the Environment (RIVM), 3720 BA Bilthoven, The Netherlands.
[3]University of Ulster, Northern Ireland Centre for Food and Health (NICHE), Cromore Road, Coleraine, BT52 1SA Northern Ireland, UK.

Authors' contributions

This work was carried out in collaboration between all authors. Authors HV and SW designed the work. Author PdR conducted the literature research, analysed the data and wrote the first version. Authors HV, SW and EV were responsible for subsequent reviewing and scientific editing, while author PdR was the primary responsible for final content. All authors read and approved the final manuscript.

ABSTRACT

Extracts from *Rosmarinus officinalis* L., more commonly known as rosemary, have been approved for use in the EU as food additive E932 under Regulation 1333/2008 of the European Parliament and the Council. Rosemary extracts are currently widely used to increase shelf life of food products. Rosemary extracts are characterised by two reference antioxidant compounds, carnosol and carnosic acid. This characterization allows for differences in rosemary extracts. Four approved production methods, as described by Commission Regulation 231/2012, produce rosemary extracts with different compositions and antioxidant activity. This results in difficulties to compare scientific data and to assess the safety of approved rosemary extracts. Based on unpublished studies for each of the four approved extract types, EFSA concluded that "the proposed uses and use levels would not be of safety concern". Yet, gaps in knowledge still exist for the approved extracts as many different rosemary extracts are used in scientific research.

**Corresponding author: Email: Hans.verhagen@rivm.nl*

Keywords: Food additives; E932; extracts of rosemary; legal status; toxicological data.

1. INTRODUCTION

Oxidative degradation by free radicals in foods results in a reduction of quality, odour, and texture, but this process can be delayed by the addition of antioxidants. In the food industry, both synthetic and natural antioxidants are being used to improve the storage stability of foods. The benefit of synthetic antioxidants originates from the high level of activity in comparison with natural anti-oxidants [1]. However, some synthetic antioxidants have shown potential adverse effects. For example, butylated hydroxytolunene (BHT) and butylated hydroxyanisole (BHA) have been extensively used as synthetic antioxidants in foods. However, after research showed a potential carcinogenic effect [2,3] regulations regarding these synthetic antioxidants became restricted. Recent research however contradicts this by showing a potential anti-carcinogenic effect [4,5]. As a consequence of the contradiction and the general consumer request for natural products, the use of natural antioxidants has increased in recent years.

Rosmarinus officinalis L., more commonly known as rosemary, belongs to the Lamiaceae family of herbs and is a popular spice originating from Mediterranean areas. Traditionally it was used for flavour and aromatic reasons, but after the discovery of the antioxidant properties of rosemary extracts the European Rosemary Extract Manufactures Group (EREMG) applied for the authorization by the European Commission of rosemary extracts as a food additive. Subsequently, the European Food Safety Authority (EFSA) evaluated the safety of rosemary extracts in 2008. After the assessment of EFSA rosemary extracts were approved as a food additive and specified for use in the EU. Henceforth the extracts were formally labelled"E392 extracts of rosemary". However, some gaps in knowledge were present during the assessment. To assess the current state of knowledge in relation with EU legislation a literature search was performed using PubMed with "rosemary extracts" "extracts of rosemary", and "*Rosmarinus officinalis* L extracts" as search terms in combination with section specific search terms.

2. CHEMICAL STRUCTURE

Commission Regulation 231/2012 specifies approved food additive in Regulation 1333/2008, and defines rosemary extracts as: "Extracts of rosemary contain several components, which have been proven to exert antioxidative functions. These components belong mainly to the classes of phenolic acids, flavonoids, diterpenoids. Besides the antioxidant compounds the extract can also contain triterpenes and organic solvents extractable material specifically defined in the following specification". More specifically, carnosol (CAS No 5957-80-2) and carnosic acid (CAS No 3650-09-7) (Fig. 1) are identified as the reference antioxidant compounds in extracts of rosemary, and should not comprise less than 90% of the total phenolic diterpenes in the extract.

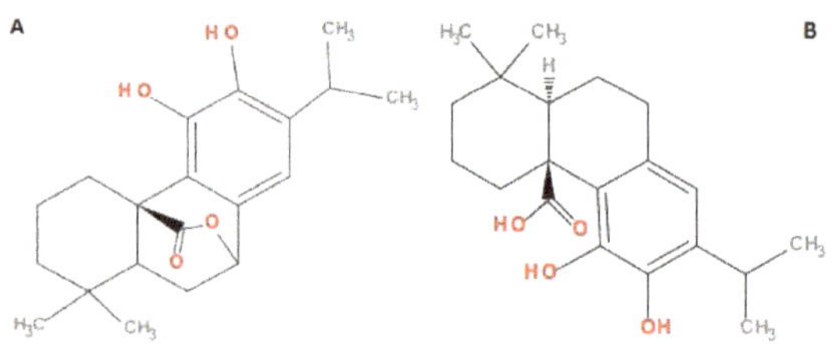

Fig. 1. Chemical structure of carnosol $C_{20}H_{28}O_4$ (A) and carnosic acid $C_{20}H_{28}O_4$ (B). Based on Senorans et al. [7]

Commission Regulation 231/2012 only mentions carnosol and carnosic acid specifically as reference compounds, as a consequence all other antioxidant compounds in rosemary extracts can vary. Table 1 illustrates the findings from Almela et al. [6] on the compounds present in rosemary extracts. Although, many compounds have been identified, there still remain unidentified substances. Senorans et al. [7] found several non-identified compounds using a liquid chromatographic and mass spectrometric analysis. Furthermore, both studies concluded that the exact composition for rosemary extracts was influenced by the extraction and purification process used. Additionally, the composition of active components in the final extract was found to be dependent on which parts of the rosemary plant were used in the extraction process, and whether or not they were dried [8]. Furthermore, evidence exists that the composition of rosemary essential oil is influenced by the harvesting time [9]. Although, rosemary essential oil and rosemary extracts are different, both are produced from the leaves. Therefore rosemary extracts composition might also be influenced by harvest time, however there is no evidence to either support or contradict this.

3. MANUFACTURING

Within the EU, food additives, in accordance with Regulation 1333/2008, need authorization prior to their use in foodstuffs. Article 12 of this Regulation limits the production of a food additive to the specified production methods as included in the Community list of approved food additives. Significant changes in the production method, starting materials or a change in particle size requires additional authorization.

In accordance with Commission Regulation 231/2012 extracts of rosemary can be produced from dried leaves of rosemary using four methods. The first method utilizes acetone to dissolve the active components from rosemary leaves. After filtration the solvent is evaporated and the residue is dried and sieved by conventional food processing methods. The second method is based on supercritical extraction [10]. A supercritical fluid is a substance at a temperature and pressure above its critical point. As a result the supercritical fluid can effuse through solids like a gas and dissolve materials from this solid similarly to a liquid. In this manufacturing method supercritical CO_2 is used to dissolve the active components. In the second step of the manufacturing process the extract is separated from the supercritical CO_2 [10]. Ethanol extracts of rosemary constitute the third manufacturing method. This method is similar to the first mentioned method, but ethanol is used as a solvent and deodorizer [10]. The fourth and last approved method for the manufacturing of rosemary extracts is also based on a two-step process. Extracts of rosemary are produced in accordance with the third method; subsequently the deodorized ethanol rosemary extracts are decolorized with a hexane extraction [10]. Regardless of the method used, the extracts can be combined with, by Regulation 1333/2008 approved, food grade carriers or standardized to a specific carnosol plus carnosic acid content.

The first two manufacturing methods produce rosemary extracts with similar compositions. However, supercritical fluid extraction produces extracts with a higher antioxidant concentration than organic solvents [11]. Furthermore, supercritical fluids have a higher diffusion coefficient, lower viscosity, lower surface tension for rapid penetration, and are more selective in extraction substances [12]. Additionally, the supercritical extraction method provides the possibility to remove flavour from the rosemary extract, although this characteristic of the method is not specifically mentioned in Regulation 231/2012 [13]. Aside from a different composition the third method produces an odourless extract and the fourth method produces both an odourless and colourless extract.

Table 1. Main compounds identified in rosemary extracts. Based on Almela et al. [6]

Type of compound	*Chemical name*	*Related compounds*
Phenolic acids	Rosmarinic acid	
Diterpenes	Carnosic Acid	Methyl Carnosate (C_{20}:$COOCH_3$)
		Carnosol (δ-lactone bridge C_{20}-C_7)
		Epiisorosmanol (δ-lactone bridge C_{20}-C_7:C_6-OH)
		Rosmanol (γ-lactone bridge C_{20}-C_6:C_7:-OH)
		Epirosmanol (γ-lactone bridge C_{20}-C_6:C_7:-OH
		Epirosmanol-methylether (γ-lactone bridge C_{20}-C_6:C_7:-OH)
		Rosmadial (lactone bridge C_{20}-C_{11}:C_6=C_7=CHO)
Flavones	Lueolin	6-hydroxyluteolin 7-glucoside
	Apigenin	Genkwanin (7-methyl ether)
		4"-methoxytectochrysin (4":7-dimethyl ether)
		Homoplataginin (6-methyl ether, 7-glucoside)
		Scutellarein (6-hydroxy)
		4',5,7,8-Tetrahydroxyflavone (8-hydroxyapigenin)
		Cirsimaritin (6,7-dimethyl ether)

In 2002, Ibanez et al. [14] reported on the use of subcritical water for the extraction for rosemary extracts. In short, subcritical water (H_2O at a temperature and pressure below its critical point) is used as a solvent, and in a second step the extracts of rosemary are separated from the subcritical water. It was concluded that this extraction method produces extracts with similar antioxidant activity as supercritical fluid extraction, however the extract composition was different. The petitioner specifically excluded water extracts within its application. Subsequently, water based extracts have not been assessed by EFSA and are not incorporated in Regulation 321/2012.

4. USES IN FOODS

4.1 Function in Foods

Unless intentionally removed, the extracts of rosemary still contain flavourings and can be used to improve the flavour of products, however rosemary extracts are currently widely used to increase shelf life of food products due to the high antioxidant activity of its main components [15,16]. Research on the antioxidant activity of the compounds in rosemary extract in foods is primarily focussed on carnosic acid. The potency of carnosic acid in comparison to the other compounds in rosemary extracts in soybean oil was found to be more than twice as high based on Rancimat measurements [17]. Furthermore, the same study also found higher antioxidant activity for carnosic acid in comparison tothe synthetic antioxidants BHT and BHA, whereas tertiary butylhydroquinone (TBHQ) was found to surpass rosemary extracts in antioxidant activity. Zhang et al. [18] found the similar results when comparing BHT, BHA, and TBHQ with carnosic acid based on peroxide values, TBARS assays, free fatty acid measurements, and p-anisidine values in sunflower oil. However, antioxidant activity of rosemary extracts seems to depend on the medium. Frankel et al. [19] measured the antioxidant activity of both carnosol and carnosic acid in a corn oil emulsion and bulk oil using hydroperoxide and hexanal formations and found reduced antioxidant activity in the emulsion system. Furthermore, antioxidant activity at pH 7 was lower compared to activity at pH 4 and 5. Koleva et al. [20] found a similar difference between bulk oil measurements and emulsion oil measurements for rosemaric acid.

The aforementioned findings suggest that rosemary extracts might not increase shelf life in all products, but evidence exist for several food categories. Using the TBARS assay, Sebranek et al. [21] found similar antioxidant activity for an unspecified rosemary extracts as BHA and BHT in precooked frozen pork sausages, but in raw frozen pork sausages rosemary extracts was found to be more effective than BHA and BHT in preventing increased TBARS values, and in preventing loss of red colour. Moreover, Nassu et al. [22] investigated the efficacy of a commercial albeit unspecified rosemary extract in fermented goat sausage using a TBARS assay and found a significant difference in oxidation with 0.050% w/w extract compared to controls. Using peroxide values according to the AOCS(1989) method, Frutos et al. [23] found significant protection against oxidation in bread with an oil, garlic and parsley dressing when using a 4 g/l concentration of rosemary extracts(carnosic acid (20–30%), rosmarinic acid (0–1%) and rosmanol (0.5–1.5%)). Despite the possible differences in the extract composition, these studies depict the efficacy of rosemary extracts in prolonging oxidative degradation in actual food products, however no evidence was found for the approved rosemary extracts.

4.2 Dietary Exposure

In 2008 EFSA published its opinion on the use of extracts of rosemary as a food additive [24]. As part of this opinion, expected dietary exposure was assessed based on consumption in the UK. The UK National Dietary and Nutrition Survey for adults [25] and pre-school children [26] were used as the main data sources. Despite the exclusion of rosemary as flavouring agent, a conservative estimate, based on the assumption of maximum usage levels in all proposed foods, was made. The potential mean exposure to carnosol plus carnosic acid from all proposed uses was estimated at 0.04 mg/kg bw/day for adults and 0.11 mg/kg bw/day for pre-school children. At the 90th, 95th and 97,5th percentile the expected intake was estimated at 0.08, 0.10 and 0.12 mg/kg bw/day for adults. For pre-school children this was estimated at 0.18, 0.20 and 0.23 mg/kg bw/day respectively.

5. LEGAL STATUS IN EU AND BEYOND

With the introduction of the legislative Package on Food Improvement Agents in December 2008 also the regulation of food additives in the European Union was reformed. Regulation 1333/2008 of the European Parliament and the Council, introduced one regime for the use of food additives in the Union, food colours sweeteners and the remaining food additives that until then were regulated in separate Directives (respectively Directive 94/36/EC, Directive 94/35/EC and Directive 95/2/EC).In accordance with Regulation 1333/2008 rosemary extracts can only be used in the approved food categories, in food additive preparations, and in food grade carriers. For rosemary extracts the maximum levels of use for each approved food category ranges from 30 mg/l or mg/kg for fats and oils essentially free from water till 400 mg/l or mg/kg for food supplements. A more detailed description is shown in Table 2. For colour, beta-carotene, and lycopene preparationsa maximum level of 1000 mg/kg is allowed in the preparation and 5 mg/kg for the final product. Additionally, rosemary extracts may be added to all food flavourings with a maximum of 1000 mg/kg. These levels are expressed as the sum of carnosic acid and carnosol rather than the dosage of the whole rosemary extract. Regulation 1333/2008 also specifies labelling requirements, as a result extracts of rosemary can either be labelled with the E-number "E392" or "antioxidant: rosemary extract".

Besides a maximum level of use, the purity of food additives is also specified by EU legislation (Commission Regulation 231/2012). Acetone extracts of rosemary should contain more than 10% w/w of carnosol plus carnosic acid and should not contain more than 500 mg/kg acetone. Supercritical extracts should contain more than 13% w/w and no more than 2% ethanol. Deodorized ethanolic extracts should contain more than 5% w/w and no more than 200 mg/kg ethanol. Decolorized and deodorized hexane and ethanol extracts should contain more than 5% w/w, and not more than 25 mg/kg hexane and/or 500 mg/kg ethanol. Moreover, all rosemary extracts should have an antioxidant/volatiles ratio higher than 15.

In its assessment, EFSA concluded that "the proposed uses and use levels would not be of safety concern" even though EFSA could not establish a numerical acceptable daily intake due to the lack of reproductive and development toxicity data [24]. On a global scale, rosemary extracts have not yet been evaluated by the Joint FAO/WHO Expert Committee on Food Additives (JECFA). In March 2013 the Codex Alimentarius Commission expressed its intention to include rosemary extracts as an antioxidant in the standard for fish oils. However, to date rosemary extracts (internationally labelled as INS 392) has not been included in the Codex General Standard for Food Additives (GSFA) [27].

Within the Unites States (US) food additives can be used if one of three conditions are met. The Federal Code of Regulations 21 U.S.C. 321 – Definitions: Generally states that food additives as defined by that code are allowed when the additive has either received premarket approval from the Food and Drug Administration (FDA), has a Generally Recognized As Safe (GRAS) status, or was consumed in significant quantities prior to January, 1, 1958. Natural extracts of rosemary have received a GRAS status (Federal Code of Regulations 21 U.S.C. 182.20), which has been affirmed by the FDA [29]. Subsequently, rosemary extracts is listed in the 'Everything added to food in the United States (EAFUS)' database. As a result the use of rosemary extracts as a food additive is allowed in the US based on the general recognition of safety, rather than a safety assessment. Furthermore, no clear characterisation of natural extract of rosemary is described by US legislation.

6. BIOLOGICAL AND TOXICOLOGICAL DATA

6.1 Absorption, Distribution, Metabolism, and Excretion

Providing a complete description of the absorption, distribution, metabolism and excretion is difficult due to the complex nature of rosemary extracts. However, Vaquero et al. [30] attempted to establish a complete metabolic profile for all major bioactive components of rosemary extracts in Zucker rats. A commercial rosemary extract was used containing carnosic acid (38.9±1.7%), carnosol (6.5±0.1%), and methyl carnosate (6.9±0.6%). Other compounds or the production method were not specified. This study reported that after oral administration of 100 mg rosemary extract traces of both carnosol and carnosic acid were found in the small intestine after 25 minutes. After 13.3 hour large quantities of both carnosol and carnosic acid were present in the small intestine, caecum and colon. In total twenty-seven metabolites were found following the intake of rosemary extract by Vaquero et al. [30].

Table 2. Detailed description of the approved food categories, maximum levels, and restriction for the use of extracts of rosemary as stated in EU regulation 1333/2008 [28]

Food category	*Maximum level (mg/l or mg/kg) **	*Restrictions*
Dehydrated milk as defined by Directive 2001/114/EC	200	Only milk powder for vending machines
	30	Only dried milk for manufacturing of ice cream
Fats and oils essentially free from water (excluding anhydrous milkfat)	30	Only vegetable oils (excluding virgin oils and olive oils) and fat where content of polyunsaturated fatty acids is higher than 15% w/w of the total fatty acid, for the use in non-heat- treated food products
	50	Only fish oil and algal oil; lard, beef, poultry sheep and porcine fat; fat and oils for the professional manufacture of heat-treated foods; frying oils and frying fat, excluding olive oil and pomace oil
Vegetable oil pan spray	50	Only fats and oils for the professional manufacture of heat-treated foods
Nut butters and nut spreads	200	
Processed potato products	200	Only dehydrated potatoes products
Chewing gum	200	
Decorations, coatings and fillings, except fruit-based fillings	100	Only sauces
Fillings of stuffed pasta (ravioli and similar)	250	Only in fillings of stuffed dry pasta Period of application: From 25 December 2012
Fine bakery wares	200	
Non-heat-treated processed meat	100	Only dried sausages
	150	Excluding dried sausages
	150	Only dehydrated meat
Heat-treated processed meat	100	Only dried sausages
	150	Excluding dried sausages
	150	Only dehydrated meat
Processed fish and fishery products including molluscs and crustaceans	150	
Processed eggs and egg products	200	
Seasonings and condiments	200	
Mustard	100	
Soups and broths	50	
Sauces	100	
Potato-, cereal-, flour- or starch-based snacks	50	
Processed nuts	200	
Food supplements supplied in a solid form including capsules and tablets and similar forms, excluding chewable forms	400	
Food supplements supplied in a liquid form	400	
Food supplements supplied in a syrup-type or chewable form	400	

** Extracts of rosemary as the sum of carnosol and carnosic acid*

Furthermore, nine metabolites, including carnosol and carnosic acid, were found in varying concentrations in the plasma, liver and brain. Vaquero et al. [30] also reported onthe time course of carnosic acid and carnosol in plasma. Carnosic acid reached the maximum peak concentration of 26.6 μM after 0.4 hour and the last measurable concentration was found after 1.7 hour. The clearance of carnosol had not started at the end of the measurements, as such no maximum peak concentration was found (after 13.3 hours the concentration was 18.2 μM). For both carnosic acid and carnosol the limit of detection was 1 μM (0.3 ppm) and the limit of quantification was 3 μM (1 ppm).In contrast, Doolaege et al. [31] found a maximum peak concentration for carnosic acid of 0.035 mg/ml after 2.3 hour using a lower dosage (limit of detection and limit of quantification were both 0.005 mg/ml).Additionally, traces of carnosic acid were found in the liver andmuscle tissue.

Only the metabolism of rosmarinic acid has been evaluated in humans. Baba et al. [32] reports a peak concentration in plasma after 0.5 hour, which is followed by an increase of methyl-rosmarinic acid with a peak concentration after 2 hour. Furthermore, 75% of the ingested rosmarinic acid is excreted in the urine as methylated rosmarinic acid, caffeic acid, ferulic acid, and m-coumaric acid after 6 hours.

6.2 Acute, Short Term, Subchronic Toxicity, and Chronic Toxicity

Based on unpublished data from the petitioner EFSA concluded that two versions of ethanol extraction extracts can be considered to be of low acute toxicity [24]. This assessment was based on the acute oral administration by gavage of 8.5/10 g/kg bw and 24/28.5 g/kg bw for female and male rats respectively and the daily administration by gavage of 4.5/5 g/kg bw and 11.8/14.1 g/kg bw for females and males respectively during five days. No mortality was reported in any of the experiments, however both extracts showed an increase in liver weight for both sexes and increase in fatty liver.

In lower concentrations however, no adverse effects have been reported. In the acute oral safety study of Anadon et al. [33], Winstarrats were orally given a single dose of 2000 mg/kg bw of rosemary extract. For this study a supercritical fluid extraction method was used to obtain the rosemary extract. During the two-week observation period no abnormal signs, changes in behaviour, or body weight were observed. After two weeks no histological, hematological, or organ weight changes were found. Moreover, Fahim et al. [34] found no adverse effect using a daily oral dose of 0.15 g / 100 g bw ethanol extract for three weeks, and even reports a potential hepatoprotective effect for the extract compared to a reference compound. However, no acute toxicity studies have been performed with acetone or hexane rosemary extracts.

For the assessment of short-term and sub-chronic toxicity ESFA used unpublished data from five13-week experiments provided by the petitioner [24]. For two ethanol extracts and a hexane extract administered dosages ranged from 500 or 1000 till 5000 mg/kg diet, for which the latter is equal to 400 mg/kg bw/day rosemary extract. Furthermore, an acetone extract was administered at 3800 mg/kgdiet, and a supercritical CO_2 extract was administered at 300,600 & 2400 mg/kgdiet. Aside from some palatability issues, no mortality or any clinical effects were found in any of the studies. The only notable effect was a dose dependent increase in liver weight, which was only significant at the highest dose groups for all types of rosemary extracts. Upon further investigation, using a supercritical CO_2 extract at dose of 2400 mg/kgdiet,a1.5-fold increase in liver enzymes cytochrome P450 andseveral CYP2 enzymes were found. However, these effects were shown to be reversible after a 28 day recovery period. Using an ethanol extract, a similar increase was found, however, this change was not significant. Moreover, the increase in liver weight was accompanied by minimal centrilobular hypertrophy and microsomal enzyme induction. Furthermore, using a hexane extract a reduction in hepatic DNA content was found for the 5000mg/kg diet group suggesting hypertrophy rather than hyperplasia. The changes in liver weight, caused by centrilobular hypertrophy and microsomal enzyme induction, are generally not considered a toxic response. Additionally, the low magnitude, reversibility, and no increases in plasma levels of aspartate aminotransferase, alanine aminotransferase, and alkaline phosphatase lead the EFSA panel to conclude that the increase in liver weight represents an adaptive response rather than hepatoxicity [24]. No further effects were reported for any of the studies and the NOAEL range was determined at 180 till 400 mg extract/kg bw/day, which is equivalent to 20-60 mg/kg bw/day of carnosol plus carnosic acid. Currently, no published data is present to support or contradict these findings.

6.3 Genotoxicity, Carcinogenicity, Reproductive and Developmental Toxicity

At the time of evaluation by EFSA unpublished studies were used to evaluate the mutagenicity of rosemary extracts [24]. Several extracts were tested with the Ames test using different strains of *S. typhimurium*. Based on these tests ESFA concluded that ethanol and hexane-ethanol rosemary extracts were not mutagenic. Another study showed no changes in mutation frequency for a hexane-ethanol rosemary extract in a human lymphoblastoid cell line. Additionally, it was concluded, based on another study, that hexane-ethanol rosemary extracts do not increase mutagenicity *in vivo* in mice. Genotoxicity of an ethanol rosemary extract was further assessed with an *in vitro* chromosome aberration test using human lymphocytes. No significant damage occurred in the chromosomes and it was concluded that no genotoxic effect was found. No studies have investigated the genotoxicity of supercritical CO_2 and acetone rosemary extracts.

Currently, there are no published studies to report a genotoxic effect for rosemary extract, however several studies suggest rosemary extracts to affect enzymes related to carcinogenicity. Offord et al. [35,36] and Debersac et al. [37] reported that several different rosemary extracts affect several phase I and II enzymes involved with carcinogenicity/genotoxicity, like cytochrome P450 1A1 (decreased) and glutathione-S-transferase (increased). However, the extracts used differ from the approved extracts. Mohebati et al. [38] replicated these findings with one of the main components, carnosol, present in the approved extracts. Furthermore, several other authors reported on potential protective properties against carcinogenicity/genotoxicity of different rosemary extracts through different pathways [39-44]. Due to the difference in composition between rosemary extract it is unclear whether similar effect are also present for the approved extracts.

A rosemary extract (carnosic acid 88%, carnosol 12%) was also identified as anti-mutagenic by Minnunni et al. [45] by preventing reactive oxygen species from reacting with genetic information in the Ames test with strain TA102. Furthermore, Slamenova et al. [46] reported on the protective properties of an ethanol rosemary extract to DNA by scavenging for both OH radicals and singlet oxygen. No long term studies for carcinogenicity have been performed for any of the rosemary extracts.

Only two published studies investigated reproductive and developmental toxicity. The effect of rosemary extract on fertility after 63 days was examined in rats by Nusier et al. [47]. Two ingestion levels were used, 250 and 500 mg/kg/day, of which the 500 mg/kg/day group showed a significant decline in spermatogenesis, and a decrease in sperm motility and density. However, these results were found using an extract made by a 70% ethanol and water 30% extraction resulting in an extract of a polar nature. Therefore EFSA concluded that these results were not applicable to the proposed extracts under assessment. Lemonica et al. [48] comparedadaily dosage of 26 mg of a 30% w/w rosemary extract with a saline solution on the incidence of abortion and/or interference with normal development in Winstarrats. In contrast to most other studies, a water based rosemary extracts was made from leaves, flowers and stems, which was administered by gavage prior and during pregnancy. Due to the production method of the rosemary extract it suffers from the same problem as the aforementioned study. Nevertheless, no differences were found post-implantation or in the development of the foetuses, however a non-significant difference was found in implantation loss.

Based on the histopathology conducted in previously mentioned 13-week studies EFSA was able to conclude that at the tested dose levels do not affect the male reproductive system. As mentioned before, due to the lack of reproductive and developmental toxicity or long-term studies EFSA was not able to establish an acceptable daily intake.

7. OTHER DATA

Rosemary was also found to have potential antimicrobial, antifungal, and antiviral effects albeit mainly as essential oil not as extract. De Melo et al. [49] investigated the antimicrobial effects of rosemary essential oils for active packaging. A significant difference in microbial activity was found for 50% w/w rosemary essential oil. Similarly to the antioxidant activity of rosemary extracts, the antimicrobial effects were reported to be strongly dependent on the composition of the medium [50]. Rosemary oil and rosemary extracts have several compounds in common, however no evidence exist to support an antimicrobial effect of rosemary extracts. Furthermore, significant differences in the biological control of fungi in foodstuffs were found for low concentrations of an unspecified rosemary extract by Centeno et al. [51], this was also reported for rosemary essential oils [52].

8. DISCUSSION AND CONCLUSION

Rosemary is a spice that was traditionally used as a flavour and aromatic agent. In 2008 EFSA published its opinion regarding the safety as a food additive, stating: "the proposed uses and use levels would not be of safety concern" [24], and subsequently the European Commission approved the use of rosemary extracts in 2010 [28]. Despite the approval under Regulation 1333/2008 and further specification under Commission Regulation 231/2012, extracts of rosemary are not yet fully characterised. The characterisation based on the reference compounds leaves room for different compositions. These differences in composition are partially restricted by Commission Regulation 231/2013 that only allows production of rosemary extracts from dried rosemary leaves. However, the different approved production methods still result in different composition and antioxidant activity [10]. Moreover, the characterisation of rosemary extracts is different outside the EU and research using rosemary extracts are not bound to these restrictions. This results in difficult to compare scientific evidence and difficulty to assess the safety of approved rosemary extracts.

Despite the differences in composition, the use of rosemary extracts has shown antioxidant properties in several food categories. However, this effect might not be present in all foods, since the antioxidant activity of one of the major components of rosemary extracts, carnosic acid, was found to be medium dependent [19,20]. From a functional perspective, the room for different compositions might allow the optimization of antioxidant activity of the rosemary extract for a specific medium or food. However, this depends on the properties of the other components of the extract, which are currently unknown.

No new evidence is present to suggest adverse effects, but some discrepancies still exist. The unpublished data suggests an increase in total cytochrome P450 content, however Offord et al. [35,36] reports the inhibition of cytochrome P4501A1 mRNA expression and decreased enzyme activity in human bronchial epithelial cells. In contrast, Debersac et al. [37] did not find any changes for cytochrome p450 in hepatic tissue of Wistar rats. The reason for this contradiction is difficult to identify due to differences in test materials and extract composition. Additionally, both studies report a strong induction of phase II enzymes, particularly glutathione-S-transferase, and quinone reductase. These studies and others that show similar protective effects tend to show a large diversity in extract compositions, potential mechanism, and effect magnitude [39-44]. Many studies also focus solely on carnosol or carnosic acid. While there is no reason for concern, the lack of knowledge about other components than carnosol and carnosic acid and possible combined effects of extract components are clear.

The increase in liver weight was not seen as an adverse effect;as such the NOAEL range for all five extracts was determined at 20-60 mg/kg bw/day of carnosol plus carnosic acid. Based on the estimated dietary exposure, this results in a safety margin for the 95th percentile between 200-600 for adults, and 100-300 for pre-school children. Furthermore, no lowest observed adverse effect level was determined. However, the effect of the rosemary extract from Nusier et al. [47] on spermatogenesis and sperm motility and density raise some concern. Although the extracts are considered to be different, both are produced from the leaves of rosemary and could contain many of the same compounds. However even if it is considered as an adverse effect, it was only present above the intake of 60mg/kg bw/day carnosol plus carnosic acid. Leaving a greater margin of safety than for the increase in liver weight effect. Therefore there is no reason for concern, however further research is need to ensure this adverse effect, and any other effect on reproduction, development or otherwise are not present for the approved rosemary extracts. The indications for other potential functional properties of rosemary extracts only strengthens the need for further investigation. Additionally, further specification, characterisation of rosemary extracts might also help fill the knowledge gaps.

Despite the uncertainty created by the compositional differences between rosemary extracts and the gaps in knowledge, there is no evidence to suggest a direct need for re-evaluation of approval. Moreover, since rosemary extracts have been approved by Directive 2010/69/EU in 2010, it will not be re-evaluated under the process started by

Regulation 1333/2008 to re-evaluate all food additives authorized for use in the Union prior to 20 January 2009 (see Commission Regulation No 257/2010).

ACKNOWLEDGEMENTS

The authors declare no conflicts of interest and this research received no grant from any funding agency.

COMPETING INTERESTS

Authors have declared that no competing interests exist.

REFERENCES

1. Pokorný J. Are natural antioxidants better – and safer – than synthetic antioxidants? European Journal of Lipid Science and Technology. 2007;109(6):629-642.
2. Ito N, Fukushima S, Tsuda H. Carcinogenicity and modification of the carcinogenic response by bha, Bht, and other antioxidants. Critical Reviews in Toxicology. 1985;15(2):109-150.
3. Williams G. Epigenetic promoting effects of butylated hydroxyanisole. Food and Chemical Toxicology. 1986;24(10):1163-1166.
4. Botterweck A, et al. Intake of butylated hydroxyanisole and butylated hydroxytoluene and stomach cancer risk: Results from analyses in the Netherlands cohort study. Food and Chemical Toxicology. 2000;38(7):599-605.
5. Williams G, Iatropoulos M, Whysner J. Safety assessment of butylated hydroxyanisole and butylated hydroxytoluene as antioxidant food additives. Food and Chemical Toxicology. 1999;37(9):1027-1038.
6. Almela L, et al. Liquid chromatograpic–mass spectrometric analysis of phenolics and free radical scavenging activity of rosemary extract from different raw material. Journal of Chromatography A. 2006;1120(1–2):221-229.
7. Señoráns FJ, et al. Liquid chromatographic–mass spectrometric analysis of supercritical-fluid extracts of rosemary plants. Journal of Chromatography A. 2000;870(1–2):491-499.
8. del Baño MJ, et al. Phenolic diterpenes, flavones, and rosmarinic acid distribution during the development of leaves, flowers, stems, and roots of *Rosmarinus officinalis*. Antioxidant Activity. Journal of Agricultural and Food Chemistry. 2003;51(15):4247-4253.
9. Singh M, Guleria N. Influence of harvesting stage and inorganic and organic fertilizers on yield and oil composition of rosemary (*Rosmarinus officinalis* L.) in a semi-arid tropical climate. Industrial Crops and Products. 2013;42(0):37-40.
10. Commission Regulation (EU) No 231/2012 of 9 March 2012 laying down specifications for food additives listed in Annexes II and III to Regulation (EC) No 1333/2008 of the European Parliament and of the Council Text with EEA relevance.
11. Carvalho Jr, RN, et al. Supercritical fluid extraction from rosemary (*Rosmarinus officinalis*): Kinetic data, extract's global yield, composition, and antioxidant activity. The Journal of Supercritical Fluids. 2005;35(3):197-204.
12. Rozzi NL, Singh RK. Supercritical fluids and the food industry. Comprehensive Reviews in Food Science and Food Safety. 2002;1(1):33-44.
13. López-Sebastián S, et al. Dearomatization of antioxidant rosemary extracts by treatment with supercritical carbon dioxide. Journal of Agricultural and Food Chemistry. 1998;46(1):13-19.
14. Ibañez E, et al. Subcritical water extraction of antioxidant compounds from rosemary plants. Journal of Agricultural and Food Chemistry. 2002;51(2):375-382.
15. Hernández-Hernández E, et al. Antioxidant effect rosemary (*Rosmarinus officinalis* L.) and oregano (*Origanum vulgare* L.) extracts on TBARS and colour of model raw pork batters. Meat Science. 2009;81(2):410-417.
16. Erkan N, Ayranci G, Ayranci E. Antioxidant activities of rosemary (*Rosmarinus Officinalis* L.) extract, blackseed (*Nigella sativa* L.) essential oil, carnosic acid, rosmarinic acid and sesamol. Food Chemistry. 2008;110(1):76-82.
17. Richheimer S, et al. Antioxidant activity of lipid-soluble phenolic diterpenes from rosemary. Journal of the American Oil Chemists' Society. 1996;73(4):507-514.
18. Zhang Y, et al. Oxidative stability of sunflower oil supplemented with carnosic

acid compared with synthetic antioxidants during accelerated storage. Food Chemistry. 2010;118(3):656-662.
19. Frankel EN, et al. Antioxidant activity of a rosemary extract and its constituents, carnosic acid, carnosol, and rosmarinic acid, in bulk oil and oil-in-water emulsion. Journal of Agricultural and Food Chemistry. 1996;44(1):131-135.
20. Koleva II, et al. Screening of plant extracts for antioxidant activity: A comparative study on three testing methods. Phytochemical Analysis. 2002;13(1):8-17.
21. Sebranek JG, et al. Comparison of a natural rosemary extract and BHA/BHT for relative antioxidant effectiveness in pork sausage. Meat Science. 2005;69(2):289-296.
22. Nassu RT, et al. Oxidative stability of fermented goat meat sausage with different levels of natural antioxidant. Meat Science. 2003;63(1):43-49.
23. Frutos MJ, Hernández-Herrero JA. Effects of rosemary extract (*Rosmarinus officinalis*) on the stability of bread with an oil, garlic and parsley dressing. LWT - Food Science and Technology. 2005;38(6):651-655.
24. Aguilar F, et al. Use of rosemary extracts as a food additive–scientific opinion of the panel on food additives, flavourings, processing aids and materials in contact with food. EFSA J. 2008;721:1-29.
25. Henderson L, et al. The national diet & nutrition survey: Adults aged 19 to 64 years. HM Stationery Office; 2002.
26. Gregory JR, et al. National diet and nutrition survey: Children aged. 1½ to 4½ years, report of the diet and nutrition survey. London: HMSO. 1995;1.
27. FAO J. Report of the twenty-third session of the codex committee on fats and oils. CL 2013/5-FO; 2013.
28. Regulation (EC) No 1333/2008 of the European Parliament and of the Council on food additives; 2008.
29. Administration, U.F.a.D. Everything Added to Food in the United States (EAFUS). 2011 Last Update Date: 03/13/2013 Access Date: 11/26/2013]; Available:http://www.fda.gov/Food/IngredientsPackagingLabeling/FoodAdditivesIngredients/ucm115326.htm
30. Romo Vaquero M, et al. Bioavailability of the major bioactive diterpenoids in a rosemary extract: Metabolic profile in the intestine, liver, plasma, and brain of Zucker rats. Molecular Nutrition & Food Research. 2013;57(10):1834-1846.
31. Doolaege EA, et al. Absorption, distribution and elimination of carnosic acid, a natural antioxidant from *Rosmarinus officinalis*, in rats. Plant Foods for Human Nutrition. 2011;66(2):196-202.
32. Baba S, et al. Absorption, metabolism, degradation and urinary excretion of rosmarinic acid after intake of Perilla frutescens extract in humans. European Journal of Nutrition. 2005;44(1):1-9.
33. Anadón A, et al. Acute oral safety study of rosemary extracts in rats. Journal of Food Protection. 2008;71(4):790-795.
34. Fahim A, AYE, Hoda M. Fadel, Khaled F. S. Hassan, Fawzia. Allied studies on the effect of *Rosmarinus officinalis* L. on experimental hepatotoxicity and mutagenesis. International Journal of Food Sciences and Nutrition. 1999;50(6):413-427.
35. Offord EA, et al. Rosemary components inhibit benzo[a]pyrene-induced genotoxicity in human bronchial cells. Carcinogenesis. 1995;16(9):2057-62.
36. Offord EA, et al. Mechanisms involved in the chemoprotective effects of rosemary extract studied in human liver and bronchial cells. Cancer Letters. 1997;114(1–2):275-281.
37. Debersac P, et al. Induction of cytochrome P450 and/or detoxication enzymes by various extracts of rosemary: Description of specific patterns. Food Chem Toxicol. 2001;39(9):907-18.
38. Mohebati A, et al. Carnosol, a constituent of Zyflamend, inhibits aryl hydrocarbon receptor-mediated activation of CYP1A1 and CYP1B1 transcription and mutagenesis. Cancer Prev Res (Phila). 2012;5(4):593-602.
39. Sancheti G, Goyal P. Modulatory influence of *Rosemarinus officinalis* on DMBA-induced mouse skin tumorigenesis. Asian Pac J Cancer Prev. 2006;7(2):331-5.
40. Lo AH, et al. Carnosol, an antioxidant in rosemary, suppresses inducible nitric oxide synthase through down-regulating nuclear factor-kappaB in mouse macrophages. Carcinogenesis. 2002;23(6):983-91.
41. Singletary K, MacDonald C, Wallig M. Inhibition by rosemary and carnosol of 7,12-dimethylbenz[a]anthracene (DMBA)-induced rat mammary tumorigenesis and in vivo DMBA-DNA adduct formation. Cancer Lett. 1996;104(1):43-8.

42. Mace K, et al. Development of *In vitro* models for cellular and molecular studies in toxicology and chemoprevention. Arch Toxicol Suppl. 1998;20:227-36.
43. Choi HR, et al. Peroxynitrite scavenging activity of herb extracts. Phytother Res, 2002;16(4):364-7.
44. Yesil-Celiktas O, et al. Inhibitory effects of rosemary extracts, carnosic acid and rosmarinic acid on the growth of various human cancer cell lines. Plant Foods Hum Nutr. 2010;65(2):158-63.
45. Minnunni M, et al. Natural antioxidants as inhibitors of oxygen species induced mutagenicity. Mutat Res. 1992;269(2):193-200.
46. Slamenova D, et al. Rosemary-stimulated reduction of DNA strand breaks and FPG-sensitive sites in mammalian cells treated with H_2O_2 or visible light-excited Methylene Blue. Cancer Lett. 2002;177(2):145-53.
47. Nusier MK, Bataineh HN, Daradkah HM. Adverse effects of rosemary (*Rosmarinus officinalis* L.) on reproductive function in adult male rats. Exp Biol Med (Maywood), 2007;232(6):809-13.
48. Lemonica IP, Damasceno DC, Di-Stasi LC. Study of the embryotoxic effects of an extract of rosemary *(Rosmarinus officinalis* L.). Braz J Med Biol Res. 1996;29(2):223-7.
49. de Melo AA, et al. Microbiological quality and other characteristics of refrigerated chicken meat in contact with cellulose acetate-based film incorporated with rosemary essential oil. Braz J Microbiol. 2012;43(4):1419-27.
50. Del Campo J, Amiot MJ, Nguyen-The C. Antimicrobial effect of rosemary extracts. J Food Prot. 2000;63(10):1359-68.
51. Centeno S, et al. Antifungal activity of extracts of *Rosmarinus officinalis* and *Thymus vulgaris* against Aspergillus flavus and A. ochraceus. Pak J Biol Sci. 2010;13(9):452-5.
52. Rasooli I, et al. Antimycotoxigenic characteristics of *Rosmarinus officinalis* and *Trachyspermum copticum* L. essential oils. Int J Food Microbiol. 2008;122(1-2):135-9.

16

Dietary Diversity of Rural Households in North Central Nigeria

Mary O. Agada[1*] and Edwin M. Igbokwe[2]

[1]Institute of Food Security, University of Agriculture, P.M.B. 2373 Makurdi, Nigeria.
[2]Department of Agricultural Extension, University of Nigeria, Nsukka, Enugu State, Nigeria.

Authors' contributions

This work was carried out in collaboration between both authors. Author MOA designed the study, wrote the protocol, managed the literature searches and wrote the first draft of the manuscript. Both authors read and approved the final manuscript.

ABSTRACT

Aims: To determine the dietary diversity score and diversity within food groups consumed and the probability of food security among rural households in North Central Nigeria.
Study Design: A cross sectional survey design was used.
Place and Duration of Study: The study was conducted among three ethnic groups in North Central Nigeria, between November and December, 2011. The ethnic groups were Tiv, Igala and Eggon in Benue, Kogi and Nasarawa States respectively.
Methodology: A sample of 340 households was interviewed to identify the variety of foods consumed over the previous 24 hours. Data from the food groups was analyzed using percentages. Since national dietary data is not available on mean Dietary Diversity Score, mean terciles were used to classify households into low, medium and high dietary diversity.
Results: The findings indicated that beside oil and fat (97%) and spices and condiments (96%), root and tuber crops (86%) constituted the food group consumed by most households. Most Tiv (96.7%) and Igala (94.4%) households reported use of root and tuber crops whereas most Eggon (78.6%) households reported use of cereals. Mean dietary diversity terciles were 2.90, 4.53 and 6.37 for low, medium and high dietary diversity respectively, with a mean of 4.60 for the study area.
Conclusion: Rural households in North Central Nigeria consumed an average of four to five food groups per day. However, the diet consumed is low in dietary variety. Consequently, access to nutritious foods remained a challenge in the region. Thus, farmers should be encouraged to

Corresponding author: Email: maryagada59@gmail.com

produce and consume food of increased quality and diversity for improved nutrition and food security.

Keywords: Dietary diversity; diet quality; food security; households; terciles.

1. INTRODUCTION

Food security is defined as a situation in which all household members have both economic and physical access to enough, safe and nutritious foods to meet their needs, so that they do not live in hunger or fear of starvation. Food access refers specifically to adequate availability of quality foods for health and wellbeing [1]. Worldwide, around 925 million people are chronically hungry due to extreme poverty, while up to two billion people lack food security intermittently due to varying degrees of poverty [2]. Household food insecurity is directly influenced by low access to food. Consequently, households that do not have sufficient access to food have low dietary diversity, reflecting poor nutrient adequacy.

Dietary diversity is the number of different foods or food groups consumed by the household over a reference period not regarding the frequency of consumption [3]. It is an outcome measure of food security at the individual or household level [4]. In the present study dietary diversity refers to the number of food groups consumed over a 24 hours period.

Household dietary diversity is influenced by accessibility to food. It is estimated that close to 1 billion people in the world suffer from hunger and food insecurity, defined as not having enough calories to live a healthy life [5]. Nearly the entire undernourished people are in developing countries with worst scenarios in Asia (578 million) and sub-Saharan Africa (239 million) [6]. In the case of Nigeria, an estimated 61 percent of the people are malnourished [7]. The number of people in the world with poor access to nutritious foods rich in essential micronutrients such as fruits and vegetables, meat, fish, dairy products, and biofortified staple foods is staggering [8]. This leads to deficiencies in micronutrients such as vitamin A, iron, and zinc and affects the survival, health, development, and wellbeing of billions of people. However, it has been observed that the more food groups included in a daily diet the greater the likelihood of meeting nutrient requirements while monotonous diets, based mainly on starches such as maize and bread, have been closely associated with food insecurity [9].

Findings from dietary diversity studies all suggest that food security conditions could be enhanced by consuming varieties of food [10,11,12,13]. However, this has not been extensively examined in North Central Nigeria. Therefore, the present study was carried out to assess the dietary diversity and the probability of food security among rural households in North Central Nigeria. The outcomes of this study could be used to guide the federal and state governments when formulating new approaches and interventions to address food access and ultimately food security in households in North Central Nigeria.

There are four dimensions to food security based on its definition, namely availability, stability, access and utilization. This study examined the access dimension of household food security using dietary diversity score (DDS) and mean dietary diversity terciles. The data examined in the study was part of a large questionnaire survey on the food security situation of households among selected ethnic groups in North Central Nigeria.

2. METHODOLOGY

The study was conducted in North Central Nigeria. North Central Nigeria is situated in the southern Guinea savanna agro-ecological zone and it comprises six states, namely Plateau, Nasarawa, Benue, Kogi, Niger and Kwara as well as the Federal Capital Territory, Abuja [14]. The region covers a land area of about 251,425 square kilometres [15,16] with a population of about 20,266,257 inhabitants [17]. Agriculture is the principal activity in the study area. It is, however, at a subsistence level. Farms are generally small, usually less than five hectares and rely on the use of manual labour and crude implements such as hoes and machetes. Cattle, sheep, goats, pigs, and poultry are some of the principal livestock kept by farmers. Like other regions of the country, the population is growing rapidly and urbanization is accelerating. This is leading to worsening food deficit despite government efforts to rectify the situation.

Primary data was obtained from a survey of households in the villages of three ethnic groups in North Central Nigeria between November and December, 2011. A multi-stage sampling procedure was adopted for the study. In the first stage, a purposive sampling technique was used to select three ethnic groups and one village per ethnic group based on differences in language and culture. In the second stage, a random sample of 120 Tiv, 108 Igala and 112 Eggon households were interviewed using the international standardized Household Dietary Diversity Score (HDDS) questionnaire. Culturally, females are responsible for household food preparation in North Central Nigeria and they constituted the respondents for this study. The interview was conducted in the homes of the interviewees by well trained interviewers who were knowledgeable about food preparations, including prevalent ethnic foods. The data collected covered the variety of foods and drinks from twelve food groups consumed by the households as a whole during the previous 24 hours [18].

Every food item in the meals was coded since quantities or proportions were not required. A dietary diversity score (DDS) was calculated by counting each of twelve food groups [18]. Since national dietary data is not available on mean DDS, terciles were used to classify the households into low, medium and high diversity [19]. Terciles were used because there are currently no international guidelines or recommendations on which to base cut-offs for low, average and high. A score of equal to or lower than three was considered as poor dietary diversity (and by association poor food security) while a score of six and above represents a very varied diet. The twelve food groups used in this study were: (1) cereals (maize, rice, millet, sorghum and fonio 'ácha'); (2) roots and tubers (yam, cassava, cocoyam and sweet potato); (3) vegetables (amarantus and talphairia); (4) fruits (orange, pawpaw, banana, fresh tomatoes); 5) meat and poultry (chicken, beef, goat or sheep and bush meat); (6) eggs; (7) fish (fresh fish and smoked frozen fish); (8) pulses, legumes and nuts (cowpea, groundnut and bambaranut); (9) milk and milk products; (10) oil (groundnut oil and red palm oil); (11) sugar; (12) condiments (pepper, salt and spices). The proportion of people who had consumed a food group at least once was also calculated.

3. RESULTS AND DISCUSSION

The proportion of households who consumed food from each food group in the previous 24 hours (Table 1) revealed that root and tuber crops constituted the food group consumed by most Tiv (97%) and Igala (94%) households whereas cereals constituted the food group consumed by most Eggon (79%) households. Over 50% of the households in each ethnic group consumed fish (fresh and smoked frozen fish) whereas about 30% of all the households consumed vegetables. Igala households consumed more fruits (35%) compared to Tiv (23%) and Eggon (17%) households. Conversely, Eggon households (46%) reported use of legumes compared to Igala (26%) and Tiv households (23%). The Tiv consumed meat the most (36%) and Eggon the least (7%). Overall, beside spices and condiments (96%) and oil and fat (93%), root and tuber crops (86%) constituted the food group most consumed in the study area. Few households reported use of legumes (32%), vegetables (26%) and fruits (25%) while meat (19%), sugar (17%), milk (10%) and egg (5%) were delicacies. These findings are consistent with the observation that lack of diversity is a particularly severe problem among populations in the developing world where diets are based predominantly on staples and often include only a few animal products at most and only seasonal fruits and vegetables [20].

The role of root and tuber crops in the diets of the study populations could result in a high prevalence of protein inadequacy. The diminishing role of pulses and animal products, two valuable sources of protein and micronutrients, suggests a low quality diet in North Central Nigeria. These findings concur with those of other researchers that poor households subsist on monotonous staple-based diets and lack access to nutritious foods such as fruits, vegetables, animal-source foods (such as fish, meat, eggs, and dairy products), or wild foods of high nutrient content [21,22,23]. Lack of diversity in the diet is strongly associated with inadequate intake and risks of deficiencies of essential micronutrients such as vitamin A, iron, and zinc [5]. Micronutrient deficiencies have far-reaching health and nutrition consequences in both the short and the long term [24,25,26]. These deficiencies affect the survival, health, development, and well-being of those afflicted. Children and women of reproductive age are especially vulnerable because they have particularly high micronutrient requirements

[25,26,27]. The limited access to nutritious food in the study area could be attributed to limited availability of nutritious foods, economic constraints and lack of knowledge and information [5].

Most households (Tiv -61%; Igala-57%; Eggon-58%) reported consuming between four and five food groups the previous day. The mean dietary diversity terciles were 2.90, 4.53 and 6.37 with a grand mean of 4.60 (Table 2). The findings revealed that more than half of the households studied fell in the medium tercile and consumed between 4 and 5 food groups the previous day. Studies in other developing countries that have evaluated their dietary diversity scores showed a mean DDS of 4.91 for Filipino children aged 24-71 months using scores of nine food groups; in Burkina Faso the mean DDS was 4.6, in Laos 5.2 and 3.3 in Northern Uganda [10]. In the diets of adult South Africans, the DDS was 4.02 [28]. It is apparent that poor dietary variety is a feature of many developing countries, and is not restricted to the North Central Nigerian population. Although the findings of this study showed that the population fell in the medium tercile, the commonly consumed food groups, other than spices and condiments and oil and fat, were roots and tubers among Tiv and Igala ethnic groups and cereals in Eggon households. Meat, eggs, vegetables, fruits, legumes, milk, sugar and honey were consumed the least often (Table 1). However, research has shown that increased dietary diversity improves the capacity of individuals to meet their daily requirements for a large number of essential nutrients [29]. This means that nutritional balance can be ensured through ensuring a variety of food. Therefore, the challenge for the local and state governments in the region is to evolve strategies that would increase access to varieties of food by the rural households.

Table 1. Percentage distribution of food groups consumed by households in North Central Nigeria

Food groups	Tiv (n=120) %	Igala (n=108) %	Eggon (n=112) %	Total (n=340) %
Cereals	28.3	7.4	78.6	38.2
Roots and tubers	96.7	94.4	67.0	86.2
Vegetables	35.8	34.3	37.5	25.9
Fruits	22.5	35.2	17.0	24.7
Meat, poultry and offal	35.8	13.9	7.1	19.4
Eggs	4.2	3.7	7.1	5.0
Fish	51.7	54.6	57.1	54.4
Pulses, legumes and nuts	23.3	25.9	46.4	31.8
Milk	4.2	9.3	16.1	9.7
Oil and fats (ground nut oil and palm oil)	96.7	94.4	86.6	92.6
Sugar and honey	7.5	17.6	25.9	16.8
Spices and condiments	100.0	97.2	91.1	96.2

Table 2. Distribution of food groups consumed by households in North Central Nigeria by dietary diversity terciles

Ethnic group	Low (≤ 3 food groups)		Medium (4 & 5 food groups)		High (6 ≥ food groups)	
	Percentage	Mean	Percentage	Mean	Percentage	Mean
Anter (n = 120)	3.33	2.75	60.83	4.43	35.83	6.35
Ikem (n = 108)	13.89	3.00	57.41	4.55	28.70	6.45
Randa (n = 112)	9.82	2.82	58.04	4.62	32.14	6.33
Total (n = 340)	8.82	2.90	58.82	4.53	32.35	6.37

The strength of this study is in the realization that in spite of the various strategies by the Federal Government of Nigeria to improve food security in the country, the situation in North Central Nigeria is still a challenge because rural households are still unable to access nutritious foods in the right quantity and quality. However, the study is limited in the sense that the use of households rather than individuals meant that only households' economic access to food was assessed and not quality of individual's diet. Furthermore, the use of 24-hour recall instead of longer days for measuring household dietary diversity made it impossible to capture individual's usual food intake. Lack of internationally acceptable cut-offs for mean dietary diversity terciles made it unsafe for comparisms across different locations.

4. CONCLUSION

The study concludes that rural households in North Central Nigeria consumed an average of four to five food groups the previous day. However, the diet consumed is low in dietary variety as the commonly consumed food groups were spices and condiments, oil and fat, roots and tubers, cereals and fish. Eggs, milk, meat, fruits, and vegetables were the least consumed. Consequently, access to nutritious foods remained a challenge in the region. The implication of the finding for the local and state governments in North Central Nigeria is to encourage farmers to produce and consume food of increased quality and diversity for improved nutrition and food security.

COMPETING INTERESTS

Authors have declared that no competing interests exist.

REFERENCES

1. Food and Agriculture Organization of the United Nations. The state of food insecurity in the world; 2003. Accessed 4 May 2010. Available: http:www.fao.org-docrep-006-J0083e-j0083e00.html
2. Food and Agriculture organization of the United Nations. The state of food insecurity in the World; 2009. Accessed 10 June 2011. Available:http://www.fao.org/docrep/013/i1683e/i1683e.pdf.
3. Hoddinott J, Yohannes Y. Dietary diversity as a food security indicator. FCMD discussion paper no. 136. International Food Policy Research Institute, Washington; 2002.
4. Hoddinott J. Choosing outcome indicators of household food security International. Food Policy Research Institute, Washington; 1999. Accessed 17 October 2012. Available:http://www2.reliefweb.int/rw/rwt.nsf/db900SID/NVEA-5ULK83/$File/IFPR_HouseholdFoodSecurity.pdf.
5. International Food Policy Research Institute. Value chain for nutrition. Leveraging agriculture for improving nutrition and health. Brief 4. IFPRI 2020 International Conference, New Delhi, India; 2011.
6. Hunger Notes. World hunger and poverty facts and statistics 2011. World Hunger Education Service, Washington D.C. Accessed 12 May 2011. Available:http://www.worldhunger.org/learn/ worldhungerfacts2002.htm.
7. Momodu AS1, Akinbami CAO, Obisanya JF. Achieving food security and climate change mitigation through entrepreneurship development in rural Nigeria: Gender perspective. Africa J. Environ. Sci. Tech. 2011;5(10):834-854.
8. Food and Agriculture Organization of the United Nations. The State of Food Insecurity in the World; 2011. Accessed 7 August 2012. Available:http://www.fao.org/docrep/013/i1683e/i1683e.pdf.
9. Kennedy GL. Evaluation of dietary diversity scores for assessment of micronutrient intake and food security in developing countries. Ph.D Thesis. University of Wagenningen, Netherlands; 2009.
10. Kennedy G, Pedro MR, Seighieri C, Nantel G, Brouwer ID. Dietary diversity score is a useful indicator of micronutrient intake in non-breast-feeding Filipino children. J. Nutr. 2007;137(2):472-477.
11. Ruel M, Graham J, Murphy S, Allen L. Validating simple indicators of dietary diversity and animal source food intake that accurately reflect nutrient adequacy in developing countries. GL-CRSP Report; 2004.
12. Steyn NP, Nel JH, Nantel G, Kennedy G, Labadarios D. Food variety and dietary diversity scores in children: are they good indicators of dietary adequacy. Public Health Nutr. 2006;9(5):644-650.

13. Savy M, Martin-Prevel Y, Sawadogo P, Kameli Y, Delpeuch F. Use of variety/diversity scores for diet quality measurement: relation with nutritional status of women in a rural area in Burkina Faso. EJCN. 2005;59:703-716.
14. National Bureau of Statistics. The Nigerian statistical fact sheets on economic and social development. National Bureau of Statistics, Abuja, Nigeria; 2005.
15. Nigeria Annual Abstract of Statistics. Federal Office of Statistics. Lagos, Nigeria; 1996.
16. Nasarawa Agricultural Development Programme. 2009 On-Farm Adaptive Research (OFAR) Report. 22nd Central Zonal REFILS Workshop held at Taraba State ADP Jalingo; 2010.
17. National Population Commission. 2006 National Population Census Official Gazette (Extraordinary). 2009;94(4):52.
18. Swindale A, Ohri-Vachaspati P. Measuring household food consumption: A technical guide. Food and Nutrition Technical Assistance Project (FANTA), Academy for Educational Development, Washington, DC; 2005.
19. Food and Agriculture Organization. Guidelines for measuring household and individual dietary diversity, version 4, Rome; 2008.
20. Arimond M, Ruel MT. Dietary diversity is associated with child nutritional status: Evidence from 11 demographic and health surveys. J. Nutr. 2004;134:2579-2585.
21. Ruel MT. Operationalizing dietary diversity: A review of measurement issues and research priorities. J. Nutr. 2003;133(11)2:3911–3926.
22. Arimond M, Wiesmann D, Becquey E, Carriquiry A, Daniels MC, Deitchler M, et al. Simple food group diversity indicators predict micronutrient adequacy of women's diets in 5 diverse, resource-poor settings. J. Nutr. 2010;140(11):2059–2069.
23. Arimond MC, Hawkes C, Ruel MT, Sifri Z, Berti PR, Leroy JL, et al. Agricultural interventions and nutrition: lessons from the past and new evidence. In: Thompson B, Amoroso L. (eds.). Combating micronutrient deficiencies: Food-based approaches. Food and Agriculture Organization and CAB International, Rome; 2010.
24. United Nations Children's Fund and Micronutrient Initiative. Vitamin and mineral deficiency: A global progress report. Ottawa: Micronutrient Initiative; 2004. Accessed November 29 2011. Available:http://www.micronutrient.org/CMFiles/PubLib/VMd-GPR-English1KWW-3242008-681.pdf.
25. Black RE, Allen LH, Bhuttam ZA, Caulfield LE, De Onis M, Ezatti M. Maternal and child undernutrition: Global and regional exposures and health consequences. Lancet. 2008;371:243–260.
26. Micronutrient Initiative. Investing in the future: A united call to action on vitamin and mineral deficiencies: Global Report. Micronutrient Initiative; 2009. Accessed 22 May 2011. Available:http://www.unitedcalltoaction.org/documents/Investing_in_the_future.pdf
27. United Nations System Standing Committee on Nutrition. 6th Report on the World Nutrition Situation: Progress in Nutrition; 2010. Accessed November 29 2011. Available:http://www.unscn.org/files/Publications/RWNS6/report/SCN_report.pdf.
28. Labadarios L, Steyn NP, Nel J. How diverse is the diet of adult South Africans. Nutr. J. 2011;10:33 DOI: 10. 1186/1495-2891-10-33.
29. Arimond M, Ruel MT. A report on the investigation into recurrent epidemics of pellagra in Kuito-Angola. Food and Nutrition Technical Assistance (FANTA) project, Academy for Educational Development (AED) and United States Agency for International Development (USAID). Washington D.C.; 2002.

17

From Dietary Guidelines to Dietary Guidance?

Egbert Clevers[1*] and Miriam Urlings[2]

[1]*Maastricht University, P.O.Box 616, 6200 MD Maastricht, The Netherlands.*
[2]*Department of Complex Genetics, NUTRIM School of Nutrition and Translational Research in Metabolism, Maastricht University, P.O.Box 616, 6200 MD Maastricht, The Netherlands.*

Authors' contributions

This work was carried out in collaboration between both authors. Author EC conducted the literature research, analysed the data and wrote the first version. Author MU designed the work. Author MU was responsible for subsequent reviewing and scientific editing, while author EC was the primary responsible for final content. Both authors have read and approved the final manuscript.

ABSTRACT

One challenge is to make food consumption healthful and sustainable. Tools to achieve this include science-based documented dietary advice (here: 'dietary guidelines'), and policy-based initiatives and legislation (here: 'dietary guidance'). These two scientific and political tools operate independently, which may be counterproductive, especially if the future gives rise to high-impact scenarios of food crisis. This paper analyses, based on empirical trend data, the potential of dietary guidelines and guidance, and provides suggestions for these to align.

Trend-analyses and empirical data point out that dietary guidelines, although relevant, achieve limited uptake by the consumer. Scientists apparently lack the management tools and communication channels to effectively reach the consumer, and thus major changes are not expected. By contrast, emerging issues such as the obesogenic environment and climate change are likely to call for greater involvement of policy-makers.

This paper therefore advocates alignment of science and food policy. First, dietary guidelines, used for individual health promoting purposes, should be calibrated on global considerations. Guidelines on red meat consumption are recommended, as red meat consumption is a burden on both health and sustainability. Second, dietary guidance, used for global health and sustainability promoting purposes, should be calibrated on empowering the individual. Guidance in creating self-sufficient local food networks is recommended, as self-sufficient local food networks address both health and sustainability.

**Corresponding author: Email: erw.clevers@alumni.maastrichtuniversity.nl*

Keywords: Diet; health; sustainability; dietary guidelines; recommendations; red meat; local food networks; self-sufficiency.

ABBREVIATIONS

EGCG: Epigallocatechin Gallate; FDA: Food and Drug Administration; GMO: Genetically Modified Organisms; GSTM: Glutathione S-transferase mu; GSTT: Glutathione S-transferase theta; NGx: Nutrigenomics/genetics; WHO: World Health Organization.

1. INTRODUCTION

One current challenge is to make food consumption healthful and sustainable [1-3]; a responsibility shared by scientists, policy-makers (including legislators), the industry, and consumers [4,5]. Tools to achieve this include science-based documented dietary advice (here: 'dietary guidelines'), and policy-based initiatives and legislation (here: 'dietary guidance').

Scientists and policy makers operate independently, causing dietary guidelines and dietary guidance to trigger opposite effects [6]. For example, scientists may recommend dairy as part of a healthy diet, whereas policy-makers may impose taxes on it for reasons unrelated to consumer health. This is a counterproductive situation, in which science and policy are in a balance that fluctuates on trends and high-impact future scenarios. Examples of these trends and scenarios are scientific advances, changes in consumer behaviour, an increase in food-related diseases, food supply chain crises, climate change, and international conflicts. High-impact scenarios require early anticipation [7] by scientists and policy-makers, preferably operating in concert [6]. Knowing how is crucial to make food consumption healthful and sustainable in the long term.

This paper will first explain the current trends listed above, i.e. how science and food policy influence our diets now, and how these trends may evolve. Second, it provides practical suggestions for dietary guidelines and guidance to anticipate high-impact future scenarios.

2. SCIENCE

Current guidelines, such as "the Eatwell Plate" [8], are a weighted consensus of recommended intakes of foods, macronutrients, and micronutrients. These recommendations are based on evidence on the population level for an adequate nutrition and reduction of chronic diseases such as obesity, cardiovascular disease, type II diabetes, and some forms of cancer. Convincing or probable protective factors are especially: fibre, vegetables, fruits, and fish [9-12]. Convincing or probable harmful factors are especially: red meat, salt-preserved foods, saturated and trans fatty acids [9-12]. The degree of evidence has changed over the years, but the common conclusions have remained valid [13].

In contrast to conventional nutrition research on the population level, future research may open the road to personalised dietary guidelines in order to reduce the risk of obtaining disease. Two scientific advances are nutrigenetics and nutrigenomics (NGx), i.e. crosstalk between the diet and gene expression. For example, it has recently been suggested that a protective effect of cruciferous vegetable intake on lung cancer only holds in individuals who are GSTM1-null and/or GSTT1-null [14]. Further, expression of colon cancer-associated genes could perhaps be altered by dietary components such as epigallocatechin gallate (EGCG), curcumin, and resveratrol [15-18].

Genetic expression differs per individual, so that NGx can lead to personalised dietary guidelines. However, it remains to be seen whether NGx and personalised nutrition will fulfil their technological potential [19,20], and whether they will be embraced by the consumer [21-23]. Thus, NGx and personalised nutrition could be one mechanism for science-based dietary guidelines to increase in efficacy, but its widespread implementation faces delay.

3. THE CONSUMER

Whereas science informs the consumer what a healthy diet entails, consumers make their choices on various grounds [24]. For example, while healthfulness of vegetables is well-recognised by half of the Dutch population [25], their median consumption has declined consistently over the past decades [26] to 120 grams by 2010, which is well under the target of 200 grams per day [27]. Meanwhile, consumption figures of unprocessed meats have shifted to processed meats, and crisps are more widely consumed [28]. Disregarding guidelines for diet and health is typically caused by short-term personal motives such as a lack of time [29-31] and self-control [25,32]. Consumers also base their choices on direct foodborne effects and complaints, as exemplified by an increasing number of people who want to cut back on gluten (e.g. ~30% of American adults [33]). In summary, short-term personal motives modify the efficacy of dietary guidelines in preventing unfavourable dietary trends in the long-term.

One could speculate if, and how the scientific community can progress towards a preeminent role in consumer decision-making. The method would be required to outweigh the 'lack of time for cooking'-scenario, which is a developing trend in itself [34], and which explains the inflated segments of highly-processed, energy-dense fast foods (e.g. [28]). Besides, dietary guidelines would have to compete with pseudo-scientific mass media slogans (e.g. 'superfoods', the AGE-less way, the Paleo diet). It is questionable how consumers may retain their credence towards established scientific dietary guidelines, since apparently, scientists, as risk assessors, lack the management tools or communication channels to effectively reach the consumer. Better management tools and communication channels may be found within policy-based initiatives and legislation as discussed below.

4. POLICY FOR HEALTH

Our western environment has been described as 'obesogenic', and 'unhealthy', implying that it promotes disease (especially by stimulating a sedentary lifestyle with quick-to-prepare energy-dense meals) [35,36]. To counter the unhealthy environment, policy-makers such as legislators have the means [37], and thus some degree of responsibility [38]. Various governments have enacted measures, such as regulations on frying oils in snack bars [39-41], offerings in school canteens [42,43], food advertising to children [44], and food labelling [45]. With some exceptions [46], the proposed regulations have led to acceptance and success. For example, trans fatty acid levels in Danish fast food were tenfold lower than those in neighbouring countries [47], likely due to the Danish regulation on trans fatty acids.

Several governments have taken responsibility to control the unhealthy environment, and their successes have not gone unnoticed. For example, based on Danish success, the European Commission and the FDA now consider proposing a regulation on trans fatty acid content in foods (e.g. [48]). This could further reduce the number of cardiovascular events, and health-care costs [49].

Non-legal options to stimulate a healthy environment include the new concept of smart city-design (e.g. parks with fruit and nut trees [50], or community kitchens [51]), and stimulatory programmes (e.g. lowered taxation on vegetables [52], or child breakfast programmes). In summary, large-scale implementation of regulations and smart city initiatives are still to occur, but potential effects are considerable.

5. POLICY FOR SUSTAINABILITY

Policy-makers face new challenges in a rapidly changing world [53-55]. Issues that were tenuous in the 20th century can evolve quickly in the future, by exponential growth [56,57], critical decline [58,59], or *de novo* emergence [60,61]. Several bodies have drawn scenario-based projections of the future. The next section will analyse, for each scenario, the potential role of policy-based dietary guidance.

The scenarios found in refs [62-64] build on current favourable or unfavourable trends, i.e. food prosperity versus food crises, and require appropriate anticipation by policy-makers.

Scenarios of prosperity are characterised by new technologies and innovations (e.g. GMO, *in vitro* meat, food pills, and functional foods), complex global trade systems (which might in turn increase the risk for food safety events), and consumer naivety about food and its effects (abundance of convenience foods, obesity). Consumer demands and industrial innovations drive dietary patterns, while sustainability problems do not come to expression due to technological advances. Thus, given a sustainable food supply, food policy-based dietary guidance is confined to tackling health-related problems.

Scenarios of crisis are characterised by food supply chain risks (antimicrobial resistance, contamination), resource depletion (phosphorus, freshwater, leading to price volatility), or climate change (decreased crop productivity, food pests). Common effects may be: loss of consumer confidence, food resource nationalism, and social unrest. Thus, given an unsustainable food supply, global food policy-based dietary guidance is required.

6. FOOD INDUSTRY

In an attempt to make food consumption healthful and sustainable, policy-makers face increasing involvement from the food industry [38]. This trend is characterised by the rise of major multinationals [65], especially those with a core business around unhealthy or unsustainable food products. The food industry uses its power to influence science [66,67], the consumer [68,69], and policy-makers [70], and thus the dynamics of dietary guidelines and dietary guidance. At the same time, food industrial entities have launched initiatives to contribute to a healthful [71] and sustainable [72] food consumption, possibly with the purpose of deflecting social and political pressure [73,74]. However, if such initiatives prove financially disadvantageous, one can not expect companies to retain them (e.g. [75,76]. It is unclear whether the food industry can reasonably be expected to be a full partner in the promotion of a healthful and sustainable food consumption. Therefore, the suggestions of the present paper focus on the interplay between science and policy.

7. SUGGESTIONS

The trends and projections analysed in this paper lead to two main conclusions. First, the role of science-based dietary guidelines is unlikely to grow, as 1) most old dietary guidelines have maintained their validity, and thus abrupt improvements are not expected; 2) consumers are prone to short-term influences, such as a lack of time, which renders dietary guidelines impractical. Second, the role of policy-based dietary guidance is likely to grow, as 1) there is a trend of regulations that successfully counter the unhealthy, obesogenic environment; 2) scenarios of crisis require preliminary action, for which policy-makers have the means.

One rigid view pictures scientists practising science, and policy-makers practising politics. Here, scientists can only minimally influence society towards healthy and sustainable nutrition, and, worse, their influence may be incompatible with that of policy-makers.

What if scientists, as well as consumers, would take on a pro-active role towards the political driver's seat, and if dietary guidelines and dietary guidance would coalesce? In that case, dietary guidelines, set up for individual-health promoting purposes, would be calibrated on global considerations. Dietary guidance, set up for global health and sustainability promoting purposes, would be calibrated on empowering the individual. This vision will be illustrated by suggestions concerning red meat, and self-sufficient local food networks.

7.1 Red Meat

Red meat is a suitable target for dietary guidelines, as it is a burden on both individual health as well as global sustainability [77], and its consumption is booming worldwide, most notably in developing countries [62,78-80]. First, red meat intake is associated with an increased risk for colorectal cancer [81,82], and probably other cancers [83,84] as well as type-II diabetes [85]. Second, producing meat is an unsustainable and inefficient process: it requires a large land surface [86], emits considerable amounts of greenhouse gases [87], and is prone to contamination [88]. On the above grounds, dietary guidelines to reduce red meat consumption can serve as super-scientific tool, with a broad rationale.

Discouraging red meat consumption will likely come with criticisms. First, guidelines on a single food or food category, may, from a merely scientific point of view, be inferior to a "total diet approach" [89,90] (e.g. food pyramids). However, the present paper emphasises, based on trends and empirical data, that considerations for dietary guidelines should not merely be scientific, but also political. The efficacy of the total diet approach is unknown [91-94], and it has been suggested that guidelines should be simple, clear, and specific [95]. Second, substituting red meat for plant-derived foods would reduce protein and fat intakes. However, in the western diet, proteins and fats constitute 16-17 and 33-37% of total energy intake respectively [96-99], which is higher than the WHO recommendations [100]. Third, the meat industry will suffer losses following a reduced demand for red meat, and red meat prices are forecast to rise to levels unaffordable for the lower income groups [6,62]. This type of considerations is inherent to food policy, and requires an assessment of the societal effects, per country. In summary, the example of red meat shows how dietary guidelines can evolve from a purely individualistic, scientific matter to a situation in which scientific (health) and political (health and sustainability) agendas are aligned.

7.2 Self-sufficient Local Food Networks

Self-sufficient local food networks, also termed 'alternative food networks' [101,102], are small-scaled [103] social communities with sustainable food supply chains (i.e. that can operate in a financially independent way, while being ecologically sound [104]), usually including community gardens.

The setup of these networks is a suitable means of dietary guidance, as it addresses both global sustainability as well as individual health. First, it protects against the risks associated with the complexity of global trade. Food safety events [105,106] are unlikely, because supply chains are short, specific, and easily-controlled. Consumer confidence is secured, because consumers are actively involved in the food supply chains as entrepreneurs [62,107-111]. International conflicts, which could jeopardise global trade, would leave local food networks unharmed. Second, self-sufficient local food networks rely largely on plant-derived foods, so that, in its purest form, the obesogenic environment may diminish.

Self-sufficient local food networks are a drastic innovation that comes with a couple of disadvantages. First, the variety of foodstuffs available will be strongly restricted, because the community relies on seasonal, local products [62]. A more homogeneous diet can, in theory, lead to nutrient deficiencies. Complementary innovations are required to diversify food access. Second, not every region is geographically capable of hosting a self-sufficient local food network, notably regions with barren terrain such as deserts, mountainous, and polar areas [103]. It is important to emphasise that such a network does not replace international trade, but rather serves as a sustainable shadow system, that, triggered by high-impact events, can materialise when deemed necessary. In summary, the example of self-sufficient local food networks shows how dietary guidance can evolve from a purely societal, political matter to a situation in which consumers and scientists are empowered.

ACKNOWLEDGEMENTS

The authors would like to thank Prof. Dr. Hans Verhagen for stimulating discussion within scope of the present article.

COMPETING INTERESTS

The authors declare no conflicts of interest and this work received no grant from any funding agency.

REFERENCES

1. The Government Office for Science. Foresight: The future of food and farming. Final Project Report. London; 2011.
2. Currie N, Etsell G. Towards a national food strategy. A framework for Securing the Future of Food; 2009.
3. Palokangas S, Andersen T, Borg P, Kurppa S, Marttila J, Parkkinen M, et al. Food for tomorrow – Proposal for Finland's National Food Strategy; 2010.
4. FAO/WHO. Assuring food safety and quality: Guidelines for strengthening national food control systems. Rome and Geneva; 2003.
5. Safe Food International (WHO, FAO). Guidelines for consumer organizations to promote national food safety systems. Washington, USA; 2003.
6. Knottnerus JA, Boot AW, Bovens MA, Engbersen GB, Hirsch-Ballin EM, de-Visser M, et al. WRR. [Naar een voedselbeleid] (Dutch); 2014.
7. Wijbenga FH, Van Witteloostuijn A. Entrepreneurial locus of control and competitive strategies – the moderating effect of environmental dynamism. J Econ Psychol. 2007;28(5):566-89.
8. Food Standards Agency. The Eatwell Plate; 2010. Available:http://tna.europarchive.org/20100929190231/http://www.food.gov.uk/multimedia/pdfs/publication/eatwellplate0210.pdf. Accessed Dec 2014.
9. Bandera EV, Kushi LH, Moore DF, Gifkins DM, McCullough ML. Consumption of animal foods and endometrial cancer risk: a systematic literature review and meta-analysis. Cancer Causes Control. 2007; 18(9):967-88.
10. Ben Q, Sun Y, Chai R, Qian A, Xu B, Yuan Y. Dietary fiber intake reduces risk for colorectal adenoma: A meta-analysis. Gastroenterology. 2014;146(3):689-99 e6.
11. Magalhaes B, Peleteiro B, Lunet N. Dietary patterns and colorectal cancer: Systematic review and meta-analysis. Eur J Cancer Prev. 2012;21(1):15-23.

12. WHO, FAO. Diet. Nutrition and the Prevention of Chronic Diseases; 2003.
13. Schneeman BO. Evolution of dietary guidelines. J Am Dent Assoc. 2003; 103(12):5-9.
14. Brennan P, Hsu CC, Moullan N, Szeszenia-Dabrowska N, Lissowska J, Zaridze D, et al. Effect of cruciferous vegetables on lung cancer in patients stratified by genetic status: A mendelian randomisation approach. Lancet. 2005; 366(9496):1558-60.
15. Shakibaei M, Buhrmann C, Kraehe P, Shayan P, Lueders C, Goel A. Curcumin chemosensitizes 5-fluorouracil resistant MMR-deficient human colon cancer cells in high density cultures. PLoS One. 2014; 9(1):e85397.
16. Sadik NA. Chemopreventive efficacy of green tea drinking against 1, 2-dimethyl hydrazine-induced rat colon carcinogenesis. Cell Biochem Funct. 2013; 31(3):196-207.
17. Liu YZ, Wu K, Huang J, Liu Y, Wang X, Meng ZJ, et al. The PTEN/PI3K/Akt and Wnt/beta-catenin signaling pathways are involved in the inhibitory effect of resveratrol on human colon cancer cell proliferation. Int J Oncol. 2014;45(1): 104-12.
18. Lim TG, Lee SY, Huang Z, Lim do Y, Chen H, Jung SK, et al. Curcumin suppresses proliferation of colon cancer cells by targeting CDK2. Cancer Prev Res (Phila). 2014;7(4):466-74.
19. de Roos B. Personalised nutrition: ready for practice? Proc Nutr Soc. 2013;72(1): 48-52.
20. Hesketh J. Personalised nutrition: How far has nutrigenomics progressed? Eur J Clin Nutr. 2013;67(5):430-5.
21. Berezowska A, Fischer AR, Ronteltap A, Kuznesof S, Macready A, Fallaize R, et al. Understanding consumer evaluations of personalised nutrition services in terms of the privacy calculus: A qualitative study. Public Health Genomics. 2014;17(3): 127-40.
22. Poinhos R, Van der Lans IA, Rankin A, Fischer AR, Bunting B, Kuznesof S, et al. Psychological determinants of consumer acceptance of personalised nutrition in 9 European countries. PLoS One. 2014; 9(10):e110614.
23. Stewart-Knox B, Kuznesof S, Robinson J, Rankin A, Orr K, Duffy M, et al. Factors influencing European consumer uptake of personalised nutrition. Results of a qualitative analysis. Appetite. 2013;66: 67-74.
24. O'Brien MC, McConnon A, Hollywood LE, Cuskelly GJ, Barnett J, Raats M, et al. Let's talk about health: Shoppers' discourse regarding health while food shopping. Public Health Nutr. 2014;3:1-10.
25. Lappalainen R, Kearney J, Gibney M. A pan EU survey of consumer attitudes to food, nutrition and health: An overview. Food Qual Prefer. 1998;9(6):467-78.
26. RIVM. [Project: Trend rapportage VCP] (Dutch). 1987-1998;2:Table 11.
27. Gezondheidsraad. Richtlijnen goede voeding. Den Haag, the Netherlands; 2006. Contract No.: 2006/21. ISBN 90-5549-627-8.
28. Geurts M. [Veranderingen in het aanbod van voedingsmiddelen en de voedselconsumptie: Resultaten gebaseerd op bijna 25 jaar voedselconsumptieonderzoek. RIVM briefrapport 090429001] (Dutch). RIVM; 2013.
29. Neumark-Sztainer D, Story M, Perry C, Casey MA. Factors influencing food choices of adolescents: Findings from focus-group discussions with adolescents. J Am Diet Assoc. 1999;99(8):929-37.
30. Galloway AT, Lee Y, Birch LL. Predictors and consequences of food neophobia and pickiness in young girls. J Am Diet Assoc. 2003;103(6):692-8.
31. Silliman K, Rodas-Fortier K, Neyman M. A survey of dietary and exercise habits and perceived barriers to following a healthy lifestyle in a college population. Californian J Health Promot. 2004;2(2):10-9.
32. Nicklas TA, Jahns L, Bogle ML, Chester DN, Giovanni M, Klurfeld DM, et al. Barriers and facilitators for consumer adherence to the dietary guidelines for Americans: The health study. J Acad Nutr Diet. 2013;113(10):1317-31.
33. Dieting monitor. 52 week data year ending January 30. The NPD Group; 2013.
34. Baxter J. Patterns of change and stability in the gender division of household labour in Australia, 1986–1997. J Sociol. 2002; 38(4):399-424.
35. Swinburn B, Egger G, Raza F. Dissecting obesogenic environments: The development and application of a

framework for identifying and prioritizing environmental interventions for obesity. Prev Med. 1999;29(6 Pt 1):563-70.
36. Poston WS, 2nd, Foreyt JP. Obesity is an environmental issue. Atherosclerosis. 1999;146(2):201-9.
37. Schippers EI. [Akkoord verbetering productsamenstelling zout, vet, suiker] (Dutch). Available:http://www.rijksoverheid.nl/bestanden/documenten-en-publicaties/convenanten/2014/01/23/akkoord-verbetering-productsamenstelling-zout-verzadigd-vet-suiker/akkoord-verbetering-productsamenstelling-zout-verzadigd-vet-suiker.pdf. Accessed Jan 2015. Den Haag, NL.
38. Moodie R, Stuckler D, Monteiro C, Sheron N, Neal B, Thamarangsi T, et al. Profits and pandemics: Prevention of harmful effects of tobacco, alcohol, and ultra-processed food and drink industries. Lancet. 2013;381(9867):670-9.
39. Bekendtgørelse om indhold af transfedtsyrer i olier og fedtstoffer m.v., §13, §55, stk. 2 og 78, stk. 3, i lov nr. 471; 1998.
40. Notice of adoption of an amendment (§81.08) to article 81 of the New York city health code; 2006.
41. Brownell KD, Pomeranz JL. The trans-fat ban--food regulation and long-term health. N Engl J Med. 2014;370(19):1773-5.
42. El Senado y la Cámara de Representantes de la República Oriental del Uruguay. 2013;19:140.
43. New Jersey Revised Statutes, 2013; §18A:33-16.
44. The Committee of Advertising Practice. The CAP code. The UK code of non-broadcast advertising, Sales Promotion and Direct Marketing. Sect. 2014;5(1-5):7.
45. Directive 2000/13/EC of the European Parliament and of the Council of 20 March 2000 on the approximation of the laws of the Member States relating to the labelling, presentation and advertising of foodstuffs. Article 3; 2000.
46. Soda Industry V. New York City Petition; 2012. Available:https://archive.org/stream/463325-soda-industry-v-new-york-city-petition/463325-soda-industry-v-new-york-city-petition_djvu.txt. Accessed Jan 2015.
47. Stender S, Dyerberg J, Astrup A. High levels of industrially produced trans fat in popular fast foods. N Engl J Med. 2006; 354(15):1650-2.
48. Mr Borg (European Commission). Written questions by Members of the European Parliament and their answers given by a European Union institution (2014/C 228/01); 2013.
49. Cohen JT. FDA's proposed ban on trans fats: How do the costs and benefits stack up? Clin Ther. 2014;36(3):322-7.
50. The local food assets task force. The local food action plan of the Vancouver park board; 2013. Available:http://vancouver.ca/files/cov/Local-food-action-plan.pdf. Accessed Jan 2015.
51. Mayer KA, Suzumura L. The family community kitchen pilot project: Impacts and Implications; 2011. Available:https://stti.confex.com/stti/bc41/webprogram/Paper49563.html Accessed Jan 2015.; Seattle.
52. European Parliament resolution on 'Promoting healthy diets and physical activity: A European dimension for the prevention of overweight, obesity and chronic diseases' (2006/2231(INI)). Official Journal of the European Union. 2007; C250E:93-9.
53. Barkema HG, Baum JA, Mannix EA. Management challenges in a new time. Acad Manage J. 2002;45(5):916-30.
54. Berliner DC. Educational Research: The Hardest Science of All. Educational Researcher. 2002;31(8):18-20.
55. Miller D. The structural and environmental correlates of business strategy. Strategic Manage J. 1987;8(1):55-76.
56. United Nations Department of Economic and Social Affairs (DESA). Population division. Probablistic population projections based on the world population prospects: The Revision; 2012.
57. World Intellectual Property Organization. WIPO patent report: Statistics on Worldwide Patent Activities; 2007. Available:http://www.wipo.int/edocs/pubdocs/en/patents/931/wipo_pub_931.pdf. Accessed Jan 2015.
58. Valero A. A prediction of the exergy loss of the world's mineral reserves in the 21st century (5th Dubrovnik Conference on Sustainable Development of Energy,

Water & Environment Systems.). Energy. 2011;36(4).

59. Jones DO, Yool A, Wei CL, Henson SA, Ruhl HA, Watson RA, et al. Global reductions in seafloor biomass in response to climate change. Glob Chang Biol. 2014; 20(6):1861-72.
60. WHO. Antimicrobial resistance. Fact Sheet N°194; 2014.
61. Bostrom, Cirkovic. Global Catastrophic Risks.: Oxford University Press; 2011.
62. European Commission Directorate General for Health and Consumers. Scoping study Delivering on EU food safety and nutrition in 2050 - Scenarios of future change and policy responses. Framework Contract for evaluation and evaluation related services - Lot 3: Food Chain. 2013. Available:http://ec.europa.eu/food/food/docs/final_report_scoping_study_en.pdf. Accessed Jan 2015.
63. Various. Special issue: Sustainable Food Consumption: Current Trends, Policy Approaches, and Future Scenarios. Sustain Sci Pract Policy. 2013;9(2).
64. Various. Future of the consumer society.; Tampere, Finland; 2009.
65. Lyson T, Raymer A. Stalking the wily multinational: Power and control in the US food system. Agr Hum Values. 2000;17(2): 199-208.
66. Lesser LI, Ebbeling CB, Goozner M, Wypij D, Ludwig DS. Relationship between funding source and conclusion among nutrition-related scientific articles. PLoS Med. 2007;4(1):e5.
67. Bes-Rastrollo M, Schulze MB, Ruiz-Canela M, Martinez-Gonzalez MA. Financial conflicts of interest and reporting bias regarding the association between sugar-sweetened beverages and weight gain: A systematic review of systematic reviews. PLoS Med. 2013;10(12):e1001578; dicsussion e.
68. Kersh R, Morone J. How the personal becomes political: Prohibitions, public health, and obesity. J Adult Dev. 2002; 16(2):162-75.
69. Kwan S. Individual versus Corporate Responsibility: Market Choice, the Food Industry, and the Pervasiveness of Moral Models of Fatness. Food Cult Soc. 2009; 12(4):477-95.
70. Boseley S. Political context of the World Health Organization: Sugar industry threatens to scupper the WHO. Int J Health Serv. 2003;33(4):831-3.
71. Webster JL, Dunford EK, Hawkes C, Neal BC. Salt reduction initiatives around the world. J Hypertens. 2011;29(6):1043-50.
72. Chkanikova O, Mont O. Overview of sustainability initiatives in European Food Retail Sector; 2011.
73. Sharma LL, Teret SP, Brownell KD. The food industry and self-regulation: standards to promote success and to avoid public health failures. Am J Public Health. 2010;100(2):240-6.
74. Ronit K, Jensen JD. Obesity and industry self-regulation of food and beverage marketing: A literature review. Eur J Clin Nutr. 2014;68(7):753-9.
75. Kelland K. Special report: a pinch of doubt over salt. 2011. Available:http://www.reboundhealth.com/cms/images/pdf/NewspaperandArticle2/a%20pinch%20a%20doubt%20over%20salt%20id%2019727.pdf. Accessed May 2015.
76. Freedhoff Y. The food industry is neither friend, nor foe, nor partner. Obes Rev. 2014;15(1):6-8.
77. Westhoek H, Lesschen JP, Rood T, Wagner S, Marco AD, Murphy-Bokern D, et al. Food choices, health and environment: Effects of cutting Europe's meat and dairy intake. Global Environ Chang. 2014;26:196-205.
78. USDA. USDA Agricultural Projections to 2023; 2014.
79. Alexandratos N, Bruinsma J. FAO. World agriculture towards 2030/2050. ESA working paper no 12-03; 2012.
80. FAO. Food Balance Sheets; 2009.
81. Chan DS, Lau R, Aune D, Vieira R, Greenwood DC, Kampman E, et al. Red and processed meat and colorectal cancer incidence: Meta-analysis of prospective studies. PLoS One. 2011;6(6):e20456.
82. Hebels DG, Sveje KM, de Kok MC, van Herwijnen MH, Kuhnle GG, Engels LG, et al. Red meat intake-induced increases in fecal water genotoxicity correlate with pro-carcinogenic gene expression changes in the human colon. Food Chem Toxicol. 2012;50(2):95-103.
83. Larsson SC, Wolk A. Red and processed meat consumption and risk of pancreatic cancer: Meta-analysis of prospective studies. Br J Cancer. 2012;106(3):603-7.

84. Song P, Lu M, Yin Q, Wu L, Zhang D, Fu B, et al. Red meat consumption and stomach cancer risk: A meta-analysis. J Cancer Res Clin Oncol. 2014;140(6): 979-92.
85. Feskens EJ, Sluik D, van Woudenbergh GJ. Meat consumption, diabetes, and its complications. Curr Diab Rep. 2013;13(2): 298-306.
86. Westhoek H, Rood T, Van-den-Berg M, Janse J, Nijdam D, Reudink M, et al. PBL. The protein puzzle: The consumption and production of meat, dairy and fish in the European Union; 2011.
87. Hajer M. Planbureau voor de Leefomgeving. [Balans van de leefomgeving] (Dutch); 2012.
88. Dalli J. Feed contamination - Dioxin in Germany. 2011. Available:http://ec.europa.eu/food/food/chemicalsafety/contaminants/speech_id_24012011_en.pdf (Accessed Jan 2015)
89. Freeland-Graves JH, Nitzke S, Academy of N, Dietetics. Position of the academy of nutrition and dietetics: Total diet approach to healthy eating. J Acad Nutr Diet. 2013; 113(2):307-17.
90. Nitzke S, Freeland-Graves J, American Dietetic A. Position of the American Dietetic Association: Total diet approach to communicating food and nutrition information. J Am Diet Assoc. 2007; 107(7):1224-32.
91. Brown KA, Timotijevic L, Barnett J, Shepherd R, Lahteenmaki L, Raats MM. A review of consumer awareness, understanding and use of food-based dietary guidelines. Br J Nutr. 2011; 106(1):15-26.
92. Guthrie JF, Smallwood DM. Evaluating the effects of the Dietary Guidelines for Americans on consumer behavior and health: Methodological challenges. J Am Diet Assoc. 2003;103(12 Suppl 2):S42-9.
93. Haack SA, Byker CJ. Recent population adherence to and knowledge of United States federal nutrition guides, 1992-2013: a systematic review. Nutr Rev. 2014; 72(10):613-26.
94. Serra-Majem L, Ribas-Barba L, Salvador G, Serra J, Castell C, Cabezas C, et al. Compliance with dietary guidelines in the Catalan population: Basis for a nutrition policy at the regional level (the PAAS strategy). Public Health Nutr. 2007; 10(11A):1406-14.
95. Boylan S, Louie JC, Gill TP. Consumer response to healthy eating, physical activity and weight-related recommendations: A systematic review. Obes Rev. 2012;13(7):606-17.
96. Austin GL, Ogden LG, Hill JO. Trends in carbohydrate, fat, and protein intakes and association with energy intake in normal-weight, overweight, and obese individuals: 1971-2006. Am J Clin Nutr. 2011; 93(4):836-43.
97. Irish Universities Nutrition Alliance. North/South Ireland Food Consumption Survey; 2001.
98. Public Health England and Food Standards Agency. National Diet and Nutrition Survey: Results from Years 1 to 4 (combined) of the rolling programme for 2008 and 2009 to 2011 and 2012. Available:https://www.gov.uk/government/uploads/system/uploads/attachment_data/file/310995/NDNS_Y1_to_4_UK_report.pdf (Accessed Jan 2015)
99. RIVM. Dutch National Food Consumption Survey 2007-2010|Part 5 Macronutrients, Version 2, based on dataset DFC_2010_core_20120822; 2010.
100. WHO. Population nutrient intake goals for preventing diet-related chronic diseases; 2003.
101. Renting H, Marsing TK, Banks J. Understanding alternative food networks: exploring the role of short food supply chains in rural development. Environ Plann A. 2003;35:393-411.
102. Tregear A. Progressing knowledge in alternative and local food networks: Critical reflections and a research agenda. J Rural Stud. 2011;27(4):419-30.
103. Pradhan P, Ludeke MK, Reusser DE, Kropp JP. Food self-sufficiency across scales: How local can we go? Environ Sci Technol. 2014;48(16):9463-70.
104. Feenstra GW. Local food systems and sustainable communities. Am J Alternative Agr. 1997;12(1):28-36.
105. Commission Decision of 28 July 1989 concerning certain protection measures relating to bovine spongiform encephalopathy in the United Kingdom., 89/469/EEC; 1989.
106. Food Safety Authority of Ireland. FSAI survey finds horse DNA in some beef burger products; 2013.

Available:https://www.fsai.ie/news_centre/press_releases/horseDNA15012013.html. (Accessed Jan 2015)

107. TNS Opinion & Social. Special Eurobarometer 368. The Common Agricultural Policy; 2011. Available:http://ec.europa.eu/public_opinion/archives/ebs/ebs_368_en.pdf (Accessed Jan 2015)

108. Sage C. Social embeddedness and relations of regard: Alternative 'good food' networks in south-west Ireland. J Rural Stud. 2003;19(1):47-60.

109. Seyfang G. Ecological citizenship and sustainable consumption: Examining local organic food networks. J Rural Stud. 2006;22(4):383-95.

110. Dubuisson-Quellier S, Lamine C. Consumer involvement in fair trade and local food system: Delegation and empowerment regimes. Geo Journal. 2008;73(1):55-65.

111. Vermeir I, Verbeke W. Sustainable food consumption: Exploring the consumer 'attitude - behavioral intention' gap. J Agr Environ Ethic. 2006;19(2):169-94.

18

Perspectives on Low Calorie Intense Sweeteners with a Focus on Aspartame and Stevia

Caomhan Logue[1], Stephan J. A. C. Peters[2], Alison M. Gallagher[1] and Hans Verhagen[1,3*]

[1]Northern Ireland Centre for Food and Health (NICHE), University of Ulster, Cromore Road, Coleraine, Northern Ireland. BT52 1SA, Northern Ireland.
[2]Netherlands Nutrition Centre (Voedingscentrum), P.O.Box 85700, 2508CK, Den Haag, The Netherlands.
[3]National Institute for Public Health and the Environment (RIVM), P.O.Box 1, 3720 BA, Bilthoven, The Netherlands.

Author's contribution

This paper is a translation into English of two earlier papers in Dutch (Verhagen H. (2012). Voeding Nu 10: 10-11; Peters S. and Verhagen H. (2012) Voeding Nu 11: 21-23), with updates. HV and SP wrote the original texts in Dutch. HV did the translation. CL integrated and updated the work. All authors read, improved and approved the final manuscript.

ABSTRACT

The safety of some food additives/E-numbers, including low calorie (intense) sweeteners (LCS), is constantly the subject of dispute and controversy. However, since LCS have been assigned an acceptable daily intake (ADI) and an E-number following extensive assessment of available safety and toxicological data, consumer safety is assured. These substances have been carefully evaluated, for example by the European Food Safety Authority (EFSA), leading to the conclusion that they are essentially safe when consumption is below ADI levels. Although, intake data indicate that general consumption of LCS is relatively low, many people appear to remain concerned about their safety, particularly aspartame (E951). More recently, stevia (steviol glycosides, E960) has been marketed as a "natural" alternative to aspartame. However, it is unclear whether stevia can live up to its promises. With regards to public health, the real risk within our diet is not the safety of food additives, but rather more likely to be the potential impacts of consuming too much energy and/or an unhealthy dietary pattern.

**Corresponding author: Email: Hans.Verhagen@rivm.nl*

Keywords: Aspartame, stevia, steviol glycosides, low calorie sweeteners, safety, acceptable daily intake.

1. INTRODUCTION

Hazard and risk are essentially different concepts. A hazard is a potential source of an adverse health effect; however it is only considered a risk when the potential for excessive exposure exists. In principle, any substance that we are exposed to, for example via the diet, can be a health risk when exposure levels become too high. This applies to everyday substances like water, oxygen and salt and therefore logically applies to food additives such as low calorie sweeteners (LCS). LCS are substances that can be added to our food and beverages in order to provide a sweet taste without additional calories and they are now widely used in a variety of products. Toxicology is concerned with establishing the level at which intake/exposure is safe and above which the risk of adverse health effects may be increased. Acceptable Daily Intake (ADI) is the term used to describe the safe level of intake for food additives, and this has been defined as, "the daily intake of a chemical, which during an entire lifetime appears to be without appreciable risk on the basis of all known facts at that time" [1]. The ADI is expressed as mg per kg body weight (bw) (Table 1).

The ADI is typically determined following assessment of available data from *in vitro* and *in vivo* studies (usually from animal studies). The highest level of intake that does not cause apparent adverse effects for the most sensitive endpoint in the most sensitive species of test animals is established and identified as the "no observed adverse effect level" (NOAEL). The ADI is then calculated after applying a large safety/uncertainty factor in order to account for inter-species differences (e.g. between rat and humans) and intra-human differences (e.g. between the most sensitive and least sensitive individuals) [1]. In the case of most food additives, including LCS, this means applying an overall safety factor of 100 to the NOAEL; a factor of 10 for inter-species differences and a further factor of 10 for intra-human differences [2] as illustrated in Figure 1. This wide margin of safety is designed to ensure that at daily consumption below the ADI over an entire lifetime, no health risks are to be expected. Any potential hazard, therefore, is not an actual risk as a result of the application of these safety factors.

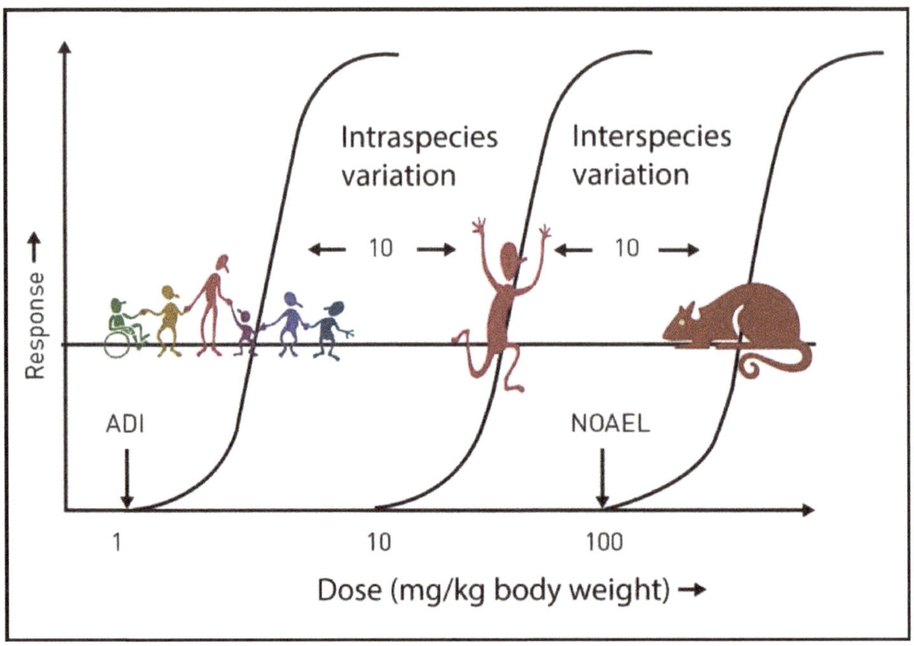

Figure 1. A safety factor of 100 is typically used to calculate and ADI from the NOAEL (10 for interspecies differences and 10 for intraspecies differences)

However, before a food additive can be approved for use in foods, a second step in the safety evaluation of the additive application is required. It is necessary to demonstrate that, under the proposed conditions of use, levels of consumption will not exceed the ADI. Hence, when a food additive has been approved and assigned an E-number, this should provide additional assurances that the ADI will not be exceeded.

2. RISK ASSESSMENT IN THE EU/EFSA

The safety of all LCS that have been permitted on the (EU) market have been extensively evaluated by the European Food Safety Authority (EFSA) or its predecessor the EU Scientific Committee on Food (SCF). "EFSA is the keystone of European Union (EU) risk assessment regarding food and feed safety. ..... EFSA was set up in January 2002, following a series of food crises in the late 1990s, as an independent source of scientific advice and communication on risks associated with the food chain. EFSA was created as part of a comprehensive programme to improve EU food safety, ensure a high level of consumer protection and restore and maintain confidence in the EU food supply" [3]. The establishment of EFSA has been laid down in Regulation EC/178/2002 [4] following the recommendations in the 2000 EU White paper on food safety [5], which also recommended that risk assessment is done independently from risk management. It is important to note that "EFSA's most critical commitment is to provide objective and independent science-based advice and clear communication grounded in the most up-to-date scientific information and knowledge. EFSA's independent scientific advice underpins the European food safety system........ EFSA's activities are guided by a set of core values. These are: excellence in science, independence, openness and transparency, and responsiveness" [3].

Over the years, a number of LCS have been approved for use as food additives in the EU; these are listed in Table 1.

Permitted food additives are periodically re-evaluated in the EU and by 2020 this work must be completed [6] under EU Regulation EU/257/2010 [7]. The use of some food additives are frequently a cause for concern within some sections of society, especially after the publication of new scientific findings which may receive unbalanced (media) attention. One of the most well-known examples of this has been aspartame, whose latest re-evaluation was brought forward in 2011 at the request of the European Commission as a result of ongoing intense interest around its safety. In order to demonstrate transparency and promote debate within the EU on the matter, a public consultation on the draft opinion was carried out [8]. EFSA concluded that " ..... aspartame was not of safety concern at the current aspartame exposure estimates or at the ADI of 40 mg/kg bw/day. Therefore, there was no reason to revise the ADI of aspartame. Current exposures to aspartame - and its degradation product DKP (diketopiperazine) - were below their respective ADIs". In 2013 EFSA published its final Opinion on the safety of aspartame, again confirming its safety at the current levels of use [9].

Table 1: overview of LCS approved for use in the EU

Sweetener	E-number	Sweetness[c]	ADI (mg/kg body weight)	Year of approval
Acesulfame K	E 950	200	9	1984
Aspartame	E 951	180-200	40	1984
Cyclamic acid and its salts[a]	E 952	30	7	1984
Saccharin and its salts[b]	E 954	300-500	5	1977
Sucralose	E 955	600	15	2000
Thaumatin	E 957	2000-3000	'not specified'	1984
Neohesperidine DC	E 959	1900	5	1988
Steviol glycosides	E 960	300	4 (expressed as steviol equivalents)	2011
Neotame	E 961	7000	2	2009
Salt of aspartame acesulfame K	E 962	350	Aspartame: 40; acesulfame K: 9	2000
Advantame	E 969	37000	5	2014

[a]Sodium and Calcium salts, [b] Sodium, Potassium and Calcium salts, [c] Relative to sucrose

3. APPLICATION OF LCS IN SOFT BEVERAGES

Food additives that have been evaluated and considered to be safe for consumption by humans under the proposed conditions of use are allocated an "E-number" (E = European). Although this is meant to signify a guarantee for safe use, E-numbers are sometimes wrongly viewed as representing non-natural ingredients or even a risk to human health even though it is possible for a food additive to be of natural origin or be nature-identical.

With regards to manufacturers' LCS use, maximum usable dosage (MUD) levels for each sweetener have been specified for relevant food and beverage products which must not be exceeded. The actual levels used in products, however, will not always equate to the MUD as this will also be influenced by additional factors such as the intrinsic sweetening properties of the sweetener and the desired taste of the final product. Additionally, it is possible for some LCS to be used in combination as they have been shown to work synergistically to enhance the sweetness intensity or mask undesirable aftertastes. Quite often, this will result in a lower level of LCS being used in products than would be required if they were being used in isolation. Table 2 illustrates the volume of low calorie sweetened beverages that would need to be consumed in a day before an individual would surpass the ADI, assuming that the MUD has been used.

It remains important to monitor actual consumption levels of LCS in order to ensure intakes remain below the ADI and, within Europe, Member states have been instructed to do this on a regular basis as per EU Directive 94/35/EC [10]. Available data from a number of European countries such as the UK, France, Belgium and the Netherlands, indicate that indeed the consumption of LCS is (far) below the ADI, with only cyclamate intake among young children potentially exceeding the ADI [11].

4. FOCUS ON ASPARTAME AND STEVIA

Aspartame is a methyl ester of two amino acids (the building blocks of proteins); aspartic acid and phenylalanine. Gram for gram, it has the same energy content as sucrose, however because it is up to 200 times sweeter, very small amounts are required to achieve the same level of sweetness hence the labels "low calorie" or "intense". As the most commonly used LCS, it can be found in around 6,000 products worldwide, however food consumption surveys have consistently shown that intake levels do not exceed the ADI [9].

Animal studies have reported no negative effects from aspartame when administered in doses of up to 4,000 mg/kg-bw (a very high dose as because tests for toxicity normally do not exceed 1,000 mg/kg-bw). On the basis of the NOAEL at 4000 mg/kg-bw the ADI was established at 40 mg/kg-bw following the application of a safety factor of 100 [9,12].

Following ingestion, aspartame is broken down into its component parts; aspartic acid, phenylalanine and methanol which can also be obtained, quite often in greater quantities, from commonly consumed foods and beverages such as meat, poultry and fruit juices. Upon absorption, all three components are metabolized in the same way as they would be if obtained from the normal diet [13]. At current estimates of aspartame intake, the amount of absorbed phenylalanine and aspartic acid does not reach physiologically significant levels within the general population. The final component of aspartame metabolism is methanol which can be very toxic at high levels of intake. Excess methanol is converted to formic acid via formaldehyde in the liver, and it is this which ultimately produces the toxic effects of methanol.

Table 2. Lifetime daily consumption of light beverages for a 25 kg child and a 70 kg adult before surpassing the ADI

Low calorie sweetener	Child (bw = 25 kg)[a]	Adult (bw = 70 kg)[a]
Aspartame	1,67 L	4,66 L
Acesulfame K	0,64 L	1,80 L
Cyclamate	0,70 L	1,96 L
Saccharine	1,56 L	4,37 L
Sucralose	1,25 L	3,50 L
Steviol Glycosides	1,25 L	3,50 L

[a]*(calculated via www.zoetstoffen.nl).*

At high enough levels of intake, it can cause metabolic acidosis, blindness and poisoning of the central nervous system [14]. Fortunately, the body is capable of dealing with a certain amount of methanol; a single bolus of 2g of methanol can be tolerated without any negative effects [15]. It has been estimated that from one litre of soft drink sweetened with aspartame, 55 mg of methanol would be formed which is less than what is naturally found in one litre of fruit juice [9] and well within the 2g shown to be safe, therefore the risk of methanol toxicity from aspartame is negligible. Furthermore, the liver naturally processes 22 mg of formaldehyde per minute (around 50 grams per day) as part of normal biochemical processes [15] so the contribution of aspartame to the level of formaldehyde within the body is also negligible. A large amount of methanol would have to be consumed at once for it to become toxic, therefore if it is assumed that 55 mg methanol would be formed per litre of an aspartame sweetened soft drink, more than 36 litres of that beverage would have to be consumed at one sitting in order to exceed the safe amount of two grams. It is clear that at this volume of intake, one would develop problems other than methanol poisoning and this provides a good illustration of the importance of making a proper distinction between the terms "hazard" and "risk".

4.1 Aspartame and Cancer

In 2006 and 2007, two studies were published by the Italian Ramazzini Institute which purported to prove that aspartame causes cancer in rats [16,17]. Substances which could potentially be genotoxic carcinogens in humans are not permitted and, although these were the only studies which provide support for such cancer claims, EFSA commissioned a panel of independent experts to examine the results of these studies. Based on the data from the first study, EFSA concluded that the tumours found were not as a result of the aspartame treatment, but of chronic inflammation of the lungs in the rats. Furthermore, it was found that the statistics appeared to be flawed and the results were not consistent. In order to assist in the assessment of the second study, EFSA made a number of requests for additional data but only some of this was subsequently made available to the panel. Another study, published by the same institute [18], claimed to demonstrate a dose response relationship between exposure to aspartame and the incidence of hepatocellular and alveolar/bronchiolar carcinomas in male Swiss mice. Once again, EFSA evaluated the results of this work and concluded that there were a number of flaws in the study design and how the study findings were reported, therefore it was decided that there was no reason to reconsider their original opinion on the use of aspartame at the permitted levels [19].

It is also worth noting that, upon examination of other publications by the Ramazzini Institute, the rats used by this group tend to develop cancer from many of the substances to which they are exposed [20-22], including Coca cola which does not contain aspartame [23]. To date, hundreds of safety studies have been carried out on aspartame and in none but the Ramazzini studies were any associations with genotoxicity or cancer reported [9]. In short, there are no indications that aspartame is carcinogenic, nor is this to be expected because the substances into which aspartame is metabolized in the body are known and occur naturally in food and in the body.

4.2 Diketopiperazine (DKP)

Diketopiperazine (DKP) is the major degradation product of aspartame and may form when a soft drink containing aspartame is stored for a long period of time at room temperature. DKP has an ADI of 7.5 mg/kg-bw. In the studies that were carried out to establish the ADI for aspartame, DKP was concurrently administered to the test animals in a ratio of 3:1 (aspartame: DKP) with no effects observed [9]. The safety of DKP has also been evaluated and confirmed in no less than thirteen toxicological studies [13]. It is also worth noting that DKPs are the most frequently occurring peptide derivatives in nature and can be found in many commonly consumed protein containing foods. EFSA included DKP in its recent Opinion on the safety of aspartame and concluded that with exposure estimates of up to 5.5mg/kg-bw at the 95th percentile for the general population, there was no reason to reconsider the current ADI [9].

4.3 Obesity and Premature Births

Based on epidemiological studies, it is regularly incorrectly reported (usually in the media) that there is a positive relationship between LCS beverage (hence, aspartame) intake and obesity. However, many of these studies do not consider the possibility of reverse causality, since people who are overweight or obese often tend to consume LCS beverages as part of a strategy to manage their weight more effectively. It is therefore necessary to view apparent associations between LCS beverage intake and the development of obesity with caution. A number of studies have also been published which have investigated a possible association between intake of LCS beverages and the risk of preterm delivery [24,25]. Overall the evidence does not conclusively support the case of negative effects ASB intake on pregnancy outcomes [26] and additional studies would be necessary to establish a causal link. It is also worth noting that most epidemiological studies investigating the health effects of LCS simply consider intakes of all LCS collectively and do not consider the intake levels of individual LCS in isolation. This is a major weakness in this type of research as it may result in erroneous associations or mask any true associations that may exist.

4.4 Phenylketonuria

Phenylalanine, an essential amino acid, is formed from aspartame metabolism. However, patients with phenylketonuria (PKU), an inborn error of metabolism which is diagnosed by means of the neonatal heal prick test, are unable to break down the amino acid properly resulting in high concentrations in the blood. High levels of phenylalanine in children can result in a number of potentially very negative health effects such as impaired brain development. For this reason, the ADI is not applicable to PKU patients and as such, products containing aspartame must carry a warning that they are a source of phenylalanine. This warning, however, does not relate to any potential danger of aspartame to those in the general population.

4.5 Stevia as Aspartame's Alternative?

Although "natural" is not synonymous with healthy or nontoxic, there is an appetite within the market for a natural alternative to aspartame. One such alternative may come in the form of the leaves of a shrub native to South America; the Stevia rebaudiana plant of the chrysanthemum family. The leaves of this plant contain a large number of sweet diterpene glycosides known as steviol glycosides [27]. These can be isolated by a process of water extraction followed by purification and crystallization and have been used as a sweetener in various countries for many years.

In 2008, an ADI of 4mg/kg-bw (expressed as steviol equivalents) was established at the 69th Joint FAO/WHO Expert Committee on Food Additives (JECFA) [28], however it was not until 2011 that the use of steviol glycosides was approved in Europe following a positive Opinion by EFSA [29]. The primary reason for this delay was due to the timing of the submission of the safety dossier by the petitioners to EFSA. As per JECFA specifications [30], the final sweetening mixture must consist of not less than 95% of nine named steviol glycosides with the most abundant being stevioside and/or rebaudioside A. Thus, following isolation, the sweetener mixture is approximately 300 times sweeter than sucrose and is therefore classified as a low calorie intense sweetener. Steviol glycosides have been allocated an E-number (E960), indicating that they fulfil the same safety requirements as aspartame and any other approved food additive.

4.6 Metabolism, Safety and Application of Steviol Glycosides

Following ingestion, both stevioside and rebaudioside A experience similar metabolic fates; both are very poorly absorbed, if at all, and instead are converted to their common aglycone, steviol, via bacterial hydrolysis in the large intestine. A large proportion of steviol is then absorbed while the remainder is excreted in the faeces. Very limited amounts of free steviol are found in the plasma as it undergoes extensive first pass metabolism in the liver and is excreted in the urine predominantly as steviol glucuronide [31].

As with all approved food additives, extensive safety and toxicological testing has been conducted via *in vitro* and *in vivo* studies in order to establish the safety of steviol glycosides. Although the safety of steviol glycosides have been established, some *in vitro* studies have shown that the aglycone, steviol, and some of its oxidative derivatives exhibit genotoxic activity. These results, however, have not been replicated *in vivo* with doses of up to 8000 mg/kg-bw [32] therefore the associated risk is negligible. The ADI for steviol glycosides was established following analysis of data from a 2 year carcinogenic study in the rat in which a NOAEL for stevioside of 967 mg/kg-bw, equating to 388 mg/kg-bw steviol. Following the application of a safety factor of 100 the ADI was therefore set at 4 mg/kg-bw, to be expressed as steviol equivalents [30].

Can the new low calorie sweetener E960 completely replace the other LCS as a natural alternative? Although this would prove to be an attractive option for some, it does not seem likely. This is mainly due to the fact that E960 has a liquorice-like aftertaste which would negatively affect the taste of products too much for it to be used in isolation. Nevertheless, two of the major soft drinks manufacturers, Coca-Cola Company and PepsiCo, have each claimed to have developed a new beverage product in which stevia has replaced 30% of the sugar without compromising on taste. Officially, these products can be marketed as light (or diet) products as they contain 30% less sugar than the regular varieties, albeit that manufacturers may not do this as consumers tend to equate the label "light" in soft drinks with "calorie-free." The "stevia drinks" may have to be marketed differently, possibly by exploiting the perceived advantages of sweetening the products with a sweetener that has been derived from a natural source rather than using artificial alternatives.

4. CONCLUSION

All approved food additives, including LCS are safe to use in the permitted quantities. The decision of which sweetener(s) to use in products will be dependent upon a number of factors including the permitted MUD and the properties of the sweetener. The impact of substituting LCS for sugar in commonly consumed food and beverage products on health, particularly weight status, has been the subject of debate in recent years. Recent research [33] has shown that children who drank a soft drink sweetened with sweeteners instead of a soft drink with sugar in a placebo controlled double blind study had a lower body weight and/or consumed fewer calories. This suggests that LCS could indeed make a positive contribution to public health, as previously calculated [34].

ACKNOWLEDGEMENTS

Dr. Le Roy Dowey and Prof Sean Strain, University of Ulster, are thanked for stimulating discussions.

COMPETING INTERESTS

Authors have declared that no competing interests exist.

REFERENCES

1. World Health Organisation. Principles for the Safety Assessment of Food Additives and Contaminants in Food. Environmental Health Criteria; 70. Accessed 24 September 2014. Available:http://www.inchem.org/documents/ehc/ehc/ehc70.htm#SectionNumber:5.5
2. Renwick AG. Acceptable daily intake and the regulation of intense sweeteners. Food Addit Contam. 1990;7(4):463-475.
3. European Food Safety Authority. About EFSA. Accessed 15 September 2014. Available:http://www.efsa.europa.eu/en/aboutefsa.htm
4. European Commission. Regulation (EC) No 178/2002 of the European Parliament and of the Council of 28 January 2002 laying down the general principles and requirements of food law, establishing the European Food Safety Authority and laying down procedures in matters of food safety. OJ. 2002;L31/1:1-24. Accessed 14 September 2014. Available: http://www.eur-lex.europa.eu/legal-content/EN/TXT/PDF/?uri=CELEX:32002R0178&from=EN
5. European Commission. White Paper on Food Safety. 2000. Accessed: 14 September 2014. Available: http://www.ec.europa.eu/dgs/health_consumer/library/pub/pub06_en.pdf
6. European Commission. Re-evaluation of food additives. Accessed 14 September 2014. Available:http://ec.europa.eu/food/food/fAEF/additives/re_evaluation_fA_en.htm
7. European Commission. Commission Regulation (EU) No 257/2010 of 25 March 2010 setting up a programme for the re-evaluation of approved food additives in accordance with Regulation (EC) No 1333/2008 of the European Parliament and of the Council on food additives. OJ. 2010;L80/19:19-27. Accessed 14 September 2014. Available: http://eur-lex.europa.eu/LexUriServ/LexUriServ.do?uri=OJ:L:2010:080:0019:0027:EN:PDF
8. European Food Safety Authority. Public Consultation: Draft Opinion on the re-evaluation of aspartame (E951) as a food additive. 2013. Accessed 13 September 2014.

Available:http://www.efsa.europa.eu/en/consultationsclosed/call/130108.pdf
9. European Food Safety Authority. Scientific Opinion on the re-evaluation of aspartame (E951) for the proposed uses as a food additive. EFSA Journal. 2013;11(12):3496. Available:http://www.efsa.europa.eu/en/efsajournal/pub/3496.htm
10. European Commission. European Parliament and Council Directive 94/35/EC of 30 June 1994 on sweeteners for use in foodstuffs. OJ. 1994;L237/3:3-12. Accessed 14 September 2014. Available: http://www.cur-lex.europa.eu/legalcontent/EN/TXT/PDF/?uri=CELEX:31994L0035&from=EN
11. Renwick AG. The intake of intense sweeteners: an update review. Food Addit Contam. 2006;23(4):327-338.
12. Scientific Committee on Food. Sweeteners. Reports of the Scientific Committee for Food (Sixteenth Series). 1985. EUR 10210 EN, Commission of the European Communities, Luxembourg.
13. Butchko HH, Stargel WW, Comer CP, Mayhew DA, Benninger C, Blackburn GL, et al. Aspartame: review of safety. Regul Toxicol Pharmacol. 2002;35(2 Pt2):S1-93.
14. Liesivuori J and Savolainen H. Methanol and formic acid toxicity: biochemical mechanisms. Pharmacol Toxicol. 1991;69(3):157163.
15. 15. Magnuson BA, Burdock GA, Doull J, Kroes RM, Marsh GM, Pariza MW, et al. Aspartame: A Safety Evaluation Based on Current Use Levels, Regulations, and Toxicological and Epidemiological Studies. Crit Rev Toxicol. 2007;37:629-727.
16. Soffritti M, Belpoggi F, Degli Esposti D, Lambertini L, Tibaldi E, Rigano. A First experimental demonstration of the multipotential carcinogenic effects of aspartame administered in the feed to Sprague-Dawley rats. Environ Health Persp. 2006;114:379-385.
17. Soffritti M, Belpoggi F, Tibaldi E, Degli Esposti D, Lauriola M. Life-span exposure to low doses of aspartame beginning during pre-natal life increases cancer effects in rats. Environ Health Persp. 2007;115:1293-1297.
18. Soffritti M, Belpoggi F, Manservigi M, Tibaldi E, Lauriola M, Falcioni L, et al. Aspartame administered in feed, beginning pre-natally through life span, induces cancers of the liver and lung in male Swiss mice. Am J Ind Med. 2010;53:1197-1206.
19. European Food Safety Authority. Statement of EFSA on the scientific evaluation of two studies related to the safety of artificial sweeteners. EFSA Journal. 2011;9(2):2089.
20. Minardi F, Belpoggi F, Soffritti M, Ciliberti A, Lauriola M, Cattin E et al. Results of Long-term Carcinogenicity Bioassay on Vinyl Acetate Monomer in Sprague-Dawley Rats. Ann N.Y. Acad. 2002;982:106-122.
21. Soffritti M, Belpoggi F, Padovani M, Lauriola M, Esposti DD, Minardi F. Life-time carcinogenicity bioassays of toluene given by stomach tube to Sprague-Dawley rats. Eur J Oncol 2004;9(2):91-102.
22. Soffritti M, Falcioni L, Bua L, Tibaldi E, Manservigi M, Belpoggi F. Potential Carcinogenic Effects of World Trade Center Dust After Intratracheal Instillation to Sprague-Dawley Rats: First Observation. Am J Ind Med 2013;56:155-162.
23. Belpoggi F, Soffritti M, Tibaldi E, Falcioni L, Bua L, Trabucco F. Results of Long-Term Carcinogenicity Bioassays on Coca-Cola Administered to Sprague-Dawley Rats. Ann N.Y. Acad. 2006;1076:736-752.
24. Halldorsson TI, Strom M, Petersen SB and Olsen SF. Intake of artificially sweetened soft drinks as risks of preterm delivery: a prospective cohort study in 59,334 Danish pregnant women. Am J Clin Nutr. 2010;92(3):626-633.
25. Englund-Ogge L, Brantsaeter al, Haugen M, Sengpiel V, Khatibi A, Myhre R et al. Association between intake of artificially sweetened and sugar-sweetened beverages and preterm delivery: a large prospective cohort study. Am J Clin Nutr. 2012; 96(3):552-559.
26. Marinovich M, Galli CL, Bosetti C, Gallus S, La Vecchia C. Aspartame, low-calorie sweeteners and disease: regulatory safety and epidemiological issues. Food Chem Toxicol. 2013; 60:109-115.
27. Ceunen S, Geuns JMC. Steviol Glycosides: Chemical Diversity, Metabolism and Function. J Nat Prod. 2013;76:1201-1228.
28. Joint FAO/WHO Expert Committee on Food Additives. Safety evaluation of certain food additives. Prepared by the 69th meeting of the Joint FAO/WHO Expert Committee on Food Additives. WHO Food Additives Series. 2009;66:183-220.
29. European Food Safety Authority. Scientific Opinion on the safety of steviol glycosides

for the proposed uses as a food additive. EFSA Journal. 2010;8(4):1537.

30. Joint FAO/WHO Expert Committee on Food Additives. Steviol Glycosides. Monographs 10. 2010. Accessed 1 September 2014. Available: http://www.fao.org/ag/agn/jecfa-additives/specs/monograph10/additive-442-m10.pdf
31. Wheeler A, Boileau A, Winkler P, Compton J, Prakash I, Jiang X, et al. Pharmacokinetics of rebaudioside A and stevioside after single oral doses in healthy men. Food Chem Toxicol. 2008;46:S54-S60.
32. Temcharoen P, Klopanichpah S, Glinsukon T, Suwannatrai M, Apibal S, Toskulkao C. Evaluation of the effect of steviol on chromosomal damage using micronucleus test in three laboratory animal species. J Med Assoc Thailand. 2000;83:s101-s108.
33. de Ruyter JC, Olthof MR, Seidell JC, Katan MB. A Trial of Sugar-free or Sugar-Sweetened Beverages and Body Weight in Children. New Engl J Med. 2012;367:1397-1406.
34. Hendriksen MA, Tijhuis MJ, Fransen HP, Verhagen H, Hoekstra J. Impact of substituting added sugar in carbonated soft drinks by intense sweeteners in young adults in the Netherlands: example of a benefit-risk approach. Eur J Nutr 2010;50:41-51.

19

Dietary Patterns and Metabolic Syndrome in Adults

Maria Stella R. de Oliveira[1,2], Natielen J. Schuch[1,3], Vivian C. Garcia[1,2] and Lígia A. Martini[1,2*]

[1]Nutrition Department, School of Public Health, University of São Paulo, Av Dr Arnaldo, 715.Zip code, 01246-904, Sao Paulo, SP, Brazil.
[2]University of São Paulo Food and Nutrition Research Center, NAPAN, USP, Brazil.
[3]Franciscan University Center, Rua dos Andradas, 1614.Zip code 97010032, Santa Maria, RS, Brazil.

Authors' contributions

All authors have significantly contributed, read and approved the final version of the paper and declare no conflict of interest.

ABSTRACT

Aims: Considering that diet is a potential factor contributing to the development of Metabolic Syndrome (MetS) and it is increasingly prevalent among population worldwide, the aim of the present study was to correlate dietary patterns with MetS and its main risk factors in adults.
Study Design: Cross-sectional.
Place and Duration of Study: School of Public Health, University of São Paulo, between August 2007 and January 2010.
Methodology: This study was conducted with 267 adults (189 women and 78 men) submitted to evaluation procedures which included body composition (waist circumference (WC), height and weight) and blood pressure measurements, biochemical analysis from a single blood sample after a 12-h fasting (serum triglycerides (TG), total cholesterol (TC), low-density lipoprotein cholesterol (LDL-c), high-density lipoprotein cholesterol (HDL-c) and fasting glucose) and assessment of dietary intake using a 24-hour recall. Dietary patterns (DPs) were identified using principal components factor analysis with varimax orthogonal rotation.
Results: Three distinct DPs were identified from the principal component factor analysis: "Traditional", "Healthy" and "Western". Among individuals with MetS, there was a positive correlation between Healthy pattern and HDL-c (P = .03), as well as between Western pattern and WC, TG and LDL-c (P = .001, P = .04 and P = .047). In people without MetS, the negative correlation was observed between Traditional pattern and LDL-c (P = .049), and positive correlation between Healthy pattern and age, TC and, LDL-c (P = .01, P = .03 and P = .03).
Conclusion: Our assessment offers information concerning food combinations that may increase the risk and prevalence of MetS. However, more studies are required to confirm these findings and

**Corresponding author: Email: lmartini@usp.br*

to assist in the prevention and development of specific nutritional recommendations for this syndrome.

Keywords: Nutritional assessment; metabolic diseases; food habits.

ABBREVIATIONS

BMI : body mass index
BP : blood pressure
DPs : dietary patterns
FG : fasting glucose
HDL-c : high-density lipoprotein cholesterol
IDF : International Diabetes Federation
LDL-c : low-density lipoprotein cholesterol
MetS : metabolic syndrome
TC : total cholesterol
TG : triglycerides
WC : waist circumference

1. INTRODUCTION

Metabolic syndrome (MetS) is a cluster of metabolic and physical changes which increases the risk of cardiovascular disease and are closely related to central obesity and insulin resistance [1,2].

Epidemiological studies suggest highly variable frequencies of MetS between populations. However, it is estimated that its prevalence is somewhere between 20 and 25% of the general people and reaches 42% among elderly individuals [3,4]. In Brazil there are no studies with representative data of the frequency of MetS, but it is known that it has been rising and increasing overall mortality about 1.5 times and cardiovascular about 2.5 times [5].

This scenario emerged from the processes of industrialization, urbanization, economic development and globalization of markets, that led to economic, political, social and cultural changes in the way individuals and communities organize their lifestyles [6]. These changes led to a dietary transition worldwide, which resulted in substantial changes in nutrition, with the enlargement in the supply of industrialized foods (rich in fats, sugars and sodium), increased consumption of energy-dense diets (high in saturated fat and low in unrefined carbohydrates) and reduced consumption of healthier foods, such as cereals, fruits and vegetables in addition to the general decrease in physical activity [7].

Nevertheless, the relationship between nutrients and cardiovascular diseases, such as MetS is not yet well understood, and the questions around this issue may be due to the traditional methods often carried out to examine the relationship between diet and the risk of chronic diseases which analyzes the effects of a single nutrient or food on these disorders [8].

Recently, dietary pattern analysis has emerged as an alternative approach to the assessment of food consumption [9] enabling to explore the effects of overall diet and, unlike the analysis of individual nutrients or foods, allows to determine the effects of diet on disease development due to the complexity of the food and the interaction between diet components [10].

In this study we aimed to investigate the relations between dietary patterns (DPs) derived from a 24-hour recall and the presence of MetS and its main risk factors, including body mass index (BMI), blood pressure (BP), waist circumference (WC), serum lipids and fasting glucose (FG) in adults living in São Paolo, Brazil.

2. MATERIALS AND METHODS

2.1 Research Design and Study Population

This cross-sectional study was conducted with adults of both sexes who attend the primary health care center at the School of Public Health, University of São Paulo, and from the Adult Health Survey of the City of São Paulo (ISA-SP), described elsewhere (11) . This research is part of the project entitled "Relationship between nutritional status of vitamin D and metabolic syndrome in adults living in the metropolitan region of São Paulo" (12) which was carried out from August 2007 to January 2010.

Initially, 462 individuals were invited to participate. Individuals under 18 years of age, pregnant and lactating women, African-descendants and those with chronic diseases that potentially alter the metabolism of vitamin D as such osteoporosis, cancer, renal failure and immune diseases, as well individuals who were using vitamin D and/or calcium supplementation or multivitamins were not included. The present study consisted of a subsample (n = 267) of the primary study, including only individuals with dietary assessment data.

2.2 Measurements

The participants were submitted to several evaluation procedures at the cited center for primary health care which included measurements of body composition and blood pressure, biochemical analysis from the blood samples collected and assessment of dietary intake. All data were gathered by previously trained field researchers.

The measurements of height and weight were performed using a calibrated TanitaTM electronic scale (Tanita Corporation of America Inc, Illinois, USA) and a SecaTM stadiometer (Seca, Hamburg, Germany). BMI was calculated as weight in kilograms divided by the square of height in meters. WC was measured by horizontally positioning a measuring tape at the midpoint level between the lower intercostals border and the anterior superior iliac supine at the end of the normal expiration phase.

Blood samples were obtained by means of venipuncture from each participant after a 12-h fasting. Samples were immediately centrifuged, aliquoted and frozen at -80°C until analysis. Serum triglycerides (TG), total cholesterol (TC), low-density lipoprotein cholesterol (LDL-c), high-density lipoprotein cholesterol (HDL-c) and FG were determined by enzymatic colorimetric assay (Celm™, Barueri, SP, Brazil). For these parameters, the cutoff points were based on those established by the International Diabetes Federation (IDF) [13].

Dietary intake was assessed by trained nutritionists using a standardized protocol of a 24-hour recall method. To minimize reporting errors and ensure proper record of the quantities described by patients, a food photograph album was used at the interviews [14]. Data from food intake were entered and processed using Nutrition Data System for Research (NDSR) software. To identify the DPs, food items were categorized into groups based on food characteristics and nutritional composition. It is noteworthy that the foods or food groups consumed by less than 10% of the sample were excluded from the analysis. Thus, the twenty-two food groups that represent the DPs of the study are presented in Table 1.

2.3 Identifying Individuals with Metabolic Syndrome

The criteria for classifying adults with MetS were based on those established by the IDF [13]: abdominal obesity (WC >90 cm in males and >80 cm in females) and >2 of the following risk factors (i) high serum TG (>150 mg/dl); (ii) low serum HDL-c (<40 mg/dl in males or <50 mg/dl in females); (iii) high BP (>130 or >85 mmHg); and (iv) high FG (>100 mg/dl).

Table 1. Composition of food groups

Food Groups	Food Items
Fresh spice	Onion, garlic, culinary herbs
Oils	Olive and vegetable oils, margarine
Fruits	All fresh fruits and fruit salad
Vegetables	Leafy green vegetables, light- or dark- colored vegetables, cruciferous vegetables
Potatoes and tubers	Potatoes and root vegetables
Cereals and whole grains	Oat, bran, wheat germ, quinoa, flaxseed, corn flakes
Pasta	Noodles, spaghetti, gnocchi and other pasta preparations
Rice	White rice
Beans	Beans, peas, green beans, chick peas, lentils
White meat	All chicken preparations
Red meat	All beef preparations
Processed meat	Bacon, ham, salami, sausages, hamburgers
Dairy	Whole milk, full-cream milk, full fat cheese, full fat yoghurt,
Low fat dairy	Reduced-fat milk, skim milk, ricotta, low-fat cheese, cottage cheese, low-fat yoghurt
White bread	White breads, crackers, toasts, salted biscuits
Whole grain bread	Mixed grain or high-fiber breads, crackers and toasts
Sweets	Cakes, pies, muffins, cookies, biscuits, chocolates, marmalades, jams, jellies, chocolate covered bars, ice cream
Salty & savoury snacks	Sandwiches, pizzas, pastries, potato and corn chips
Juices	Fruit juice, fresh fruit juice
Soft drinks	Sodas, diet sodas, light sodas
Coffee and teas	Coffee, decaffeinated coffee, black tea, green tea and other herbal tea
Sugar	Sugar and honey

2.4 Statistical Analysis

Principal components factor analysis with varimax orthogonal rotation was performed to obtain the DPs from the twenty-two food groups, after verifying the adequacy of the sample by the Kaiser-Meyer-Olkin test (considering acceptable values above 0.50). PCA was chosen since it is a multivariate technique that evaluates the intercorrelations between the initial food variables and reduces food groups into a smaller number of factors (patterns) that can explain variations in dietary intake [15].

The number of patterns retained was based on a combination of food group components with an eigenvalue >1.5, the percentage of variance criteria (sufficient factors to meet at least 60% of variance explained), and examination of the break-point in the scree plot, resulting in three factors retained for further analyses. Each dietary pattern was interpreted and named based on the food groups with loadings above 0.25 or below -0.25, which were considered as significantly contributing to a pattern.

Each variable on this study was analyzed using the Kolmogorov-Smirnov test to determine whether the distribution was normal. Spearman correlation coefficients were calculated to examine the correlation between DPs and the main risk factors for MetS.

Scores were derived from the patterns obtained, and the average score was calculated for each pattern according to the presence or absence of metabolic syndrome. Student t test was used to compare means between groups with and without MetS.

The analyses were conducted using the Statistical Package for the Social Science (SPSS) software for Windows, version 18 (SPSS Inc., Chicago, IL, USA) and the Stata - Data Analysis and Statistical Software, version 11 (StataCorp., Texas, USA). Results were taken to be statistically significant when *P* values were < 0.05.

3. RESULTS AND DISCUSSION

As presented in Table 2, a total of 267 adults (189 women and 78 men) provided biochemical and dietary pattern data. Of these, 153 (106 women and 47 men) were in the "With MetS" group and 114 (83 women and 31 men) were in the "Without MetS" group. As expected, stratifying the individuals according to the presence of MetS, the mean age as well as BMI, WC, serum TG, TC, LDL-c and FG, systolic and diastolic BP pressure were significantly higher among individuals with MetS.

Three distinct dietary patterns were identified from the principal component factor analysis. The first factor was named "Traditional Dietary Pattern" because it represented the traditional Brazilian food consumption of rice and beans and also high intakes of fresh spice, vegetables, white meat, juices, coffee & teas and sugar; and low intake of pasta. The second factor was characterized by high intakes of fresh spice, fruits, vegetables, low-fat dairy and whole grain bread and low intake of potatoes & tubers, red meat, coffee & teas and sugar and thus it was named "Healthy Dietary Pattern".

Table 2. Clinical characteristics of the studied sample

Variables	Total sample n=267	With MetS n=153	Without MetS n=114	*P* value
Gender				
Female	189 (70.79%)	106 (69.28%)	83 (72.81%)	
Male	78 (29.21%)	47 (30.72%)	31 (27.19%)	0.531
Age (years)	51.48 (14.25)	55.29 (12.26)	46.36 (15.15)	0.000
BMI (kg/m^2)	29.74 (6.01)	31.10 (5.86)	27.91 (5.74)	0.000
WC (cm)	98.06 (13.61)	102.01 (12.25)	92.61 (13.56)	0.000
TG (mg/dl)	130.95 (68.58)	158.02 (74.62)	94,61 (35.47)	0.000
TC (mg/dl)	193.85 (43.70)	200.88 (45.15)	184.43 (39.97)	0.002
LDL-c (mg/dl)	122.05 (38.86)	126.74 (40.13)	115.87 (36.38)	0.024
HDL-c (mg/dl)	44.89 (12.33)	41.70 (10.62)	49.17 (13.18)	0.000
FG (mg/dl)	98.95 (18.60)	104.58 (18,99)	91.39 (15.13)	0.000
SBP (mmHg)	130.34 (19.40)	137.57 (19.13)	120.69 (15.15)	0.000
DBP (mmHg)	79.88 (10.40)	83.25 (10.00)	75.39 (9.18)	0.000

Abbreviations: MetS – metabolic syndrome; BMI – body mass index; WC – waist circumference; TG – triglycerides; TC - total cholesterol; LDL-c – LDL-cholesterol; HDL-c – HDL-cholesterol; FG – fasting glucose; SBP – systolic blood pressure; DBP – diastolic blood pressure.

Data are expressed as means (SD), except gender n (%). T Test.

The third factor represented high intakes of olive & vegetable oils, processed meat, dairy, white bread, juices, sodas and low intake of low-fat dairy and whole grain bread and the pattern was named "Western Dietary Pattern". These patterns explained 25.59% of the dietary intake variance. Factor loadings of food groups across these DPs and the variance explained by each dietary pattern are presented in Table 3.

Table 4 shows Spearman correlation coefficients of each DP and categories of risk factors for MetS according to its presence or absence within the study participants. Among individuals with MetS, there was a positive correlation between Healthy DP and serum HDL-c (*P* = .03), as well as between Western DP and WC (*P* = .001), TG (*P* = .04) and serum TG (*P* = .047). The same did not applied to subjects without MetS, in which a positive correlation was seen between Healthy DP and age (P = .01), serum TC (P = .03) and serum LDL-c (P = .03). In addition, a negative correlation was found between Traditional DP and serum LDL-c (P = .049).

The major objective of this research was to examine the relation between patterns of food consumption and the occurrence of MetS and its main risk factors in our study population.

Table 3. Dietary patterns and factor loadings

Food groups	Factor loadings of each dietary pattern		
	Traditional	Healthy	Western
Fresh spice	***0.4264***	***0.3958***	0.1214
Olive and vegetable oils	-0.0234	0.1678	***0.5773***
Fruits	0.0420	***0.4387***	0.0087
Vegetables	***0.3399***	***0.3008***	-0.0735
Potatoes and tubers	-0.0623	***-0.3894***	-0.0745
Cereals and whole grains	-0.1952	0.2332	-0.0176
Pasta	***-0.2679***	0.2137	0.0047
Rice	***0.7677***	-0.1281	-0.0178
Beans	***0.6065***	-0.1948	0.0040
White meat	***0.5758***	0.2015	-0.0489
Red meat	0.2191	***-0.3894***	-0.0607
Processed meat	0.0347	-0.0073	***0.3129***
Dairy	0.2499	-0.1707	***0.4643***
Low fat dairy	-0.1083	***0.4254***	***-0.2728***
White bread	-0.0156	-0.0422	***0.6897***
Whole grain bread	-0.0621	***0.3105***	***-0.2966***
Sweets and snacks	0.1238	0.0245	-0.0019
Salty & savoury snacks	-0.2038	-0.1831	0.0918
Juices	***0.3637***	0.1009	***0.2607***
Sodas	-0.1308	-0.0157	***0.5608***
Coffee & teas	***0.2688***	***-0.3390***	-0.0948
Sugar	***0.2901***	***-0.5350***	0.0209
% Total variance explained	***9.96***	***17.78***	***25.59***

Table 4. Correlation between each dietary pattern and MetS components

	With MetS			Without MetS		
	Traditional	Healthy	Western	Traditional	Healthy	Western
Age	-0.099	0.088	-0,044	-0.019	**0.213**[f]	-0.033
BMI	-0.038	0.067	0.147	-0.134	0.113	0.111
WC	0.003	-0.086	**0.288**[b]	0.011	0.135	0.168
FG	0.042	0.031	0.047	-0.013	0.004	-0.006
TG	0.063	0.166	**0.183**[c]	0.115	0.116	0.001
TC	-0.026	0.072	0.149	-0.153	**0.192**[a]	-0.079
LDL-c	-0.029	0.052	**0.172**[d]	**-0.171**[e]	**0.185**[a]	-0.099
HDL-c	-0.122	**0.184**[a]	-0.161	-0.038	0.077	-0.076
SBP	0.040	0.019	0.133	0.134	0.083	-0.082
DBP	0.073	-0.061	-0.005	0.006	0.012	-0.086

Abbreviations: MetS – metabolic syndrome; BMI – body mass index; WC – waist circumference; FG – fasting glucose; TG – triglycerides; TC - total cholesterol; LDL-c – LDL-cholesterol; HDL-c – HDL-cholesterol; SBP – systolic blood pressure; DBP – diastolic blood pressure.

Spearman correlation: [a]P = .03; [b]P = .001; [c]P = .04; [d]P = .047; [e]P = .049; [f]P = .01.

Our results showed no direct association between DPs and the presence or absence of MetS. Other studies [16,17] have found similar results, in which DPs were not associated with overall prevalence of the syndrome. Differently from these findings, a Western DP was associated with MetS in Iran women, Americans, Mexicans and Chinese adults [18-21].

However looking separately to the risk factors for MetS, it was possible to note that some were associated with the DPs. For instance, the correlation between Western pattern and WC and serum TG was positive among those with MetS. Similar results were found in two different researches with Samoans adults in which a modern pattern of eating (characterized by high intake of processed foods and low intake of local traditional fresh foods) was associated with higher serum TG levels [22], BMI and WC [23]. Also, a study with German adults pointed that a high adherence to a "processed foods" pattern (composed of refined grains, processed meat, red meat, high-sugar beverages, eggs, potatoes, beer, sweets and cakes, snacks and butter) increased the occurrence of abdominal obesity, hypertension and hypertriacylglycerolaemia [24].

The positive correlation found between Healthy DP and serum HDL-c was similar to the findings of a cross-sectional study with 19 750 Dutch men and women, in which a dietary pattern characterized by greater intakes of vegetables, salad, rice, chicken, fish, and wine was significantly associated with lower blood pressure and higher HDL-c concentrations [25].

We also found some unexpected results, such as the positive correlation between Healthy DP and serum TC and serum LDL-c among those without MetS. These findings are in discordance to other studies that confirm that vegetables, fruits and whole grains have been associated with lower incidence of risk factors for MetS [26-30]. However, this divergence detected in our results could be explained by the positive correlation also found between this pattern and age. Other studies also pointed out the direct association between these last two variables [31,32].

The Traditional DP was negatively correlated to serum LDL-c among individuals without MetS. This finding could be consistent with the results of another study [33] which indicated that the consumption of a traditional Brazilian DP was inversely associated with LDL-c and TC values among men and women [34].

4. CONCLUSION

The present study demonstrated that DPs had some influence on the components of MetS, indicating that Western pattern was involved in metabolic deregulation. However, prospective and intervention studies could confirm that nutritional advice in form of food combinations could prevent and treat the MetS.

4.1 Limitations

The present study had some limitations. First, was its cross-sectional design, wich does not permit to examine the causal link between dietary patterns and MetS risk, because the temporal relation of these events could not be clearly established. Second, the findings from this study could be specific to southeastern Brazilian adults, and did not represent national findings. Finally ,the sample size and the use of only one 24h dietary recall could underestimated the individual variability of food intakes which implicates that the dietary patterns identified by principal were plausible but they did not reveal all possible patterns.

ACKNOWLEDGEMENTS

This study was supported by São Paulo Research Foundation (FAPESP) under the protocol n° 2009/52636-0 and 2007/52420-2. MSR Oliveira acknowledges receipt of scholarship funds from Coordination for the Improvement of Higher Level Personnel (CAPES).

ETHICAL APPROVAL

The Ethics Committees of University of São Paulo approved the study protocol and the consent form was obtained from all participants.

COMPETING INTERESTS

Authors have declared that no competing interests exist.

REFERENCES

1. Von Bernhardi R, Zanlungo S, Arrese M, Arteaga A, Rigotti A. El síndrome metabólico: de factor agravante a principal factor de riesgo patogénico en diversas enfermedades crônicas. Rev Med Chil. 2010;138(8):1012-9. Spanish.

2. Kassi E, Pervanidou P, Kaltsas G, Chrousos G. Metabolic syndrome: definitions and controversies. BMC Med. 2011;9:48.
3. Ribeiro Filho FF, Mariosa LS, Ferreira SRG, Zanella MT. Gordura visceral e síndrome metabólica: mais que uma simples associação. Arq Bras Endocrinol Metabol. 2006;50(2):230-8. Spanish.
4. Kac G, Sichieri R, Gigante DP. Epidemiologia nutricional. Rio de Janeiro: Fiocruz/Atheneu; 2007.
5. Sociedade Brasileira de Hipertensão; Sociedade Brasileira de Cardiologia; Sociedade Brasileira de Endocrinologia e Metabologia; Sociedade Brasileira de Diabetes; Associação Brasileira para o Estudo sobre a Obesidade. I Diretriz Brasileira de Diagnóstico e Tratamento da Síndrome Metabólica. Arq Bras Cardiol. 2005;84(Supl 1):3-28. Portuguese.
6. World Health Organization. Diet, nutrition, and the prevention of chronic diseases. Geneva; 2002. (WHO Technical Report Series, 797).
7. Ministério da Saúde, Secretaria de Vigilância à Saúde, Secretaria de Atenção à Saúde. Diretrizes e recomendações para o cuidado integral de doenças crônicas não-transmissíveis: promoção da saúde, vigilância, prevenção e assistência. Brasília (DF); 2008. (Série B. Textos Básicos de Atenção à Saúde). Portuguese.
8. Kim JH, Lee JE, Jung IK. Dietary Pattern Classifications and the Association with General Obesity and Abdominal Obesity in Korean Women. J Acad Nutr Diet 2012;112(10):1550-9.
9. Hu FB. Dietary pattern analysis: a new direction in nutritional epidemiology. Curr Opin Lipidol. 2002;13:3–9.
10. Alves AL, Olinto MT, Costa JS, Bairros FS, Balbinotti MA. Padrões alimentares de mulheres adultas residentes em área urbana no Sul do Brasil. Rev Saude Publica. 2006;40(5):865-73.
11. Martini LA, Verly E Jr, Marchioni DM, Fisberg RM Prevalence and correlates of calcium and vitamin D status adequacy in adolescents, adults, and elderly from the Health Survey-São Paulo. Nutrition. 2013; 29(6):845-50.
12. Garcez MR, Pereira JL, Fontanelli Mde M, Marchioni DM, Fisberg RM. Prevalence of dyslipidemia according to the nutritional status in a representative sample of São Paulo. Arq Bras Cardiol. 2014;103(6):476-84.
13. IDF-International Diabetes Federation. The IDF consensus worldwide definition of the metabolic syndrome. Brussels; 2006.
14. Zaboto CB, Viana RPT, Gil MF. Registro fotográfico para inquéritos alimentares. NEPA – UNICAMP; 1996. Portuguese.
15. Hair Jr, Black WC, Babin BJ et al. Multivariate data analysis, Pearson Prentice Hall; 2006.
16. Kim JA, Kim SM, Lee JS, Oh HJ, Han JH, Song Y et al. Dietary patterns and the metabolic syndrome in Korean adolescents: 2001 Korean National Health and Nutrition Survey. Diabetes Care. 2007; 30(7):1904–5.
17. Fonseca MJ, Gaio R, Lopes C, Santos AC. Association between dietary patterns and metabolic syndrome in a sample of portuguese adults. Nutr J. 2012;11:64.
18. Esmaillzadeh A, Kimiagar M, Mehrabi Y, Azadbakht L, Hu FB, Willett WC. Dietary patterns, insulin resistance, and prevalence of the metabolic syndrome in women. Am J Clin Nutr. 2007;85(3):910–8.
19. Lutsey PL, Steffen LM, Stevens J. Dietary intake and the development of the metabolic syndrome: the Atherosclerosis Risk in Communities study. Circulation 2008;117(6):754–61.
20. Denova-Gutiérrez E, Castañón S, Talavera JO, Gallegos-Carrillo K, Flores M, Dosamantes-Carrasco D et al. Dietary patterns are associated with metabolic syndrome in an urban Mexican population. J Nutr. 2010;140(10):1855–1863.
21. He Y, Li Y, Lai J, Wang D, Zhang J, Fu P et al. Dietary patterns as compared with physical activity in relation to metabolic syndrome among Chinese adults. Nutr Metab Cardiovasc Dis. 2012;S0939-4753(12):00216-5
22. DiBello JR, McGarvey ST, Kraft P, Goldberg R, Campos H, Quested C et al. Dietary patterns are associated with metabolic syndrome in adult Samoans. J Nutr. 2009;139(10):1933-43.
23. Baylin A, Deka R, Tuitele J, Viali S, Weeks DE, McGarvey ST. INSIG2 variants, dietary patterns and metabolic risk in Samoa. Eur J Clin Nutr. 2013;67(1):101-7.
24. Heidemann C, Scheidt-Nave C, Richter A, Mensink GB. Dietary patterns are associated with cardiometabolic risk factors in a representative study population

of German adults. Br J Nutr. 2011;106(8):1253-62.
25. van Dam RM, Grievink L, Ocké MC, Feskens EJ. Patterns of food consumption and risk factors for cardiovascular disease in the general Dutch population. Am J Clin Nutr. 2003;77(5):1156-63.
26. Williams DE, Prevost AT, Whichelow MJ, Cox BD, Day NE, Wareham NJ. A cross-sectional study of dietary patterns with glucose intolerance and other features of the metabolic syndrome. Br J Nutr. 2000;83:257–66.
27. Azadbakht L, Mirmiran P, Esmaillzadeh A, Azizi T, Azizi F. Beneficial effects of a dietary approaches to stop hypertension eating plan on features of the metabolic syndrome. Diabetes Care. 2005;28:2823-31.
28. Esmaillzadeh A, Mirmiran P, Azizi F. Whole-grain consumption and the metabolic syndrome: favourable association in Tehranian adults. Eur J Clin Nutr. 2005;59:353–62.
29. Sahyoun NR, Jacques PF, Zhang XL, Juan W, McKeown NM. Wholegrain intake is inversely associated with the metabolic syndrome and mortality in older adults. Am J Clin Nutr. 2006;83:124–31.
30. Esmaillzadeh A, Kimiagar M, Mehrabi Y, Azadbakht L, Hu FB, Willett WC. Fruit and vegetables intakes, C-reactive protein, and the metabolic syndrome. Am J Clin Nutr. 2006;84:1489–97.
31. Gimeno SG, Mondini L, Moraes SA, Freitas IC. Padrões de consumo de alimentos e fatores associados em adultos de Ribeirão Preto, São Paulo, Brasil: Projeto OBEDIARP. Cad Saude Publica. 2011;27(3):533-45. Portuguese.
32. Mohammadifard N, Sarrafzadegan N, Nouri F, Sajjadi F, Alikhasi H, Maghroun M et al. Using factor analysis to identify dietary patterns in Iranian adults: Isfahan Healthy Heart Program. Int J Public Health. 2012;57(1):235-41.
33. Mattei J, Hu FB, Campos H. A higher ratio of beans to white rice is associated with lower cardiometabolic risk factors in Costa Rican adults. Am J Clin Nutr 2011;94(3):869-76.
34. Olinto MT, Gigante DP, Horta B, Silveira V, Oliveira I, Willett W. Major dietary patterns and cardiovascular risk factors among young Brazilian adults. Eur J Nutr 2012;51(3):281-91.

20

Customers Purchasing Organic Food - Do They Live Healthier? Results of the German National Nutrition Survey II

Marianne Eisinger-Watzl[1*], Friederike Wittig[1], Thorsten Heuer[1] and Ingrid Hoffmann[1]

[1]*Department of Nutritional Behaviour, Max Rubner-Institute, Federal Research Institute of Nutrition and Food, Haid-und-Neu-Str. 9, 76131 Karlsruhe, Germany.*

Author's contribution

This work was carried out in collaboration between all authors. Author IH headed the project. Author MEW conducted the analysis and interpretation and drafted the manuscript. Author FW was involved in the early work and assisted with the interpretation of the results. Author TH was in charge of data management. All authors were involved in the critical revision of the manuscript for important intellectual content. All authors read and approved the final manuscript.

ABSTRACT

Aims: Using national food consumption data this paper addresses issues whether buyers of organic food make healthier food choices and pursue a healthy lifestyle concerning smoking, physical exercise and body weight compared to non-buyers.

Study Design: The German National Nutrition Survey II (NVS II) is a nationwide food consumption study providing additional information on social demographics, health, and lifestyle aspects as well as anthropometric measurements. Using data of several assessment tools, a comparison was conducted between buyers and non-buyers of organic food.

Place and Duration of the Study: From November 2005 to November 2006, data collection took place in about 500 randomly chosen sample points across Germany.

Methodology: 13,074 participants aged 18-80 years were divided into groups of buyers (44.9%) and non-buyers (55.1%) of organic food. According to their purchase frequency, the organic food buyers were further differentiated into intensive, occasional or infrequent purchase groups. A diet history method was applied to assess food consumption, trained staff measured BMI while questionnaires were used for social demographic description and healthy lifestyle factors.

Results: Buyers of organic food consumed 17% more fruit and 23% more vegetables as well as

**Corresponding author: Email: marianne.eisinger-watzl@mri.bund.de*

less meat/sausages (25%) and soft drinks (58%) than non-buyers did (P< .001, resp.). These results are more pronounced for women and for intensive buyers. Additionally, buyers of organic food exhibit healthier lifestyle characteristics in respect to smoking behaviour, physical activity, and body weight compared to non-buyers.

Conclusion: German buyers of organic food demonstrate health behaviours complying better with the recommendations for a healthy lifestyle compared with non-buyers. Independent of the discussion whether organically produced food exerts additional health effects, buyers of organic food make healthier food choices than non-buyers, thereby benefiting for their overall health.

Keywords: Food consumption survey; organic food; food choice; healthy lifestyle.

1. INTRODUCTION

Data on market statistics exhibit raising growth of the global organic market although the numbers vary a lot between countries and regions [1]. The enormous increase of organic markets in many countries prompted numerous studies to gain more insights in the supply as well as the demand sector. In the latter case, a multitude of studies concentrate on customers' perception toward organic food and the factors explaining customers' decision-making process [2-13]. Summarizing these studies reveals that customer motivation for buying organic food is not homogenous. However, some motives are consistent, such as a high concern regarding personal and environmental health, including the aspect of food safety, and animal welfare [6,8,14-16].

In the focus of marketing research, these customer attitudes and motivations are relevant in order to understand and manage developments in the organic markets. From a public health point of view, interests concentrate on the connection of this health driven customer decision and other healthy lifestyle aspects. Emerging questions are whether organic food buyers compared to non-buyers make healthier food choices and whether health consciousness of organic buyers stated as purchase motivation is related to smoking, physical activity and body weight. These lifestyle factors are fundamental behaviour connected to risk management of major chronic diseases and therefore important indicators of overall health behaviour [17,18].

So far, there are only few studies investigating whether and how the stated interest in health of organic food buyers reflects in lifestyle behaviour. Regarding food choice, a study in Great Britain showed that, compared to official census data, 274 regular customers of organic food, who completed a diet diary, consumed more vegetables, fruit and cereals as well as less animal products with the exception of fish [19]. A Norwegian study obtained similar results studying 63,808 pregnant women by means of a food frequency questionnaire. Women with frequent organic food consumption were more in line with public recommendations for healthy eating [20].

Two other studies, conducted in Germany respectively Denmark, investigated eating habits using household panel data and expenditures on organic or conventional food. Households with heavy expenditure on organic food complied closer with the recommendations for a healthy diet [21,22]. However, the data resulted from food purchase of entire households rather than food consumption of individuals. The terms 'consumer' and 'purchaser' are often used synonymously in literature. Hughner et al. [10] emphasise the incorrect terminology in their overview on studies regarding the decision-making process for buying organic food. They mention that in all studies the term' consumer' is used relying on purchase data. In consequence, they demand the accurate differentiation between 'purchasers' (or 'customers') and 'consumers' of organic food as well as additional research on food consumption.

The French study of the Nutrinet-Santé cohort is the only study published so far comparing organic food consumer and non-consumer based on the assessment of food consumption of a large general population (n=54,311) [23]. For Germany, the National Nutrition Survey II (NVS II), a representative food consumption study, offers the required data for a sample of more than 13.000 participants. Besides the data on food consumption, the survey provides information on social demographics, health and lifestyle aspects as well as anthropometric measurements [24,25] and therefore allows the examination of differences in food choice as well as lifestyle characteristics between buyers and non-buyers of organic food in Germany.

2. METHODOLOGY

2.1 Data Collection and Description of Variables

The baseline study NVS II was commissioned by the Ministry of Food, Agriculture and Consumer Protection and the dietary assessment referred to in the present study was undertaken from 2005 until 2006 [24]. At about 500 randomly chosen sample points across Germany, the local register offices randomly selected 14-80 year old participants of which 18-80 year old participants were integrated in this evaluation. Computer assisted personal interviews were conducted at study centers to obtain basic socio-demographic information followed by a computer assisted diet history interview. The diet history method is a dietary assessment instrument asking directly for the habitual food consumption [26,27]. As software DISHES (Diet Interview Software for Health Examination Studies) was used, a program developed by the Robert Koch-Institute, Berlin [28,29]. The standardized interviews lead participants through the sequence of meals of a day followed by questions on the frequencies of consumed food. Participants also completed a questionnaire on health-related aspects and leisure time activities. Additionally, anthropometric measurements were conducted at the study centers [24]. During data assessment, participants and interviewers were not aware of the future research question regarding organic purchase. This diminishes a possible response bias.

There was no specific differentiation for the consumption of organically or conventionally produced food. By means of a questionnaire, participants were classified into buyers and non-buyers based on the question: Do you buy organic food? (yes/no). Furthermore, buyers were asked to give detailed information on the frequency of their purchase of twelve organic food items. The frequencies were coded with points and a ratio was achieved by dividing the sums of points by the number of given answers to correct for the possibility that no answer was given or a person did not eat/drink a food item. Based on the ratios buyers were divided in either the intensive, occasional or infrequent purchase group.

Social class was defined as an interrelation of several factors, so an index was generated based on the degree of education, the employment status of the household's principle earner, and the monthly net income of the household. Individuals were classified according to these characteristics by a point system [24]. The answers to four questions, (1) arrangement of food items perceived as important for a healthy diet, (2) classification of probiotic yoghurt, (3) classification of ACE-beverages and (4) comprehension of the 5-a-day campaign, from the questionnaire were combined to a 'NVS II nutrition knowledge index'. Information about smoking (yes/no) and physical activity (yes/no) was retrieved according to statements from a questionnaire. Body Mass Index (BMI) was calculated dividing weight into kilograms by height in square meters. In order to classify the BMI, the WHO definition was employed [30].

The consumption of fruit, vegetables, meat and sausages as well as sweets and soft drinks serves as indicator for a healthy food pattern. Food consumption data were evaluated and compared with the dietary guidelines of the German Nutrition Society [31,32].

2.2 Data Analysis

For descriptive analysis, Pearson's Chi-square test was applied to 2x2 contingency tables, testing for independence of two criteria. For multiple comparisons, the Bonferroni correction was subsequently used. Food consumption was calculated as arithmetic mean (± standard error of mean). The presentation of the means instead of medians was chosen because several food groups were consumed by less than 50% of the participants. Consumption data did not show normal distribution and could not be normalized by log-transformation, thus non-parametric tests were carried out. The Mann-Whitney-Wilcoxon test was applied to compare food consumption of buyers and non-buyers of organic food. Kruskal-Wallis test was used to compare more than two groups.

Statistically significant differences were defined at the p-level of ≤ 0.05. Calculations were performed with SAS program version 9.2. (SAS Institute Inc., Cary, North Carolina, USA).

3. RESULTS

Of the 13,074 participants 44.9% stated to purchase organic food whereas, 55.1% did state not to do so. The frequency of purchasing organic food revealed a proportion of 5.1% intensive, 26.2% occasional and 13.6%

infrequent buyers summing up to the 44.9% of total buyers.

3.1 Social Demographic Description

More women (54%) than men participated in the NVS II; sex difference was even more pronounced regarding buyers of organic food with 61% being female. The more intense the organic purchase the higher the proportion of women: 67% intensive, 61% occasional and 59% infrequent buyers of organic food were female. In the non-buyer group, sex difference was almost balanced (49% female vs. 51% male).

The comparison of age categories revealed that the proportions of the youngest (age 18-24) and the senior participants (age 65-80) were bigger in the non-buyer-group than in the buyer-group and vice versa in the age groups of 35-50 and 51-64 years. There was no difference in the age group of 25-34 years (Fig. 1). The group of intensive buyers reached the highest average age of 52.4 years compared to 49.5 years for the occasional buyers, 47.9 years for the infrequent buyers and 48.4 years for the non-buyers (P< .001, resp.).

Regarding social class affiliation more buyers of organic food belong to the two upper classes (60%) compared to non-buyers (44%) (Fig. 2). The proportions of the two lowest classes sum up to 14% for buyers and to 24% for non-buyers. Every social class is represented in the three frequency groups of organic food purchase (intensive, occasional, infrequent).

3.2 Food Consumption

Analyses of the diet history interviews revealed that mean usual consumption of fruit of participants buying organic food was 17% higher than the consumption of non-buyers (P < .001) (Fig. 3, by sex). Half of the organic food buyers (50%) met the dietary guideline of the German Nutrition Society of 250 g fruit per day [31,32], while 40% of the non-buyers met the guideline (P < .001). Regarding the frequency of the organic purchase 65% of the intensive, 52% of the occasional and 42% of the infrequent buyers met the recommendation (P < .001, resp.). Differentiated by sex 54% of all female organic

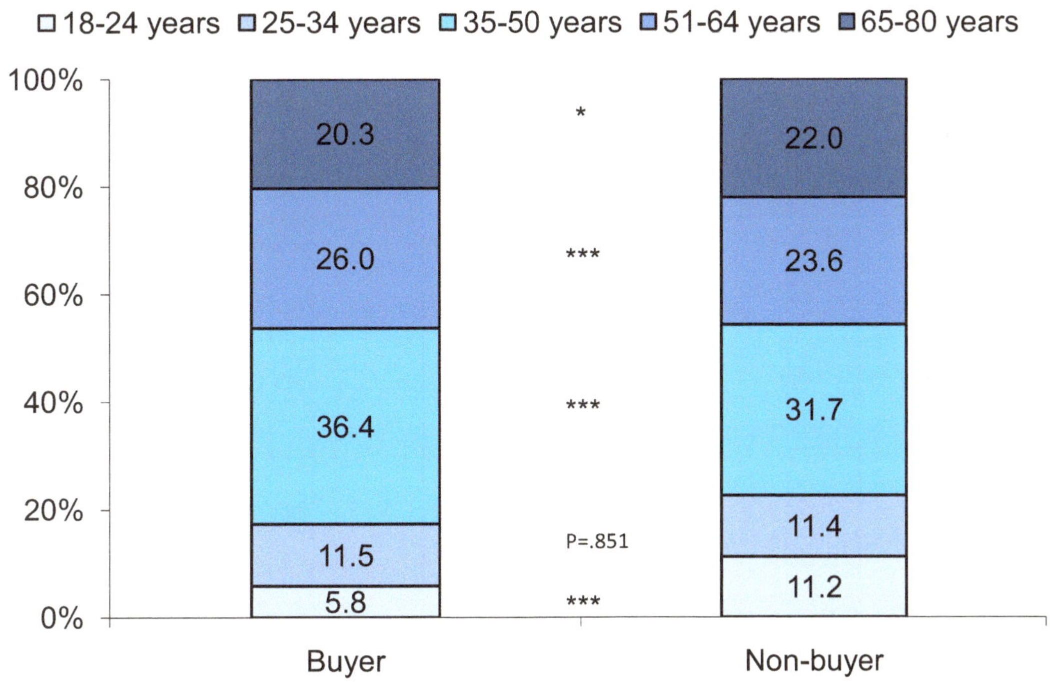

Fig. 1. Buyer (n= 5875) and non-buyer (n= 7199) of organic food by age group
**** P < .001, * P <.05*
Comparisons are based on Chi square test, Bonferroni correction

buyers and 44% of all male organic buyers consumed at least 250 g/d fruit (non-buyers: 45% for women and 36% for men). There were significant differences between female organic buyers and non-buyers as well as between the male organic buyers and non-buyers (P < .001, resp.).

For women and men, mean consumption of vegetables was 23% higher for the organic food buyer group than for the non-buyer group (P < .001) (Fig. 3, by sex). The German dietary guideline for vegetable consumption of at least 400 g per day [31,32] was reached by 18% of all organic food buyers and by 10% of all non-buyers (P < .001). The proportion of meeting the recommendation within the three frequency groups of organic food purchase was 27% for intensive, 19% for occasional and 14% for infrequent buyers (P < .001, resp.). Differentiated by sex, 20% of the women and 16% of the men of the organic food buyers met the recommendation compared to 11% female and 10% male non-buyers (differences between female buyers and non-buyers as well as male buyers and non-buyers: P < .001, resp.).

Organic food buyers consumed 25% less meat and sausages than non-buyers of organic food (P < .001) (Fig. 3, by sex). The German dietary guideline for meat and meat products is 300 to 600 g per week [31,32]. Of all organic food buyers, 48% exceed the recommendation compared to 64% of the non-buyers (P < .001). This was the case for 33% of the intensive buyers compared to 48% of the occasional and 53% of the infrequent buyers (P < .001, resp.). The proportion of women exceeding the recommendation was 35% for organic food buyers and 45% for non-buyers (P < .001) while the proportion of men exceeding the recommendation was 68% for buyers of organic food and 81% for non-buyers (P <.001).

All men regardless of buying organic food and in which frequencies, consumed comparable amounts of sweets while organic food buying women ate fewer sweets (45 g/d) than non-buying women (48 g/d) (P < .05) (Fig. 4). Female intensive buyers consume the least sweets (41 g/d), followed by female occasional buyers (45 g/d) and female infrequent buyers (48 g/d) (P < .001, resp.). Buyers of organic food consumed 58% less soft drinks (like soda and ice tea) than non-buyers (P < .001) (Fig. 4, by sex).

However, a high proportion of all participants did not drink soft drinks at all: 77% of organic food buyers and 65% of non-buyers (P <.001). Considering only the consumers of soft drinks,

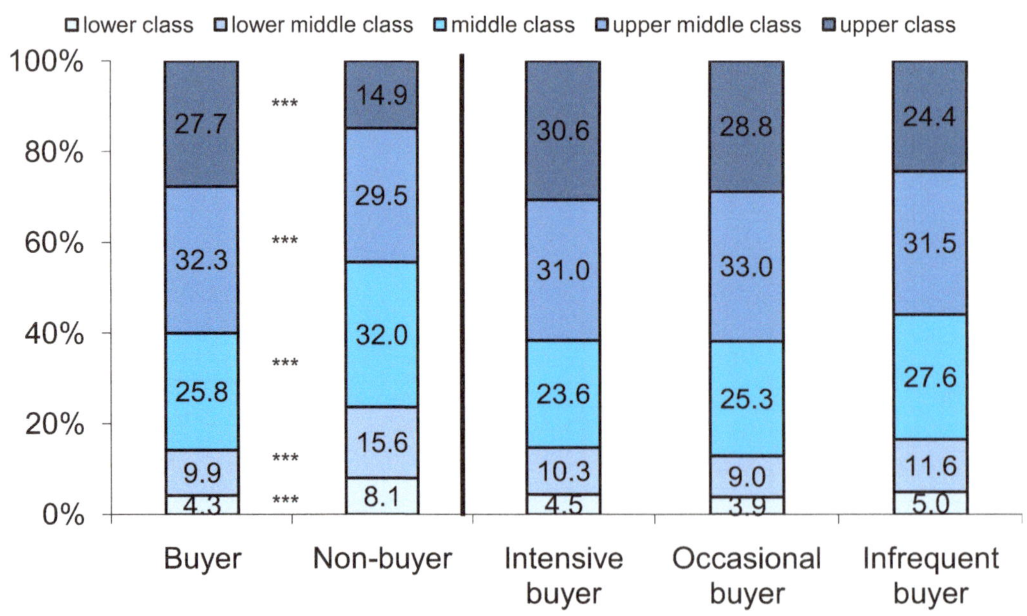

Fig. 2. Buyer (incl. sub-groups) and non-buyer of organic food by socialclass, intensive (n= 661), occasional (n= 3428), infrequent (n=1786)

**** P < .001; Comparisons are based on Chi square test, Bonferroni correction.*

the recorded quantities of the buyers were still 37% less than the quantities of the non-buyers of organic food (P < .001). The average soft drink consumption of women buying intensive, occasional and infrequent organic food were 163 ml/d, 226 ml/d and 260 ml/d, while women belonging to the non-buyer group consumed 313 ml/d (P< .001, resp.). The corresponding consumption in men was 345 ml/d (intensive), 310 ml/d (occasional), 329 ml/d (infrequent) and 505 ml/d for non-buyers (P < .001, resp.).

Altogether, organic food buyers more often met the food-based dietary guidelines of the German Nutrition Society [31]. Consequently, the absolute nutrient intake as well as nutrient densities of buyers of organic food corresponded better to the recommendations than those of the non-buyers [33] (data not presented).

3.3 Health Aspects and Anthropometric Measurements

Buyers of organic food more often were non-smokers and vegetarians and indicated more often to be physically active (Table 1). In addition, organic food buyers more often were interested in nutrition information and achieved a higher score in the NVS II nutrition knowledge index. More frequently, they stated to be satisfied with their personal health status. Within the buyer group, the proportion of normal body weight was higher and the proportions of overweight and of obese participants lower than in the non-buyer group. This was also the case for women and men respectively (with the exception of the male overweight participants) (Fig. 5).

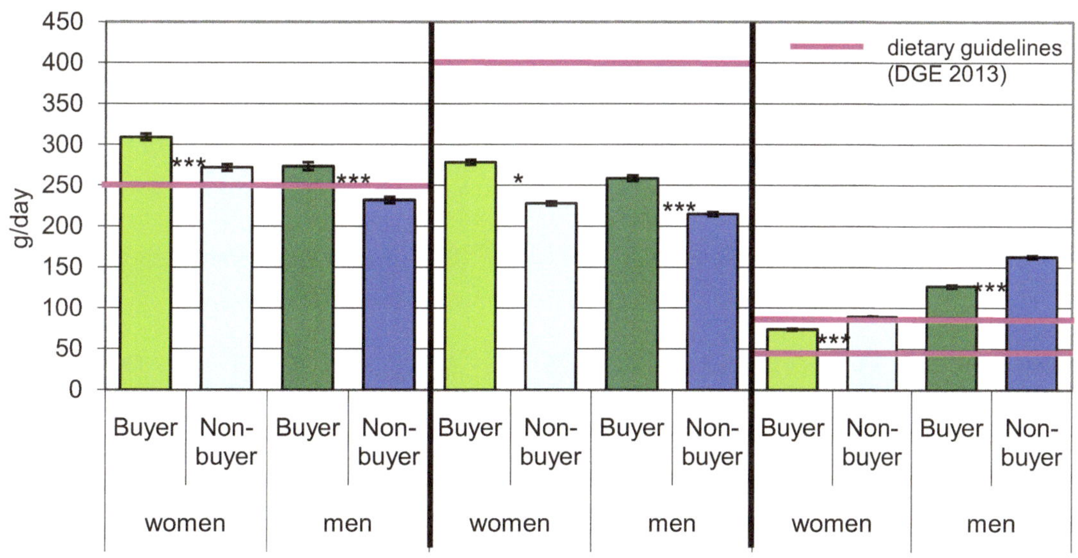

Fig. 3. Mean consumption (± SE) of fruit, vegetables and meat/sausages of buyer and non-buyer of organic food, by sex Buyer: women (n=3577), men (n= 2298); Non-buyer: women (n= 3536), men (n= 5961)

**** P < .001; Comparisons are based on Mann-Whitney-Wilcoxon test.*

Table 1. Lifestyle factors of buyers and non-buyers of organic food

Percentage of purchase group[1]	Buyers n=5875	Non-buyers n=7199
Non smoker	80,0	71,0
Being physical active (yes/no)	66,0	52,0
Vegetarian	1,8	0,3
Interest in nutrition information	84,0	64,0
High score in the NVS II nutrition knowledge index	39,0	23,0
High satisfaction with personal health status	78,0	74,0

[1]differences between the purchase groups in every case: P < .001; Comparisons are based on Chi square test

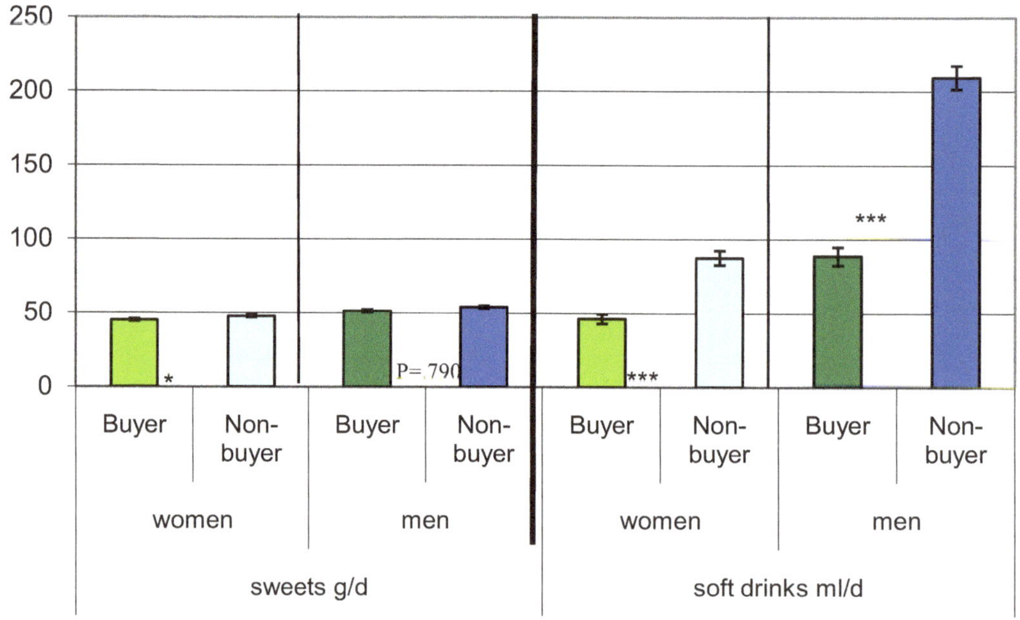

Fig. 4. Mean consumption (± SE) of sweets and soft drinks of buyer and non-buyer of organic food, by sex buyer: women (n=3577), men (n= 2298); non-buyer: women (n= 3536), men (n= 5961)

**** P < .001, * P <.05; Comparisons are based on Mann-Whitney Wilcoxon test.*

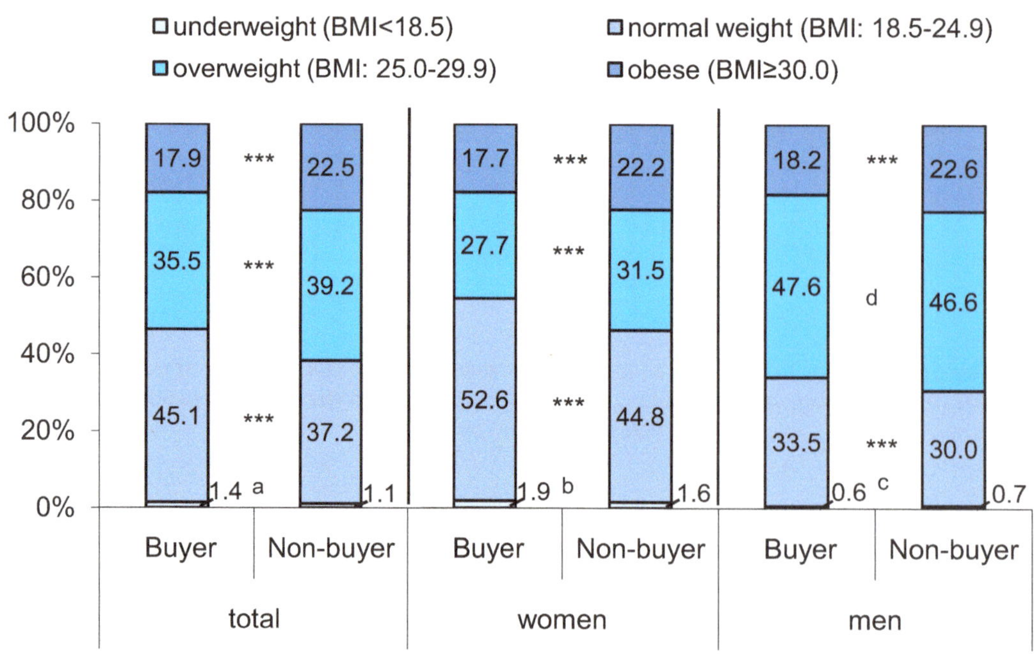

Fig. 5. Organic food buyer and non-buyer by body weight (BMI) and sex organic food buyer: women (n=3577), men (n= 2298);Non-buyer: women (n= 3536), men (n= 5961)

**** P < .001, [a]P=.178, [b] P=.277, [c]P=.675, [d]P=.473;*

Comparisons are based on Chi square test, Bonferroni correction

4. DISCUSSION AND CONCLUSION

With the data of the NVS II, the nationwide representative nutrition survey for Germany, a comparison between buyers and non-buyers of organic food was conducted focusing on food consumption and lifestyle factors. The results show that, compared to non-buyers, German buyers of organic food eat more fruit and vegetables, less meat and sausages and therefore demonstrate a more favorable food choice. This food pattern is more in line with national and international recommendations [31,34-38]. These results apply to both, women and men, while women do even better. The more intense the purchase of organic food the more favorable the food choice turns out to be. Additionally, buyers of organic food decide to lead a healthier lifestyle as indicated by lower body weight, higher level of physical activity and less smokers than non-buyers.

The French web-based cohort study from a large sample of volunteers identified five clusters, two of them comprising occasional and regular organic food consumers [23]. The comparison of food consumption, assessed with three 24-h recalls, between non-consumers and consumers of organic food revealed that food patterns of organic food consumers show a higher agreement with dietary guidelines, which gets even better with increasing frequency of consumption. In addition, the presented results regarding body weight were confirmed by the French study.

A higher percentage of normal body weight in the organic food group was also found in the Norwegian study with pregnant women [39]. In 1998, a comparison in the Netherlands between a sample of 271 customers of health food stores and a nationwide household sample (n=581) showed that organic food buyers displayed the lowest BMI [40].

Healthy food choices, normal body weight, being physically active, as well as non-smoking contribute substantially to a healthy lifestyle [18]. There is a robust understanding of links between healthy lifestyle factors and a lower risk of developing diabetes, heart disease, and cancer. In the EPIC-study, the lifestyle factors healthy diet, non-smoking, high level of physical activity and normal body weight are associated with a reduced risk of developing diabetes (93%), myocardial infarction (81%), stroke (50%) and cancer (36%) compared to participants who do not indicate any health factors [41]. Data of the NHANES study in the United States show the impact of three lifestyle factors (healthy diet, non-smoking and being physically active) with a risk reduction of 65% for cardiovascular disease and 83% for cancer [17]. Other studies also confirm these findings but are even more specific for special diseases like heart disease [42] and diabetes mellitus [43]. The four lifestyle factors (healthy diet, non-smoking, high level of physical activity and normal body weight) also proved responsible for a risk reduction of 66% regarding the endpoint all-cause mortality [44]. Ford et al. [17], found strong relationships for three lifestyle factors (healthy diet, non-smoking and being physically active) and all-cause mortality with an 82% risk reduction compared to participants without an indication of health factors.

The outcome is not to be confused with the question whether health effects of organically produced food compared to conventionally produced food are evident. Related research concentrates on differences in health effects of foods from the two farming systems and is discussed elsewhere [45-51]. Instead, the data set of the NVS II allows the comparison of food consumption and lifestyle factors of buyers and non-buyers of organic food. The results show overall better health behaviour of buyers of organic food.

The presented finding that for the German population from 2005-2006 45% are buyers and 55% were non-buyers differs from results of the 'Ökobarometer'. The 'Ökobarometer' is an annual assessment initiated by the Federal Ministry of Food, Agriculture and Consumer Protection of Germany. For the years 2005 and 2007 the data collection indicated about 75% of the interviewed participants (about 1.000 participants starting at age 14) buying organic food in different frequencies [52,53]. In contrast, the German household study with a sample of 13,000 totaled 50% as buyers for 2008 [21] showing an increase from 36% of the same household panel in 2005 [54]. Thus, the finding of the present study that 45% of for the German population bought organic food in the year 2006 is in a comparable range with the household study 2008 regarding sample size and percentage of organic purchase.

Socio-demographics are used to test how the profile of the organic buyers of the NVS II compares to other international and German studies. The finding that women are generally more engaged in health and nutrition has often been demonstrated [10,14,23,55,56]. Participants with a higher social status, i.e. higher level of education and income, have also more often been linked to intensive organic purchase behavior [2,23,54,57,58], while age is a factor with no clear association [10,14,39,54]. In summary, concerning socio-demographic variables the findings of the present study are in concordance with other studies on organic buyers.

All 13,074 participants underwent a personal interview regarding their usual food consumption of the last four weeks using the diet history method. The advantage of this dietary assessment method is the focus on long-term dietary habits while limitations comprise the cognitive effort to remember all consumed food and beverages of the last four weeks [26,59]. Especially the answers for inhomogeneous food groups, like vegetables and pastries, may be affected by either social desirability or incapability of memorizing complexity [60]. As the limitation of this assessment method applies to both study groups, it should not affect the comparison between buyers and non-buyers of organic food.

As strengths of the presented study has to be seen the sample size of 13,074 participants and the representative database for Germany. Together with the French cohort [23], two studies with the general population are available being in accordance regarding the main results that consumers buying organic food display dietary patterns more in line with dietary guidelines. Additionally, with both studies body weight was assessed demonstrating that buyers of organic food display lower body weight than non-buyers. A limitation of both studies is the cross-sectional design. The data only represent a situation for a specific moment. Therefore, causes and effects cannot be differentiated. Whether the effort for a healthier nutrition leads to the purchase of organic food or buying organic food leads to healthy choices cannot be determined and causalities cannot be deduced.

The results of the presented study demonstrate for the German population that buyers of organic food make healthier food choices and their lifestyle is closer in line with recommendations for a healthy living. In doing so, subjects probably reduce their personal risk regarding several major diseases. Whether consumption of organic food has an additional positive influence on health is still under scientific debate. From a public health point of view, organic food buyers lead a way of life closer to the recommendations for a healthy lifestyle and thereby profit by scientifically well-established health benefits and risk reductions.

ACKNOWLEDGMENTS

Franziska Lindner (KIT, Dept. Mathematics, Karlsruhe) was a consultant for methods of analysis.

The authors acknowledge the work of many colleagues involved in the NVS II.

FINANCIAL SUPPORT

The Federal Ministry of Food, Agriculture and Customer Protection of Germany funded the German National Nutrition Survey II.

The Federal Ministry of Food, Agriculture and Customer Protection of Germany financed the present study in the framework of the Federal Scheme of Organic Farming (BÖL, now BÖLN; grant number: 28080E056). The project was granted in collaboration with Prof. A. Spiller, Marketing for Food and Agricultural Products, University of Göttingen.

COMPETING INTERESTS

The authors have declared that no competing interests exist.

REFERENCES

1. Willer H, Lernoud J. Organic Agriculture Worldwide: Current Statistics. BioFach, 12.2.2014, Messezentrum, Nürnberg 2014.Accessed 4 March 2014. Available:http://www.fibl.org/fileadmin/documents/de/news/2014/willer-2014-global-data.pdf
2. Dimitri C, Dettmann RL. Organic food consumers: What do we really know about them? British Food Journal. 2012;114(8):1157–83.
3. Aertsens J, Mondelaers K, Verbeke W, Buysse J, Van Huylenbroeck G. The influence of subjective and objective knowledge on attitude, motivations and consumption of organic food. British Food Journal. 2011;113(10-11):1353–78.
4. Hjelmar U. Consumers' purchase of organic food products. A matter of convenience and reflexive practices. Appetite. 2011;56(2):336–44.

5. Nie C, Zepeda L. Lifestyle segmentation of US food shoppers to examine organic and local food consumption. Appetite. 2011;57(1):28–37.
6. Thøgersen J. Country differences in sustainable consumption: The case of organic food. Journal of Macromarketing. 2010;30(2):171–85.
7. Chen M-F. Attitude toward organic foods among Taiwanese as related to health consciousness, environmental attitudes, and the mediating effects of a healthy lifestyle. British Food Journal. 2009;111(2-3):165–78.
8. Pellegrini G, Farinello F. Organic consumers and new lifestyles An Italian country survey on consumption patterns. British Food Journal. 2009;111(9):948–74.
9. de Magistris T, Gracia A. The decision to buy organic food products in Southern Italy. British Food Journal. 2008;110(8-9):929–47.
10. Hughner RS, McDonagh P, Prothero A, Shultz CJ, Stanton J. Who are organic food consumers? A compilation and review of why people purchase organic food. Journal of Consumer Behaviour. 2007;6(2-3):94-110.
11. The Quality LowInput Food. Consumer attitudes to quality and safety of organic and low input foods: a review. 2005; Midmore P, Naspetti S, Sherwood A-M, Vairo D, Wier M, Zanoli R. QLIF. Accessed 16 October 2013. Available:http://www.qlif.org/research/sub1/QLIF_Review_Reanalysis_%200509.pdf
12. Zanoli R, Naspetti S. Consumer motivations in the purchase of organic food: A means-end approach. British Food Journal. 2002;104(8/9):643–53.
13. Institut für Sozialökologische Forschung. Ernährungswende- Ernährungsstile im Alltag. Ergebnisse einer repräsentativen Untersuchung. 2005; ISOE. Accessed 2 November 2011. Available:http://www.ernaehrungswende.de/fr_aktu.html
14. Aertsens J, Verbeke W, Mondelaers K, Van Huylenbroeck G. Personal determinants of organic food consumption: A review. British Food Journal. 2009;111(10):1140–67.
15. Mondelaers K, Verbeke W, Van Huylenbroeck G. Importance of health and environment as quality traits in the buying decision of organic products. British Food Journal. 2009;111(10):1120–39.
16. Aarset B, Beckmann S, Bigne E, Beveridge M, Bjorndal T, Bunting J, et al. The European consumers' understanding and perceptions of the "organic" food regime: the case of aquaculture. British Food Journal. 2004;106(2/3):93–105.
17. Ford ES, Bergmann MM, Boeing H, Li C, Capewell S. Healthy lifestyle behaviors and all-cause mortality among adults in the United States. Prev Med. 2012;55(1):23–27.
18. World Health Organization. Diet, nutrition and the prevention of chronic diseases. WHO technical report series. 2003; WHO. Geneva, Switzerland, Accessed 13 July 2012. Available:http://whqlibdoc.who.int/trs/who_trs_916.pdf
19. Holt GC. Ecological eating: Food ideology and food choice. An analysis of the changing British diet with reference to the consumption of meat and organically produced food. Dissertation: Department of Biomedical Sciences University of Bradford; 1993.
20. Torjusen H, Lieblein G, Naes T, Haugen M, Meltzer HM, Brantsaeter AL. Food patterns and dietary quality associated with organic food consumption during pregnancy; data from a large cohort of pregnant women in Norway. BMC Public Health. 2012;12:612. Accessed 4 Febraury 2014. Available:http://www.ncbi.nlm.nih.gov/pmc/articles/PMC3490940/pdf/1471-2458-12-612.pdf
21. Buder F, Hamm U. Verbrauchsstrukturen von Öko-Intensivkäufern im Fokus. Ernährungs Umschau. 2009;56:527–33.
22. Krarup S, Christensen T, Denver S. Are Organic Consumers Healthier than Others? Vortrag at: Cultivating the Future Based on Science: 2nd Conference of the International Society of Organic Agriculture Research ISOFAR, Modena, Italy; 2008.
23. Kesse-Guyot E, Peneau S, Mejean C, Szabo de Edelenyi F, Galan P, Hercberg S, et al. Profiles of organic food consumers in a large sample of French adults: results from the Nutrinet-Sante cohort study. PLoS One. 2013;8:10. Accessed 8 October 2014. Available:http://www.ncbi.nlm.nih.gov/pubmed/24204721
24. Max Rubner-Institut, Federal Research Institute of Nutrition and Food. National Nutrition Survey II. Ergebnisbericht, Teil 1.

Die bundesweite Befragung zur Ernährung von Jugendlichen und Erwachsenen. 2008; MRI. Karlsruhe. Accessed 28 November 2013. Available:http://www.mri.bund.de/fileadmin/Institute/EV/NVS_II_Abschlussbericht_Teil_1_mit_Ergaenzungsbericht.pdf

25. Max Rubner-Institut, Federal Research Institute of Nutrition and Food. National Nutrition Survey II. Result Report, Part 2. Die bundesweite Befragung zur Ernährung von Jugendlichen und Erwachsenen. 2008; MRI. Karlsruhe. Accessed 28 November 2013. Available:http://www.mri.bund.de/fileadmin/Institute/EV/NVSII_Abschlussbericht_Teil_2.pdf
26. van Staveren WA, Ocké MC, de Vries JHM. Estimation of dietary intake. In: Erdman JW, Macdonald IA, Zeisel SH, editors. Present Knowledge in Nutrition. 10th ed. Ames: Wiley-Blackwell; 2012.
27. Merten C, Ferrari P, Bakker M, Boss A, Hearty A, Leclercq C, et al. Methodological characteristics of the national dietary surveys carried out in the European Union as included in the European Food Safety Authority (EFSA) comprehensive european food consumption database. Food Addit Contam Part A Chem Anal Control Expo Risk Assess. 2011;28(8):975–95.
28. Mensink GBM, Bauch A, Vohmann C, Stahl A, Six J, Kohler S, et al. EsKiMo - The nutrition module in Child and Adolescent Health Survey (KiGGS). Federal Health Journal - Health Research - Health. 2007;50:902–08.
29. Mensink GBM, Haftenberger M, Thamm M. Validity of DISHES 98, a computerised dietary history interview: Energy and macronutrient intake. Eur J Clin Nutr. 2001;55(6):409–17.
30. World Health Organization. Obesity: Preventing and managing the global epidemic. Report of a WHO Consultation. 2000; WHO. Geneva, Accessed 8 March 2012. Available:http://www.who.int/nutrition/publications/obesity/WHO_TRS_894/en/index.html
31. German Nutrition Society. 10 guidelines of the German Nutrition Society (DGE) for a wholesome diet. 2011; DGE, Accessed 25 February 2012. Available: http://80.237.197.43/pdf/10-guidelines-wholesome-diet-DGE-en.pdf
32. Oberritter H, Schäbethal K, von Ruesten A, Boeing H. The DGE Nutrition circle – Presentation and basis of the food-related recommendations from the German Nutrition Society (DGE). Ernaehrungs Umschau international. 2013;60(2):24–29.
33. German Nutrition Society, Austrian Nutrition Society, Swiss Society for Nutrition Research, Swiss Association for Nutrition, editors. Reference values for nutrient intake. 1. Auflage, 5. korrigierter Nachdruck ed. Neustadt a. d. Weinstraße: Umschau Buchverlag; 2013.
34. Mozaffarian D, Ludwig DS. Dietary guidelines in the 21st century-a time for food. Jama-Journal of the American Medical Association. 2010;304(6):681–82.
35. United States Department of Agriculture, United States Department of Health & Human Services. Dietary Guidelines for americans, 2010. 2010; USDA, HHS, Accessed 5 February 2013. Available:http://www.cnpp.usda.gov/Publications/DietaryGuidelines/2010/PolicyDoc/PolicyDoc.pdf
36. Heidemann C, Schulze MB, Franco OH, van Dam RM, Mantzoros CS, Hu FB. Dietary patterns and risk of mortality from cardiovascular disease, cancer, and all causes in a prospective cohort of women. Circulation. 2008;118(3):230–37.
37. World Cancer Research Fund International, American Institute for Cancer Research. Food, nutrition, physical activity, and the prevention of cancer: a global perspective. 2007; WCRF, AICR. Washington DC, Accessed 24 August 2011. Available:http://www.dietandcancerreport.org/
38. Deutsche Gessellschaft für Ernährung e.V. Vegetables and fruit in the prevention of selected chronic diseases. Opinion of the German Nutrition Society. 2012; DGE. Bonn, Accessed 14 June 2012. Available: http://www.dge.de/pdf/ws/DGE-Stellungnahme-Gemuese-Obst-2012.pdf
39. Torjusen H, Brantsaeter AL, Haugen M, Lieblein G, Stigum H, Roos G, et al. Characteristics associated with organic food consumption during pregnancy; data from a large cohort of pregnant women in Norway. BMC Public Health. 2010;10:775. Accessed 4 February 2014. Available:http://www.biomedcentral.com/1471-2458/10/775

40. Schifferstein HNJ, Ophuis P. Health-related determinants of organic food consumption in the Netherlands. Food Qual Prefer. 1998;9(3):119–33.
41. Ford ES, Bergmann MM, Kroger J, Schienkiewitz A, Weikert C, Boeing H. Healthy living Is the best revenge findings from the European Prospective Investigation Into Cancer and Nutrition-Potsdam Study. Arch Intern Med. 2009;169(15):1355–62.
42. Lichtenstein AH, Appel LJ, Brands M, Carnethon M, Daniels S, Franch HA, et al. Diet and lifestyle recommendations revision 2006: A scientific statement from the American Heart Association Nutrition Committee Circulation. 2006;114(1):82–96.
43. Franz MJ, Bantle JP, Beebe CA, Brunzell JD, Chiasson JL, Garg A, et al. Evidence-based nutrition principles and recommendations for the treatment and prevention of diabetes and related complications. Diabetes Care. 2002;25(1):148–98.
44. Loef M, Walach H. The combined effects of healthy lifestyle behaviors on all cause mortality: A systematic review and meta-analysis. Prev Med. 2012;55(3):163–70.
45. Forman J, Silverstein J, Comm N, Council Environm H. Organic foods: Health and environmental advantages and disadvantages. Pediatrics. 2012;130(5):E1406–E15.
46. Smith-Spangler C, Brandeau ML, Hunter GE, Bavinger C, Pearson M, Eschbach PJ, et al. Are organic foods safer or healthier than conventional alternatives? Ann Intern Med. 2012;157(5):348–66.
47. Brandt K, Leifert C, Sanderson R, Seal CJ. Agroecosystem management and nutritional quality of plant foods: The case of organic fruits and vegetables. Critical Reviews in Plant Sciences. 2011;30(1-2):177–97.
48. Huber M, Rembialkowska E, Srednicka D, Bugel S, van de Vijver LPL. Organic food and impact on human health: Assessing the status quo and prospects of research. Njas-Wageningen Journal of Life Sciences. 2011;58(3-4):103–09.
49. Dangour AD, Lock K, Hayter A, Aikenhead A, Allen E, Uauy R. Nutrition-related health effects of organic foods: A systematic review. Am J Clin Nutr. 2010;92(1):203–10.
50. Dangour AD, Dodhia SK, Hayter A, Allen E, Lock K, Uauy R. Nutritional quality of organic foods: A systematic review. Am J Clin Nutr. 2009;90(3):680–85.
51. Hoefkens C, Vandekinderen I, De Meulenaer B, Devlieghere F, Baert K, Sioen I, et al. A literature-based comparison of nutrient and contaminant contents between organic and conventional vegetables and potatoes. British Food Journal. 2009;111(10):1078–97.
52. Federal Organic Farming Scheme. Ökobarometer 2007: Bio-be regular customers from occasional buyers. Supermarkets make new customers on Bio attention. 2007; BÖL. Bonn, Accessed 13 February 2014. Available:http://www.oekolandbau.de/fileadmin/redaktion/dokumente/journalisten/Oekobarometer2007.pdf
53. Pleon Kohtes Klewes. Ökobarometer 2005. Representative population survey commissioned by the Federal Ministry of Consumer. 2005; Pleon, Accessed 13 February 2014. Available:http://www.oekolandbau.de/fileadmin/redaktion/dokumente/journalisten/Ergebnisstudie_oekobarometer2005.pdf
54. Federal Organic Farming Scheme. Dynamics of purchasing behavior in the organic range. Final Report for the project 09OE014 2010; BÖL. Witzenhausen. Accessed 8.October 2013. Available:http://orgprints.org/16983/1/16983-09OE014-uni_kassel-hamm-2010-kaufverhalten.pdf
55. Institute for Social-Ecological Research. Result Presentation "bio + per" - target groups for the organic food market. 2003; ISOE. Frankfurt am Main, Accessed 16 January 2013. Available: http://orgprints.org/4554/2/4554-02OE330-2003-biopro-praesentation.pdf
56. Lockie S, Lyons K, Lawrence G, Grice J. Choosing organics: A path analysis of factors underlying the selection of organic food among Australian consumers. Appetite. 2004;43(2):135-46.
57. Zhang F, Huang CL, Lin B-H, Epperson JE. Modeling fresh organic produce consumption with scanner data: A generalized double hurdle model approach. Agribusiness. 2008;24(4):510–22.
58. Denver S, Christensen BT. Organic consumption in three European countries. 3rd QLIF Congress, Hohenheim, Germany, 20-23 March; 2007.

59. Thompson FE, Subar AF. Dietary assessment methodology. In: Coulston AM, Boushey CJ, Ferruzzi MG, editors. Nutrition in the prevention and treatment of disease. 3rd ed. ed. Oxford Elsevier; 2013.
60. Eisinger-Watzl M, Strassburg A, Ramunke J, Krems C, Heuer T, Hoffmann I. Comparison of two dietary assessment methods by food consumption: results of the German National Nutrition Survey II. Eur J Nutr. 2014. DOI: 10.1007/s00394-014-0714-z.

Permissions

All chapters in this book were first published in EJNFS, by SCIENCE DOMAIN international; hereby published with permission under the Creative Commons Attribution License or equivalent. Every chapter published in this book has been scrutinized by our experts. Their significance has been extensively debated. The topics covered herein carry significant findings which will fuel the growth of the discipline. They may even be implemented as practical applications or may be referred to as a beginning point for another development.

The contributors of this book come from diverse backgrounds, making this book a truly international effort. This book will bring forth new frontiers with its revolutionizing research information and detailed analysis of the nascent developments around the world.

We would like to thank all the contributing authors for lending their expertise to make the book truly unique. They have played a crucial role in the development of this book. Without their invaluable contributions this book wouldn't have been possible. They have made vital efforts to compile up to date information on the varied aspects of this subject to make this book a valuable addition to the collection of many professionals and students.

This book was conceptualized with the vision of imparting up-to-date information and advanced data in this field. To ensure the same, a matchless editorial board was set up. Every individual on the board went through rigorous rounds of assessment to prove their worth. After which they invested a large part of their time researching and compiling the most relevant data for our readers.

The editorial board has been involved in producing this book since its inception. They have spent rigorous hours researching and exploring the diverse topics which have resulted in the successful publishing of this book. They have passed on their knowledge of decades through this book. To expedite this challenging task, the publisher supported the team at every step. A small team of assistant editors was also appointed to further simplify the editing procedure and attain best results for the readers.

Apart from the editorial board, the designing team has also invested a significant amount of their time in understanding the subject and creating the most relevant covers. They scrutinized every image to scout for the most suitable representation of the subject and create an appropriate cover for the book.

The publishing team has been an ardent support to the editorial, designing and production team. Their endless efforts to recruit the best for this project, has resulted in the accomplishment of this book. They are a veteran in the field of academics and their pool of knowledge is as vast as their experience in printing. Their expertise and guidance has proved useful at every step. Their uncompromising quality standards have made this book an exceptional effort. Their encouragement from time to time has been an inspiration for everyone.

The publisher and the editorial board hope that this book will prove to be a valuable piece of knowledge for researchers, students, practitioners and scholars across the globe.

List of Contributors

Nkiru N. Ezeama
Department of Community Medicine, Nnamdi Azikiwe University Teaching Hospital, Nnewi, Nigeria

Prosper O. U. Adogu, Christian C. Ibeh and Echendu D. Adinma
Department of Community Medicine / Consultant Public Health Physician, Nnamdi Azikiwe University / Nnamdi Azikiwe University Teaching Hospital, Nnewi, Nigeria

Jeffrey B. Blumberg and C. Y. Oliver Chen
Friedman School of Nutrition Science and Policy, Tufts University, 711 Washington Street, Boston, MA 02111, USA

Bradley W. Bolling
Department of Food Science, University of Wisconsin-Madison, 1605 Linden Drive, Madison, WI 53706, USA

Hang Xiao
Department of Food Science, University of Massachusetts Amherst, 338 Chenoweth Laboratory, 100 Holdsworth Way, Amherst, MA 01003, USA

Géraldine Boué, Sandrine Guillou and Jeanne-Marie Membré
INRA, Secalim UMR 1014, F-44026 Nantes, France
LUNAM University, Oniris, Nantes, France

Bruno Le Bizec
LUNAM University, Oniris, Laboratory Study of Residues and Contaminants in Food (LABERCA), F-44307 Nantes, France

Jean-Philippe Antignac
INRA, F-44307 Nantes, France

Ingrid Bielderman, Klaus Kraemer and Jane Badham
Sight and Life, Basel, Switzerland

Maaike J. Bruins
DSM Biotechnology Center, Delft, The Netherlands

Reina Engle-Stone
Department of Nutrition, University of California, Davis, CA, USA

Noel W. Solomons
Center for Studies of Sensory Impairment, Aging and Metabolism (CeSSIAM), Guatemala City, Guatemala

Hans Verhagen
National Institute for Public Health and the Environment (RIVM), Bilthoven, The Netherlands
University of Ulster, Northern Ireland Centre for Food and Health (NICHE), Coleraine, Northern Ireland, UK

Lindsay H. Allen
US Department of Agriculture (USDA), ARS Western Human Nutrition Research Center, Davis, CA, USA

Lynnette M. Neufeld
Global Alliance for Improved Nutrition (GAIN), Geneva, Switzerland

Renatus P. Shilangale
Central Veterinary Laboratory, Private Bag 13187, Windhoek, Namibia
Department of Biological Sciences, Faculty of Science, University of Namibia, Private Bag 13301, Windhoek, Namibia

Percy M. Chimwamurombe
Department of Biological Sciences, Faculty of Science, University of Namibia, Private Bag 13301, Windhoek, Namibia

Godwin P. Kaaya
School of Biological Sciences, University of Nairobi, P.O.Box 30197-00100, Nairobi, Kenya

Hurjus Bahra and Anne Mullen
Diabetes and Nutritional Sciences Division, King's College London, Franklin-Wilkins Building, 150 Stamford Street, London, UK

Angelika A. Linowska and Ewa Sobecka
Division of Hydrobiology, Ichthyology and Biotechnology of Breeding, West Pomeranian University of Technology, K. Królewicza 4, Szczecin, Poland

Esther Kalonji, Véronique Sirot, Irène Margaritis and Jean-Charles Leblanc
Risk Assessment Department, French Agency for Food, Environmental and Occupational Health & Safety (ANSES), 14 Rue Pierre et Marie Curie, F-94701 Maisons-Alfort, France

Laurent Noel and Thierry Guerin
CIME Unit, Food Safety Laboratory, French Agency for Food, Environmental and Occupational Health & Safety (ANSES), 23, Avenue du Général de Gaulle, F-94706 Maisons-Alfort, France

Lya G. Soeteman-Hernández, Eugène H. Jansen, Elly J. M. Buurma-Rethans and Jan van Benthem
National Institute for Public Health and the Environment (RIVM), P.O.Box 1, 3720 BA, Bilthoven, The Netherlands

Hans Verhagen
University of Ulster, Northern Ireland Centre for Food and Health (NICHE), Cromore Road, Coleraine, BT52 1SA, Northern Ireland

David Sanz, Pedro Razquin and Luis Mata
Department of R&D, ZEULAB S.L., Polígono PLAZA, C/ Bari, 25 dpdo, 50197, Zaragoza, Spain

Santiago Condón
Faculty of Veterinary Medicine, University of Zaragoza, C/ Miguel Servet, 177, 50013 Zaragoza, Spain

Teresa Juan
Agrifood Research and Technology Centre of Aragon (CITA), Montañana, 930, 50059 Zaragoza, Spain

Benito Herraiz
S.A.T. Num 42 Los Gonzalez, C/ Extramuros, s/N, 44300 Monreal del Campo, Spain

Irene Hatsu
Department of Human Science, Ohio State University, Columbus OH, USA

Fatma Huffman, Marianna Baum and Adriana Campa
Department of Dietetics and Nutrition, Florida International University, Miami, FL, USA

Paulette Johnson
Statistical Consulting, Florida International University, Miami, FL, USA

Barbara Thomlison
School of Social Work, Florida International University, Miami, FL, USA

Meredith G. Wagner
Department of Nutrition and Dietetics, Concordia College, 901 8th Street South, Moorhead, MN 56562, USA

Yeong Rhee, Kerrie Hert-Honrath and Donna Terbizan
Department of Health, Nutrition and Exercise Sciences, North Dakota State University, Dept. 2620, P.O.Box 6050, Fargo, ND 58108, USA

Elizabeth H. Blodgett Salafia
Department of Human Development and Family Science, North Dakota State University, Dept. 2615, P.O.Box 6050, Fargo, ND 58108, USA

Alvaro R. Garcia-Fuentes, Sabrina Wirtz and Ellen Vos
Maastricht University, P.O.Box 616, 6200 MD Maastricht, The Netherlands

Hans Verhagen
National Institute for Public Health and the Environment (RIVM); P.O.Box 1, 3720 BA Bilthoven, The Netherlands
University of Ulster, Northern Ireland Centre for Food and Health (NICHE), Cromore Road, Coleraine, BT52 1SA, Northern Ireland

Tracy L. Burrows, Kirrilly M. Pursey and Peter Stanwell
School of Health Sciences, Faculty of Health and Medicine, University of Newcastle, Callaghan, NSW, Australia

Paul de Raadt, Sabrina Wirtz and Ellen Vos
Maastricht University, Faculty of Law, 6200MD Maastricht, The Netherlands

Hans Verhagen
National Institute for Public Health and the Environment (RIVM), 3720 BA Bilthoven, The Netherlands
University of Ulster, Northern Ireland Centre for Food and Health (NICHE), Cromore Road, Coleraine, BT52 1SA Northern Ireland, UK

Mary O. Agada
Institute of Food Security, University of Agriculture, P.M.B. 2373 Makurdi, Nigeria

Edwin M. Igbokwe
Department of Agricultural Extension, University of Nigeria, Nsukka, Enugu State, Nigeria

Egbert Clevers
Maastricht University, P.O.Box 616, 6200 MD Maastricht, The Netherlands

Miriam Urlings
Department of Complex Genetics, NUTRIM School of Nutrition and Translational Research in
Metabolism, Maastricht University, P.O.Box 616, 6200 MD Maastricht, The Netherlands

Caomhan Logue and Alison M. Gallagher
Northern Ireland Centre for Food and Health (NICHE), University of Ulster, Cromore Road, Coleraine, Northern Ireland. BT52 1SA, Northern Ireland

Hans Verhagen
Northern Ireland Centre for Food and Health (NICHE), University of Ulster, Cromore Road, Coleraine, Northern Ireland. BT52 1SA, Northern Ireland
National Institute for Public Health and the Environment (RIVM), P.O.Box 1, 3720 BA, Bilthoven, The Netherlands

Stephan J. A. C. Peters
Netherlands Nutrition Centre (Voedingscentrum), P.O.Box 85700, 2508CK, Den Haag, The Netherlands

Maria Stella R. de Oliveira, Vivian C. Garcia and Lígia A. Martini
Nutrition Department, School of Public Health, University of São Paulo, Av Dr Arnaldo, 715.Zip code, 01246-904, Sao Paulo, SP, Brazil
University of São Paulo Food and Nutrition Research Center, NAPAN, USP, Brazil

Natielen J. Schuch
Franciscan University Center, Rua dos Andradas, 1614. Zip code 97010032, Santa Maria, RS, Brazil

Marianne Eisinger-Watzl, Friederike Wittig, Thorsten Heuer and Ingrid Hoffmann
Department of Nutritional Behaviour, Max Rubner-Institute, Federal Research Institute of Nutrition
and Food, Haid-und-Neu-Str. 9, 76131 Karlsruhe, Germany